Wiring Regulations in Brief

To Claire

Wiring Regulations in Brief

A complete guide to the requirements of the17th Edition of the
IEE Wiring Regulations, BS 7671: 2008 and Part P of the Building
Regulations

Second edition

Ray Tricker

AMSTERDAM • BOSTON • HEIDELBERG • LONDON
NEW YORK • OXFORD • PARIS • SAN DIEGO
SAN FRANCISCO • SINGAPORE • SYDNEY • TOKYO
Butterworth-Heinemann is an imprint of Elsevier

Butterworth-Heinemann is an imprint of Elsevier
Linacre House, Jordan Hill, Oxford OX2 8DP
30 Corporate Drive, Suite 400, Burlington, MA 01803

First edition 2007
Reprinted 2007
Second edition 2008

Notice
No responsibility is assumed by the publisher for any injury and/or damage to persons
or property as a matter of products liability, negligence or otherwise, or from any use
or operation of any methods, products, instructions or ideas contained in the material
herein. Because of rapid advances in the medical sciences, in particular, independent
verification of diagnoses and drug dosages should be made.

British Library Cataloguing in Publication Data
A catalogue record for this book is available from the British Library

Library of Congress Cataloging-in-Publication Data
A catalog record for this book is available from the Library of Congress

ISBN: 978-0-7506-8973-1

For information on all Butterworth-Heinemann publications
visit our web site at http://books.elsevier.com

Typeset by Charon Tec Ltd., A Macmillan Company. (www.macmillansolutions.com)

Printed and bound in the UK
08 09 10   10 9 8 7 6 5 4 3 2 1

**Working together to grow
libraries in developing countries**

www.elsevier.com | www.bookaid.org | www.sabre.org

ELSEVIER   BOOK AID
             International   Sabre Foundation

About the author

Ray Tricker (MSc, IEng, FIET, FCMI, FCQI, FIRSE) is the Principal Consultant of Herne European Consultancy – a company specializing in Integrated Management Systems – and an established Butterworth-Heinemann author (39 titles). He served with the Royal Corps of Signals (for a total of 37 years) during which time he held various managerial posts culminating in his being appointed as the Chief Engineer of NATO's Communication Security Agency (ACE COMSEC).

Most of Ray's work since joining Herne has centred on the European railways. He has held a number of posts with the Union International des Chemins de fer (UIC), for example, Quality Manager of the European Train Control System (ETCS), European Union (EU) T500 Review Team Leader, European Rail Traffic Management System (ERTMS) Users' Group Project Co-ordinator and HEROE (Harmonization of European Rail Rules) Project Co-ordinator, and currently (as well as writing books for Butterworth-Heinemann!) he is busy assisting small businesses from around the world (usually on a no-cost basis) to produce their own auditable Quality Management Systems to meet the requirements of ISO 9001:2000. He is also a Consultant to the Association of American Railroads (AAR) advising them on ISO 9001:2000 compliance, was recently appointed as UKAS Technical Specialist for the assessment of Notified Bodies for the Harmonization of the trans-European high-speed rail system, and, is currently the Quality Manager for the Trinidad Rapid Rail Project – Programme Management Consultancy who are overseeing the Design Stage of a brand-new, multi-billion dollar Trinidadian railway system.

Contents

Preface

The Industrial Revolution during the 1800s was responsible for causing poor living and working conditions in ever-expanding, densely populated urban areas. Outbreaks of cholera and other serious diseases (through poor sanitation, damp conditions and lack of ventilation) forced the government to take action. Building control took on the greater role of Health and Safety through the first Public Health Act in 1875 and this eventually led to the first set of national building standards (i.e. the Building Regulations).

As is the case with most official documents, as soon as they were published, they were almost out of date and consequently needed revising. So it wasn't too much of a surprise to learn that the committee responsible for writing the Public Health Act of 1875 had overlooked the increased use of electric power for street lighting and/or domestic purposes. Electricity was beginning to become increasingly popular but, as there were no rules and regulations governing their installation at that time, the companies or person responsible simply dug up the roads and laid the cables as and where they felt like it!

From a Health and Safety point of view the Government of the day expressed extreme concern at this exceedingly dangerous situation and so in 1882 *The Electric Lighting Clauses Act* (modelled on the previous 1847 *Gas Act*) was passed by Parliament. This legislation was implemented by *Rules and Regulations for the prevention of Fire Risks Arising from Electric Lighting* and it is this document that is the forerunner of today's IEE Wiring Regulations. Since then, this document has seen a succession of amendments, new editions and new titles and has now become the 17th edition of the IEE Wiring Regulations (i.e. BS 7671:2008 'Requirements for Electrical Installations').

The current legislation for all Building Control is the Building Act 1984, which is implemented by the Building Regulations 2000 and these Building Regulations are a set of minimum requirements designed to secure the health, safety and welfare of people in and around buildings and to conserve fuel and energy in England and Wales. They are basic performance standards which are supported by a series of documents that correspond to the different areas covered by the regulations. These are called 'Approved Documents' and they contain practical and technical guidance on ways in which the requirements of the Building Act 1984 can be met.

Since the introduction of the Public Health Act in 1875 there has always, therefore, been a direct link between Electrical Installations and Building Control and parts of all of the Approved Documents have an affect on these sorts of installation. With the publication of Approved Document P for 'Electrical Safety' in 2005, however, the design, installation, inspection and testing of electrical installations has now become inextricably linked to Building Control and the purpose of this book is attempt to draw all of the various requirements together.

Over the past 120 years there have been literally hundreds of books written on the subject of electrical installations, but the aim of *Wiring Regulations in Brief* is not just to become another book on the library shelf to be occasionally looked at. The intention is that it will provide professional engineers, students and (i.e. to a lesser degree) the unqualified DIY fraternity with an easy-to-read reference source to the official requirements of BS 7671:2008 for electrical safety and electrical installations.

Although BS7671:2008 is well structured and has separate sections for all the main topics (e.g. safety protection, selection and erection of equipment and so on) it is not the easiest of standards to get to grips with for a particular situation. Occasionally it can be very confusing and requires the reader to constantly flick backwards and forwards through the book to find what it is all about.

For example, Regulation 411.4.7 states that: *'Where a circuit-breaker is used to satisfy the requirements of Regulation 411.3.2.2 or Regulation 411.3.2.3, the maximum value of earth fault loop impedance (Z_s) shall be determined by the formula in Regulation 411.4.5. Alternatively, for a nominal voltage (U_o) of 230 V and a disconnection time of 0.4 s in accordance with Regulation 411.3.2.2 or 5 s in accordance with Regulation 411.3.2.3, the values specified in Table 41.3 for the types and ratings of overcurrent devices listed may be used instead of calculation.'*

The intention of *Wiring Regulations in Brief*, therefore, is to peel away some of this confusion and 'officialise' and provide the reader with an on-the-job reference source that can be quickly used without having to delve backwards and forwards though the standard.

 Please note, however, that this is **only** the author's impression of the most important aspects of the Wiring Regulations and their association with the Building Regulations. It should, therefore, only be treated as an aide mémoire to the Regulations and electricians should **always** consult BS 7671 to satisfy compliance.

Main changes in the 2008 edition of BS 7671

Following a full review, the 17th edition of the IEE Wiring Regulations BS 7671:2008 replaces the previous 2001 16th edition.

Technical authority for this Standard is vested in the Joint IET/BSI Technical Committee JPEL/64. This Joint Technical Committee is responsible for the work previously undertaken by the IEE Wiring Regulations Technical Committee and the BSI Technical Committee PEL/64. Copyright is held jointly by the IET and BSI.

The latest edition of the Wiring Regulations (i.e. BS 7671:2008 Requirements for Electrical Installations) was issued on 1 January 2008 and came into effect on 1 July 2008. Installations designed after 30 June 2008 are to comply with BS 7671:2008.

The Regulations apply to the design, erection and verification of electrical installations, also additions and alterations to existing installations. Existing installations that have been installed in accordance with earlier editions of the Regulations may not comply with this new edition in every respect, but this does not necessarily mean that they are unsafe for continued use or require upgrading.

BS 7671:2008 includes changes necessary to maintain technical alignment with CENELEC harmonisation documents and a summary of the main changes is provided below:

- Continuity of service now requires that an assessment be made for each circuit.
- Documentation for all electrical installations must now be provided.
- FELV is recognised as a protective measure and new requirements are detailed.
- New additional requirements to ensure the safe connection of low voltage generating sets including small-scale embedded generators (SSEGs) have now been included.
- Protection against electric shock now refers to protection under normal conditions (previously referred to as protection against direct contact) and fault protection (previously referred to as protection against indirect contact).
- Protection against overvoltages of atmospheric origin or due to switching has additional regulations enabling designers to use a risk assessment approach when designing installations which may be susceptible to overvoltages of atmospheric origin.
- Protection of low-voltage installations against temporary overvoltages due to earth faults in the high-voltage system and due to faults in the low-voltage system have now been included.
- Requirements concerning the UK reduced low-voltage system are now included.
- Requirements to protect against voltage disturbances and implement measures against electromagnetic influences have now been included.
- Requirements for safety services (e.g. emergency escape lighting, fire alarm systems, installations for fire pumps, fire rescue service lifts, smoke and heat extraction equipment) now need to be observed.
- Safety services have been expanded in line with IEC standardization.
- Selection and erection of wiring systems now includes busbar trunking systems and powertrack systems.
- The requirement that a metallic pipe of a water utility supply shall **not** be used as an earth electrode is **retained** and other metallic water supply pipework (such as a privately owned water supply network) shall now

not be used as an earth electrode unless precautions are taken against its removal and it has been considered for such a use.

- There is a new series of regulations for luminaries and lighting installations.
- Inspection and testing (which was Part 7 of BS 7671:2001) has been restructured as a new Part 6 and now includes changes to the requirements for insulation resistance, when testing SELV and PELV circuits at 250 V and for systems up to and including 500 V (including FELV).
- Special installations or locations (which was previously Part 6 of BS 7671:2001) has now been restructured and expanded as a new Part 7 with the following major amendments:
  - Agricultural and horticultural premises:
  - additional requirements applicable to life support systems have now been included
  - the reduced disconnection times (0.2 s) and the 25 V equation no longer appear.
  - Construction and demolition site installations:
  - the reduced disconnection times (0.2 s) and the 25 V equation no longer appear.
  - Electrical installations in caravan/camping parks and similar locations:
  - now includes the requirement that each socket-outlet must be provided, individually, with overcurrent and RCD protection.
  - Locations containing a bath tub or shower basin:
  - each circuit in the special location must have 30 mA RCD protection
  - Supplementary bonding is no longer required providing the installation has main bonding in accordance with Chapter 41
  - socket-outlets (other than SELV and shaver supply units to BS EN 61558-2-5) may be installed in locations containing a bath or shower 3 m horizontally beyond the boundary of zone 1
  - zone 3 is no longer defined.
  - Rooms and cabins containing sauna heaters:
  - zones A, B, C and D in BS 7671:2001 are replaced by zones 1, 2 and 3 (with changed dimensions).
  - Swimming pools and other basins:
  - this special location now includes basins of fountains
  - zones A, B and C in BS 7671:2001 are replaced by zones 0, 1 and 2.

The following new sections have also been included in Part 7:

- exhibitions, shows and stands
- floor and ceiling heating systems
- marinas and similar locations
- mobile or transportable units
- solar photovoltaic (PV) power supply systems
- temporary electrical installations for structures, amusement devices and booths at fairgrounds, amusement parks and circuses.

Note: For some reason the numbering system of section 7 of BS 7671:2008 is not sequential, so you will need to be careful.

The following new appendices have also been included:

- current-carrying capacity and voltage drop for busbar trunking and power-track systems
- definitions concerning multiple source, d.c. and other systems
- effect of harmonic currents on balanced three-phase systems
- measurement of earth fault loop impedance (consideration of the increase of the resistance of conductors with increase of temperature)
- methods for measuring the insulation resistance/impedance of floors and walls to earth or to the protective conductor system
- protection of conductors in parallel against overcurrent
- ring and radial final circuit arrangements
- voltage drop in consumers' installations.

Content of this book

To reflect these changes, this second edition of *Wiring Regulations in Brief* is structured as follows:

Chapter 1 Introduction	The background to BS 7671, what it contains and a description of its unique numbering system, objectives and legal status. The effect that the Wiring Regulations have on other Regulations and how this British Standard can be implemented.
Chapter 2 Domestic buildings	The requirements of the Building Act 1984 together with the Building Regulations:2000 and their Approved Documents (which provide guidance for conformance) and how these Building Control Regulations inter-relate with the Wiring Regulations. A resumé of the responsibilities for electrical installations. The types of inspections and tests that have to be completed and the requirements for records. The contents of Approved Document P for electrical safety and other relevant Approved Documents (such as those for Fire Safety, Access and Facilities for Disabled People, Conservation of Fuel and Power, Resistance to the Passage of Sound etc.) together with a listing of all the most important requirements that directly concern electrical installations.

 Note: Whilst the requirements from the Wiring Regulations are normally prefaced by the word 'shall' (meaning that this section **is** a mandatory requirement), you will notice that the Building Regulations use the words 'should' (i.e. recommended), 'may' (i.e. permitted) or 'can' (i.e. possible).

The reason for this is that Approved Documents reproduce the actual *requirements* contained in the Building Regulations relevant to a particular subject area. This is then followed by *practical and technical guidance* (together with examples) showing how the requirements can be met in some of the more common building situations. There may, however, be alternative ways of complying with the Building Regulations 2000's requirements to those shown in the Approved Documents and you are, therefore, under no obligation to adopt any particular solution in an Approved Document – if you prefer to meet the requirement(s) in some other way – but you **must** meet the requirement!

Chapter 3 Earthing	This chapter reminds the reader about the different types of earthing systems and earthing arrangements. It then lists all the main requirements for earthing before briefly touching on the test requirements for earthing.
Chapter 4 Safety protection	Chapter 4 lists the main requirements for safety protection. Basic protection against electric shock, fault protection, protection against both direct and indirect contact, protective conductors and protective equipment and then lists the test requirements for safety protection.
Chapter 5 Electrical equipment, components, accessories and supplies	The amount of different types of equipment, components, accessories and supplies for electrical installations currently available is enormous and any attempt to cover every type, model and/or manufacture would prove an impossible task for a book such as this. The intention of this chapter, therefore, is to provide a catalogue of all the different types identified and referred to in the Wiring Regulations (e.g. luminaires, RCDs, plugs and sockets etc.) and then make a list of the specific requirements that are sprinkled throughout the Regulations. For your convenience this catalogue has been compiled in alphabetical order.
Chapter 6 Cables and conductors	Within the Wiring Regulations there is frequent reference to different types of cables (e.g. single core, multicore, fixed, flexible etc.) conductors (such as live supply, protective, bonding etc.) and conduits, cable ducting, cable trunking and so on. Unfortunately, as is the case for equipment and components, the requirements for these items is liberally sprinkled throughout the Standard. The aim of Chapter 6, therefore, is to provide a catalogue of all the different types identified and referred to in the Wiring Regulations in three main headings (namely cables, conductors and conduits/etc) and then make a list of their essential requirements.

Chapter 7 Special installations and locations	Whilst the Regulations apply to all electrical installations in buildings, there are also some indoor and out-of-doors special installations (such as floor and central heating systems) and locations (such as swimming pools) that are subject to special requirements owing to the extra dangers they pose.
	Chapter 7 considers the requirements for these special locations and installations.
Chapter 8 External influences	The new edition of BS 7671 now includes far more details of the regulations concerning external influences and (i.e. in Appendix 5 to BS 7671:2008) provides a concise list of environmental influences. Chapter 8 of this editon of the book provides guidance on all forms of external influence. Also included in this chapter are extracts from the current Regulations that have an impact on the environment.
Chapter 9 Inspection and testing	To meet the requirements for electrical safety, it is essential for any electrician engaged in inspection, testing and certification of electrical installations to have a full working knowledge of the IEE Wiring Regulations.
	The electrician must also have above-average experience and knowledge of the type of installation under test in order to carry out any inspection and testing. Without this prerequisite, it could be quite dangerous.
	Chapter 9 provides a consolidated list of how electrical installations shall be inspected and tested as well as a brief insight into some of the test equipment that may be used.
Chapter 10 Installation, maintenance and repair	This final chapter of the book provides some guidance on the requirements for installation, maintenance and repair of electrical installations. It lists the Regulations' requirements for these activities with respect to electrical installations and (in an appendix) provides an example stage audit checklist for designers and engineers to use.

These Chapters are then supported by the following appendices:

Appendix A: Symbols used in electrical installations
Appendix B: List of electrical and electromechanical symbols
Appendix C: SI units for existing technology
Appendix D: Acronyms and abbreviations

Appendix E: British Standards currently used with the Wiring Regulations (by standard and by title)
Appendix F: List of useful contacts and further information
plus a full Index

It is hoped that the following symbols will help you get the most out of this book:

Need to be careful (e.g. very necessary requirement, a potential minefield or legal/statutory requirement).

A good idea or a useful reminder.

For your assistance, I have also highlighted all the really essential and/or mandatory requirements of a particular section as shown in the following example:

An RCD shall not be used in a TN-C system.	WR-411.4.4

For your convenience (and to save you having to look backwards and for-wards through the book for the correct Requirement) quite of a lot of these Requirements have been shown more than once (i.e. in different chapters and/or sections of the book – as have a few of the figures and tables).

Note: If any reader has any thoughts about the contents of this book (such as areas where perhaps they feel I have not given sufficiently coverage, omissions and/or mistakes etc.) then please let know by emailing me at ray@herne.org.uk and I will make suitable amendments in the next edition of this book.

Enjoy!

Ray Tricker

Acknowledgements

I would like to thank the Institution of Engineering and Technology (IET) for giving me permission to reproduce the following Tables and Figures:

Tables 3.1, 3.2, 3.3, 3.4, 4.1, 6.2, 6.6, 6.7, 6.11, 8.4, 8.5, 8.18, and 9.13, and Figures 5.2, 5.3, 6.6, 6.7 and 10.2 which are taken from The IEE Wiring Regulations: BS7671: 2001 incorporating Amendments 1 & 2: 2004 (The IEE, London, UK in agreement with BSI, 2004) ISBN 0863413730.

Figures 9.2, 9.3, 9.4 and 9.6 which are taken from The IEE On-Site Guide (BS 7671: 2001 16th edition Wiring Regulations including Amendments 1 & 2: 2004) (IEE Publications, 2004) ISBN 0863413749.

I would also like to thank the following organizations for providing me with assistance in the preparation of this book and for giving me permission to use copy-righted materials for illustration purposes in the following tables and figures:

BRE Certification Ltd for use of their logo in Figure 2.3
BSI for use of their logo in Figure 2.3, for giving permission to reproduce
   Figure 5.5 and Tables 5.3, 5.4, 5.5, 6.12, 8.7 and 9.6
CORGI Competent Persons Scheme for use of their logo in Figure 2.3
ELECSA Ltd for use of their logo in Figure 2.3
NAPIT Certification Ltd for use of their logo in Figure 2.3
NICEIC for use of their logo in Figure 2.3
OFTEC for use of their logo in Figure 2.3
TrustMark for use of their logo in Figure 2.4

Note: Please see Appendix F for full contact details for these organizations.

In addition I would like to give due recognition to the following tables and figures which are reprinted by kind permission from Elsevier.

Figure 1.3 is taken from Introduction to Health and Safety at Work, Second Edition, Hughes and Ferrett, 2005, ISBN 0750666234.

Tables 4.2 and 6.4, and Figures 3.13, 3.14, 3.15, 3.17, 4.1, 4.3, 4.8, 4.10 and the inside cover diagrams are taken from 17th edition IEE Wiring Regulations: explained and Illustrated, eighth edition, Scaddan, 2008, ISBN 9780750687201.

Figures 2.3, 2.5, 2.6, 2.8, 2.9, 2.10, 2.11, 9.16, 9.17, 9.19 and 9.21 from Building Regulations in Brief, fourth edition, Tricker, 2006, ISBN 075068058X.

Note: The BSI logo, Kitemark and the Kitemark symbol are produced with permission of the British Standards Institute and are the Registered Trademarks of such in the United Kingdom, and others apply in other countries around the world.

1

Introduction

1.1 Introduction

The IEE Wiring Regulations is a 389-page document that defines the way in which all electrical installation work must be carried out. It does not matter whether the work is carried out by a professional electrician or an unqualified DIY enthusiast, the installation **must** comply with the Wiring Regulations.

The current edition of the Regulations is BS 7671:2008: *Requirements for Electrical Installations, IEE Wiring Regulations (Seventeenth Edition)*, more commonly referred to as 'The Red Book' or 'the 17th edition'.

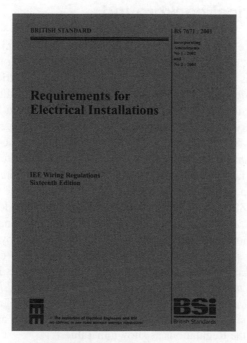

Figure 1.1 Front cover of BS 7671:2008

This British Standard is published with the full support of the BEC (i.e. the British Electrotechnical Committee – who are the UK national body responsible for formal standardisation within the electrotechnical sector) in partnership with the BSI (i.e. the British Standards Institution – who have ultimate responsibility for all British Standards produced within this sector) and The Institution of Engineering and Technology (IET) – who, with more than 135,000 members, are Europe's largest grouping of professional engineers involved in power engineering, communications, electronics, computing, software, control, informatics and manufacturing.

The technical authority for this standard is The National Committee for Electrical Installations (JPEL/64), which is a Joint IET/BSI Technical Committee responsible for all the work previously undertaken by the IEE Wiring Regulations Committee and BSI Technical Committee PEU64. Copyright is jointly held by BSI and the IET.

 Note: Please note that all references in this book to the 'Wiring Regulations' or the 'Regulation(s)', where not otherwise specifically identified, refer to BS 7671:2008, Requirements for Electrical Installations.

BS 7671:2008 was issued on 1 January 2008 and came into effect on 1 June 2008.

 All installations that were (or are) designed after 1 June 2008 **must** comply with this edition, as amended and expanded.

1.2 Historical background

The first public electricity supply in the UK was at Godalming in Surrey, in November 1881, and mainly provided street lighting. At that time, there were no existing rules and regulations available to control electrical installations and so the electricity company just dug up the roads and laid the cables in the gutters. This particular electricity supply was discontinued in 1884.

On 12 January 1882, the steam-powered Holborn Viaduct Power station opened and this facility supplied 110 V d.c. for both private consumption and street lighting. Once more, there was no one in authority to tell the electricity supplier how to lay the cables and their positioning was, therefore, dependent on the electrician responsible for that particular section of the work.

Later on in 1882, *The Electric Lighting Clauses Act* (modelled on the previous 1847 *Gas Act*) was passed by Parliament and this enabled the Board of Trade to authorise the supply of electricity in any area by a local authority, company or person and to grant powers to install this electrical supply (including breaking up the streets) through the use of the 1882 *Rules and Regulations for the prevention of Fire Risks Arising from Electric Lighting*. This document was the forerunner of today's IEE Wiring Regulations.

Historically, since 1882, there has been a succession of amendments and new editions of the Regulations as shown in Table 1.1.

By now this continual updating was seen as a bit of a problem, particularly to designers and installers who had to ensure that they were always working

Table 1.1 Succession of amendments and new editions of Wiring Regulations

1882	First edition	Entitled 'Rules and Regulations for the prevention of Fire Risks Arising from Electric Lighting'
1888	Second edition	
1897	Third edition	Entitled 'General Rules recommended for Wiring for the Supply of Electrical Energy'
1903	Fourth edition	Entitled 'Wiring Rules'
1907	Fifth edition	
1911	Sixth edition	
1916	Seventh edition	
1924	Eighth edition	Entitled 'Regulations for the Electrical Equipment of Buildings'
1927	Ninth edition	
1934	Tenth edition	
1939	Eleventh edition	Revised issue (1943), Reprinted with minor amendments (1945), Supplement issued (1946), Revised Section 8 (1948)
1950	Twelfth edition	Supplement issued (1954)
1955	Thirteenth edition	Reprinted 1958, 1961, 1962 and 1964
1966	Fourteenth edition	Reprinted 1968, 1969, 1970 (in metric units), 1972, 1973, 1974 and 1976

in compliance to the latest Regulations. With the publication of the fifteenth edition, therefore, it was decided that in future, reprints of the same edition would be contained in one of five different-coloured covers (i.e. red, green, yellow, blue and brown) and a new edition would be published when the brown-covered reprint required updating.

Table 1.2 BS 7671:2001 – Publication details

1981	Red cover	Fifteenth edition	Entitled 'Regulations for Electrical Installations'
1983	Green cover		Reprinted incorporating amendments
1984	Yellow cover		Reprinted incorporating amendments
1986	Blue cover		Reprinted incorporating amendments
1987	Brown cover		Reprinted incorporating amendments
1988	Brown cover		Reprinted with minor corrections
1991	Red cover	Sixteenth edition	Reprinted with minor corrections in 1992
			Reprinted as BS 7671 in 1992
			Amendment No 1 issued Dec 1994
1994	Green cover		Reprinted incorporating Amendment No 1
			Amendment No 2 issued Dec 1997
1997	Yellow cover		Reprinted incorporating Amendment No 2
			Amendment No 3 issued Apr 2000
2001	Blue cover		BS 7671:2001 issued Jun 2001 (see Note below)
			Amendment No 1 issued Feb 2002
			Amendment No 2 issued Mar 2004
2004	Brown cover		Reprinted incorporating Amendments No 1 and No 2
2008	Red cover	Seventeenth edition	New edition aligned with existing and new CENELEC, IEC and EN Harmonised Documents

Note: BS 7671:2008 includes some important changes that were required in order to maintain technical alignment with CENELEC harmonisation documents.

1.3 What does the Standard contain?

This Standard 'contains the rules for the design and erection of electrical installations so as to provide for safety and proper functioning for the intended use' and is based on the plan agreed internationally (i.e. through CENELEC) for the 'arrangement of safety rules for electrical installations'.

The structure of BS 7671:2008 is given in Table 1.3.

Table 1.3 BS 7671:2008 – Structure

Part 1	Sets out the scope, object and fundamental principles.
Part 2	Defines certain terms used throughout the Regulations.
Part 3	Identifies the characteristics of an installation that will need to be taken into account in choosing and applying the requirements of the subsequent Parts of the Regulations. These characteristics may vary from one part of an installation to another and need to be assessed for each location to be served by the installation.
Part 4	Describes the basic measures that are available for the protection of persons, property and livestock and against the hazards that may arise from the use of electricity.
Part 5	Describes the precautions that need to be taken in the selection and erection of the equipment of an installation.
Part 6	Covers inspection and testing.
Part 7	Identifies particular requirements for special installations or locations.

 Any intended departure from the requirements of Parts 1 to 6 requires special consideration by the installation designer and **must** be documented in the Electrical Installation Certificate specified in Part 6.

The seven parts of the Standard are then supported by the following Appendices (Table 1.4).

Table 1.4 BS 7671:2008 – Appendices

Appendix	Title	Description and remarks
1.	British and other Standards to which reference is made in the Regulations	Reproduced in the Reference section of this book.
2.	Statutory Regulations and associated memoranda	Details of all the Statutory Regulations, legislation and EU Harmonised Directives that electrical installations are required to comply with.

(*continued*)

Table 1.4 (*continued*)

Appendix	Title	Description and remarks
3.	Time/current characteristics of overcurrent protective devices and RCDs	Details of time/current characteristics for: • fuses • circuit breakers • RCDs.
4.	Current-carrying capacity and voltage drop for cables and flexible cords	Schedules of: • installation methods for conductors and cables (e.g. cleated, in conduits, on trays, in trenches) • cable specifications and current rating tables (e.g. armoured cables, mineral insulated cables, fire-resistant cables, screened cables) • correction factors (for groups of cables, mineral insulated cables, cables installed in trenches, ambient temperature where protection is against short circuits and overload) • copper conductors • aluminium conductors.
5.	Classification of external influences	Lists and schedules of external influences having an influence on electrical installations (for details see Table 1.5).
6.	Model forms for certification and reporting	Reproduced in Part 6 of this book.
7.	Harmonised cable core colours	Current details of cable core marking and colours that are to be used in all installations (for details see inside front and back cover of this book).
8.	Current-carrying capacity and voltage drop for busbar trunking and powertrack systems	Information concerning: • the basis of current-carrying capacity • rating factors for current-carrying capacity of busbar trunking systems • effective current-carrying capacity • protection against overload current • voltage drop.
9.	Definitions – multiple sources, d.c. and other systems	Examples of TN-C, TN-S, TN-C-S, TT and IT systems.
10.	Protection of conductors in parallel against overcurrent	Information concerning: • overload protection of conductors in parallel • short circuit protection of conductors in parallel.
11.	Effect of harmonic currents on balanced three-line systems	Information about rating factors for triple harmonics (examples and details of harmonic currents in line conductors)
12.	Voltage drop in consumers' installations	Information concerning the maximum allowable value of voltage drop.

(continued)

Table 1.4 (*continued*)

Appendix	Title	Description and remarks
13.	Methods for measuring the insulation resistance/ impedance of floors and walls to earth or to the protective conductor system	Test methods. Test electrodes.
14.	Measurement of earth fault loop impedance: consideration of the increase of the resistance of conductors with increase of temperature	Informative.
15.	Ring and radial final circuit arrangements	Information concerning Section 433.1.5.

Table 1.5 List of external influences relevant to electrical installations

Environment	Utilisation	Buildings
Altitude (metres)	Capability	Structure
Ambient temp. (°C)	Contact with earth	
Corrosion	Evacuation	
Electromagnetic	Materials	
Fauna	Resistance	
Flora		
Foreign bodies		
Impact		
Lightening		
Movement of air		
Other mechanical stresses		
Seismic		
Solar		
Temperature and humidity		
Vibration		
Water		
Wind		

1.3.1 What about the Standard's numbering system?

The numbering system used to identify specific requirements in BS 7671:2006 is as follows:

- The first digit signifies a Part.
- The second digit signifies a Chapter.
- The third digit signifies a Section.
- Subsequent digits signify the Regulation number.

> **Example**
>
> Section number **413** is made up as follows:
>
> - PART **4** – Protection for safety
> - Chapter **41** (first chapter of Part 4) – Protection against electric shock.
> - Section **413** (third section of Chapter 41) – Protective measure: electrical separation.

1.4 What are the objectives of the IEE Wiring Regulations?

Current legal requirements for employee competence in electrical work now call for everyone involved in certain electrical activities – for example, simply choosing the size of cable or fuse – to be aware of the regulative requirements associated with such work. BS 7671:2008 (i.e. The IEE Wiring Regulations) is the traditionally approved Code of Practice for those who are involved in (or supervise) electrical work such as electrical maintenance, control and/or instrumentation.

The stated intention of wiring safety codes is to 'provide technical, performance and material standards that will allow sufficient distribution of electrical energy and communication signals, at the same time protecting persons in the building from electric shock and preventing fire and explosion' (IET). In other words:

> **To ensure the protection of people and livestock from fire, shock or burns from any installation that complies with their requirements.**

The Regulations form the basis of safe working practice throughout the electrical industry.

1.5 What is the legal status of the IEE Wiring Regulations?

Although the IEE Wiring Regulations have always been held in high esteem throughout Europe, they had no legal status and did not require Continentals who were carrying out installation work in the UK to abide by them. This problem was overcome in October 1992 when the IEE Wiring Regulations became a British Standard, BS 7671 – thus providing them with national/international status.

Note: Although the Regulations are *non-statutory regulations* they may, however, be used as evidence in a court of law to claim compliance with a statutory requirement.

1.6 What do they cover?

As shown below, the IEE Wiring Regulations cover both electrical installations and electrical equipment.

1.6.1 Electrical installation

Definition

For the purpose of the Regulations:

Electrical installations (or *installation*) means *any assembly of associated electrical equipment supplied from a common origin to fulfil a specific purpose and having certain co-ordinated characteristics.*

The Regulations apply to the design, erection and verification of electrical installations such as those of:

- agricultural and horticultural premises;
- caravans, caravan parks and similar sites;
- commercial premises;
- construction sites, exhibitions, shows, fairgrounds and other installations for temporary purposes including professional stage and broadcast applications;
- external lighting and similar installations;
- industrial premises;
- marinas;
- mobile or transportable units;
- photovoltaic systems;
- prefabricated buildings;
- public premises;
- low-voltage generating sets;
- highway equipment and street furniture;
- residential premises.

 Note: 'Premises' covers the land and all facilities including buildings belonging to it.

The Regulations include requirements for:

- the addition to (or alteration of) installations and parts of existing installations affected by an addition or alteration;
- circuits (but not apparatus and/or equipment internal wiring) operating at voltages greater than 1000 V and derived from an installation having

a voltage not exceeding 1000 V a.c. (e.g. discharge lighting, electrostatic precipitators);

- circuits supplied at nominal voltages up to and including 1000 V a.c. or 1500 V d.c.;

  Note: although the preferred frequencies are 50 Hz, 60 Hz and 400 Hz, the use of other frequencies for special purposes is not excluded
- consumer installations external to buildings;
- fixed wiring for communication and information technology, signalling, command and control etc. (but not apparatus and/or equipment internal wiring);
- wiring systems and cables not specifically covered by the Standards for appliances.

Although the Regulations are intended as a Standard for electrical installations, in certain cases, they may need to be supplemented by the requirements and/or recommendations of other British Standards or by the requirements of the person ordering the work. Such cases could include (among others) the following:

- design and installation of temporary distribution systems delivering a.c. electrical supplies for lighting, technical services and other entertainment related purposes – BS 7909;
- electrical equipment for explosive gas atmospheres – BS EN 60079;
- electrical equipment for use in the presence of combustible dust – BS EN 50281 and BS EN 61241;
- electric signs and high-voltage luminous discharge tube installations – BS 559 and BS EN 50107;
- electric surface heating systems – BS 6351;
- electrical installations for open-cast mines and quarries – BS 6907;
- emergency lighting – BS 5266;
- fire detection and alarm systems in buildings – BS 5839;
- telecommunications systems – BS 6701.

The Regulations do **not** apply to the following installations:

- aircraft equipment;
- 'distributor's equipment' as defined in the Electricity Safety, Quality and Continuity Regulations 2002;
- electrical equipment of machines covered by BS EN 60204;
- equipment of mobile and fixed offshore installations;
- equipment on board ships;
- lightning protection systems for buildings and structures covered by BS EN 62305;
- motor vehicle equipment (except those to which the requirements of the Regulations concerning caravans are applicable);
- radio interference suppression equipment (except so far as it affects safety of the electrical installation);
- railway traction equipment, rolling stock and signalling equipment;

- those aspects of lift installations covered by relevant parts of BS 5655 and BS EN 81-1;
- those aspects of mines and quarries specifically covered by Statutory Regulations.

 Note: For installations in premises, which are subject to statutory control (e.g. via a licensing or other authority), the requirements of that authority will need to be confirmed and these requirements then complied with in the design and implementation of those installations.

1.6.2 Electrical equipment

> **Definition**
>
> For the purpose of these Regulations:
>
> *Electrical equipment* (or *Equipment*) means *any item used for generation, conversion, transmission, distribution or utilisation of electrical energy, such as machines, transformers, apparatus, measuring instruments, protective devices, wiring systems, accessories, appliances and luminaires.*

The Regulations are only applicable to the actual selection and application of items of electrical equipment **in** an electrical installation.

The Regulations do **not** deal with requirements for the construction of assemblies of electrical equipment, which are required to comply with the appropriate Standards.

1.7 What effect does using the Regulation have on other Statutory Instruments?

The requirements of the IEE Wiring Regulations also have an effect on the implementation of other Statutory Instruments such as:

- the Building Act 1984;
- the Disability Discrimination Act 1995;
- the Electricity at Work Regulations 1989;
- the Fire Precautions Act 1971;
- the Health and Safety at Work Act 1974.

1.7.1 What is the Building Act 1984?

The Building Act 1984 (as implemented by the Building Regulations 2000) is the enabling Act under which all Building Regulations have been made.

The Secretary of State (under the power given in the Building Act 1984) is required to:

- secure the health, safety, welfare and convenience of persons in or about buildings and of others who may be affected by buildings or matters connected with buildings;
- further the conservation of fuel and power;
- prevent waste, undue consumption, misuse or contamination of water

and may make regulations with respect to the design and construction of buildings and the provision of services, fittings and equipment in (or in) connection with) buildings.

 Note: The current regulations governing the Building Regulations 2000 are SI 2000/2531 (as amended) – a copy of which can be downloaded from: www.opsi.gov.uk/si/si2000/20002531.

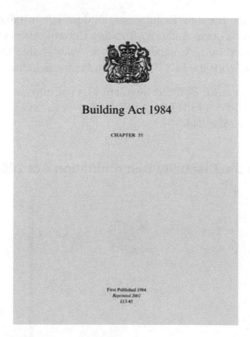

Figure 1.2 The Building Act 1984

For many years, the UK has managed to maintain a relatively high standard of electrical safety within buildings (domestic and non-domestic) based on voluntary controls centred around BS 7671. With the growing number of electrical accidents occurring in the 'home', the government has now been forced to implement a legal requirement for safety in all electrical installation work in dwellings.

 As from 1 January 2005, therefore, **all** new electrical wiring or electrical components for domestic premises (or small commercial premises linked to domestic accommodation) must be designed and installed in accordance with the Building Regulations, Part P (which is based on the fundamental principles set out in Chapter 13 of BS 7671:2008).

In addition, **all** fixed electrical installations (i.e. wiring and appliance fixed to the building fabric such as socket-outlets, switches, consumer units and ceiling fittings) **must** now be designed, installed, inspected, tested and certified to **BS 7671**.

Part P of the Building Regulations also introduces the requirement for the cable core colours of all a.c. power circuits to align with BS 7671.

 Note: Part P only applies to fixed electrical installations that are intended to operate at low-voltage or extra-low-voltage which are not controlled by the Electricity Supply Regulations 1988 as amended, or the Electricity at Work Regulations 1989 as amended.

Competent Persons Scheme

 Under Part P of the Building Regulations, all domestic installation work **must** now be inspected by Local Authority Building Control officers **unless** the work has been completed by a 'Competent Person' who is able to self-certify the work. The IEE supports the Part P Competent Person Scheme.

 For more details about the Building Regulations, visit: http://www.communities. gov.uk/index.asp?id = 1130474 or see *Building Regulations in Brief*, 5th edition (ISBN 978-0-7506-8444-6).

1.7.2 What is the Disability Discrimination Act 2005?

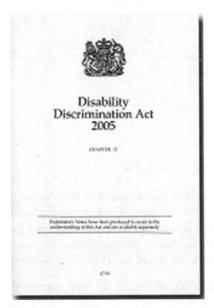

Figure 1.3 The Disability Discrimination Act 2005

The Disability Discrimination Act 1995 (DDA) is: *an Act to make it unlawful to discriminate against disabled persons in connection with employment, the provision of goods, facilities and services or the disposal or management of premises; to make provision about the employment of disabled persons; and to establish a National Disability Council.*

From the point of view of BS 7671:2001, the Disability Act 1995 as amended in 2005 makes it unlawful:

- for a trade organisation to discriminate against a disabled person;
- for a qualifications body to discriminate against a disabled person;
- for service providers to make it impossible or unreasonably difficult for disabled persons to make use of that service.

For more details about the DDA see: http://www.direct.gov.uk/en/Disabled People

1.7.3 What are the Electricity at Work Regulations 1989?

Figure 1.4 The Electricity at Work Regulations 1989

The Electricity at Work Regulations 1989 impose health and safety requirements with respect to electricity used at work. General duties are imposed to ensure that:

- all electrical systems have been properly constructed and maintained, and are used in such a way so as not to give rise to danger;

- responsibilities for safety fall with the Employer or Self-Employed Person who employs one or more individuals under a Contract of Employment;
- maintenance of fixed electrical installations and portable appliances is carried out and regular inspections are made to ensure their safety;
- persons responsible for buildings or electrical installations and appliances ensure that electrical test certificates confirming the installations and appliances have been tested are in place.

Note: The Electricity at Work Regulations 1989 also state that where an accident occurs and it is found that the systems are not covered by a valid test certificate, the Health & Safety Executive (HSE) takes a keen interest in prosecutions resulting from electrocution or death within the workplace. Reducing the risk of such an accident is a legal requirement.

Overall the Regulations require that:

- all electrical systems shall be constructed and maintained to prevent danger;
- all electrical equipment and installations are maintained in a safe condition;
- all people working with electricity are competent to do the job. Complicated tasks (i.e. equipment repairs, alterations, installation work and testing) may require a suitably qualified electrician;
- all staff are aware of your organisation's electrical safety arrangements;
- all work activities are to be carried out so as not to give rise to danger;
- equipment and procedures are safe and suitable for the working environment;
- equipment is switched off and/or unplugged before making adjustments. 'Live working' must be eliminated from work practices.

Electricity is recognised as a major hazard for not only can it kill (research has shown that the majority of electric shock fatalities occur at voltages up to 230 V), but it can cause fires and explosions. Even non-fatal shocks can cause severe and permanent injury. Most of the electrical risks can be controlled by using suitable equipment, following safe procedures when carrying out electrical work and/or ensuring that all electrical equipment and installations are properly maintained.

Additional precautions will also be required for harsh and particular conditions (i.e. wet surroundings, cramped spaces, work out of doors or near live parts of equipment). For this reason the Electricity at Work Regulations 1989 are used to impose health and safety requirements for electricity used at work.

While the majority of the Regulations concern hardware requirements, others are more generalised. For example:

- *Installations shall be of proper construction.*
- *Conductors shall be insulated.*
- *Means of cutting off the power and for electrical isolation shall be available.*

In brief, the Regulations concern the following:

Systems, work activities and protective equipment	All systems shall at all times be constructed to prevent, so far as is reasonably practicable, danger.	Regulation 4
Strength and capability of electrical equipment	No electrical equipment is to be used where its strength and capability may be exceeded so as to give rise to danger.	Regulation 5
Adverse or hazardous environments	Electrical equipment sited in adverse or hazardous environments must be suitable for those conditions.	Regulation 6
Insulation, protection and placing of conductors	Permanent safeguarding or suitable positioning of live conductors is required.	Regulation 7
Earthing and other suitable precautions	Equipment must be earthed or other suitable precautions must be taken (e.g. the use of residual current devices, double insulated equipment, reduced voltage equipment).	Regulation 8
Integrity of reference conductors	Nothing is to be placed in an earthed circuit conductor which might, without suitable precautions, give rise to danger by breaking the electrical continuity or by introducing a high impedance.	Regulation 9
Connections	All joints and connections in systems must be mechanically and electrical suitable for use.	Regulation 10
Means for protecting from excess current	Suitable protective devices should be installed in each system to ensure all parts of the system **and** users of the system are safeguarded from the effects of fault conditions.	Regulation 11

Note: Regulations 5 to 11 in effect, therefore, place a duty on the designer, installer and end user to ensure the suitability and protection of all electrical equipment.

Means of cutting off the supply and for isolation	Where necessary to prevent danger, suitable means shall be available for cutting off the electrical supply to any electrical equipment.	Regulation 12
Precautions for work on equipment made dead	Adequate precautions must be taken to prevent electrical equipment, which has been made dead in order to prevent danger, from becoming live – while any work is carried out.	Regulation 13
Work on or near live conductors	No work can be carried out on live electrical equipment unless this can be properly justified, which means that risk assessments are required. If such work is to be carried out, suitable precautions must be taken to prevent injury.	Regulation 14

(continued)

Working space, access and lighting	Adequate working space, adequate means of access and adequate lighting shall be provided at all electrical equipment on which or near which work is being done in circumstances that may give rise to danger.	Regulation 15
Competence to prevent danger and injury	No person shall engage in work that requires technical knowledge or experience to prevent danger or injury, unless he has that knowledge or experience, or is under appropriate supervision.	Regulation 16

For more information about the Electricity at Work Regulations 1989, contact:

Health and Safety Executive Local Authorities Enforcement Liaison Committee
(HELA)
www.hse.gov.uk
Tel: 020 7717 6441
Fax: 020 7717 6418
HSE Infoline – 0845 345 0055 (a 'one-stop' shop, providing rapid access to
expert advice and guidance)
e-mail: LAU.enquiries@hse.gsi.gov.uk
Or to download a copy of the Electricity at Work Regulations 1989 (Statutory
Instrument 1989 No. 635) go to:
www.opsi.gov.uk/si/si1989/Uksi_19890635_en_1.htm

1.7.4 What are the Fire Precautions (Workplace) Regulations 1997?

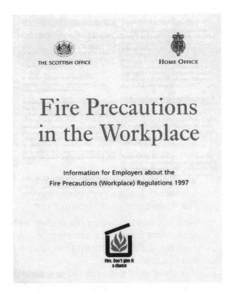

Figure 1.5 The Fire Precautions (Workplace) Regulations 1997

The Fire Precautions (Workplace) Regulations 1997 (as amended by the Fire Precautions (Workplace) (Amendment) Regulations 1999) stipulate that:

> *All offices, shops, railway premises and factories which have more than 20 persons employed in the building (or more than 10 persons employed anywhere other than on the ground floor) require a Fire Certificate.*
>
> *Any hotel or boarding house provided sleeping accommodation for more than six persons (guests or staff) or where this sleeping accommodation is above the first floor or below the ground floor, requires a Fire Certificate.*

When a Fire Certificate is issued the owner or occupier is required to provide and maintain:

- the means of escape;
- other means for ensuring that the means of escape can be safely and effectively used at all material times;
- means of fighting fire;
- means of providing warning in case of fire.

These requirements are reflected in the electrical installation.

 For further information about the Fire Precautions (Workplace) (Amendment) Regulations 1999 (Statutory Instrument 1999 No. 1877) visit www.opsi.gov.uk/si/si1999/19991877.htm or for a copy of the Act, use the following link: http://www.fire.org.uk/si/amd1840.htm

1.7.5 What is the Health and Safety at Work Act 1974?

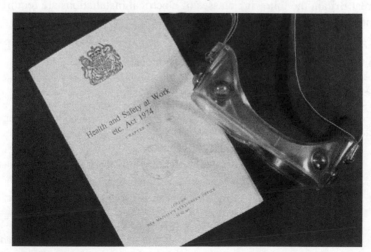

Figure 1.6 The Health and Safety at Work Act 1974

Any company with more than 5 employees is legally obliged to possess a comprehensive health and safety policy.

Over the years, the IEE Wiring Regulations have been regularly used by HSE in their guidance and installation notices, and installations which conform to BS 7671 (as amended) are regarded by HSE as likely to achieve conformity with the relevant parts of the Electricity at Work Regulations 1989. In certain instances where the Regulations have been used they may also be accompanied by Codes of Practice approved under Section 16 of the Health and Safety at Work Act 1974.

 Although some existing installations may have been designed and installed to conform to the Standards set by earlier editions of the Wiring Regulations, this does **not** necessarily mean that they will fail to achieve conformity with the relevant parts of the Electricity at Work Regulations 1989.

 For further information about the Health and Safety at Work Act 1974 visit www.hse.gov.uk/legislation/hswa.htm

1.8 How are the IEE Wiring Regulations implemented?

Although the IEE Wiring Regulations rely (primarily) on British Standards for their implementation (see Reference section for details) they do, however, include the policy decisions made in a number of Statutory Instruments and by the Council of European Communities in the relative EU Harmonised Directives.

1.8.1 Statutory Instruments

In Great Britain the following classes of electrical installations are required to comply with the Statutory Regulations.

Table 1.6 Statutory Instruments affecting Electrical Installations

Type of electrical installation	Statutory instrument
Building generally (subject to certain exemptions)	Building Regulations 2000 (as amended) (for England and Wales) ● SI 2000 No 2531 Building (Scotland) Regulations 2004 (as amended) ● Scottish SI 2004 No 406 Building Regulations (Northern Ireland) 2000 (as amended) ● Statutory Rule 2000 No 38

(continued)

Table 1.6 (*continued*)

Type of electrical installation	Statutory instrument
Cinematograph installations	Cinematograph (Safety) Regulations 1955 (as amended under the Cinematograph Act, 1909, and/or Cinematograph Act, 1952) • SI 1982 No 1856
Distributors' installations generally (subject to certain exemptions)	Electricity Safety, Quality and Continuity Regulations 2002 • SI 2002 No 2665 • SI 2006 No 1521
High-voltage luminous tube	Conditions of licence under: • in England and Wales – The Local Government (Miscellaneous provisions) Act 1982 • in Scotland – The Civic Government (Scotland) Act 1982
Machinery	The Supply of Machinery (Safety) Regulations 1992 as amended • SI 1992 No 3073 • SI 1994 No 2063
Theatres and other places licensed for public entertainment, music, dancing, etc.	Conditions of licence under: • In England and Wales – The Local Government (Miscellaneous provisions) Act 1982 • In Scotland – The Civic Government (Scotland) Act 1982
Work activity Places of work Non-domestic installations	The Electricity at Work Regulations 1989 as amended • SI 1989 No 635 • SI 1996 No 192 • SI 1997 No 1993 • SI 1999 No 2024 • The Electricity at Work Regulations (Northern Ireland) 1991 • Statutory Rule No 13

The full text of **all** Statutory Instruments that have been published since 1987 is now available from the Office of Public Sector Information (OPSI) via their website: www.hmso.gov.uk/stat.htm

With effect from July 1999, Statutory Instruments which have also been made by the National Assembly for Wales have been published via the Wales Legislation section: http://www.opsi.gov.uk/legislation/wales/w-stat.htm

The series of Scottish Statutory Instruments have been published via the Scottish Legislation: http://www.opsi.gov.uk/legislation/scotland/s-stat.htm section.

1.8.2 CENELEC Harmonised Documents

As well as British Standards, the Wiring Regulations also take account of the technical substance of agreements reached in CENELEC. In particular:

Listed by subject

Agricultural and horticultural premises	HD 60364-7-705:2007
Application of measures for protection against overcurrent	HD 384.4.473 AI:1980
Caravan parks, camping parks and similar locations	HD 384.7.708:2005
Conducting locations with restricted movement	HD 60364-7-706:2007
Construction and demolition site installations	HD 60364-7-704:2007
Earthing arrangements, protective conductors and protective bonding conductors	prHD 60364-5-54:2004
Electrical installations in caravans and motor caravans	prHD 60364-7-721:2007
Exhibitions, shows and stands	HD 384.7.711:2003
Extra-low-voltage lighting installations	HD 60364-7-715:2005
Fundamental principles, assessment of general characteristics and definitions	prHD 60364-:2007
Identification of cores in cables and flexible cords	HD 308 S2:2001
Initial verification	HD 384.6.61 S2:2003
Locations containing a bath or shower	HD 60364-7-701:2007
Marinas and similar locations	prHD 60364-7-709:2007
Mobile or transportable units	HD 60364-7-717:2004
Outdoor lighting installations	HD 384.7.714 51:2000
Protection against electric shock	HD 384.4.41 S2/AI:2002
Protection against fire where particular risks or danger exist	HD 384.4.482 51:1997
Protection against overcurrent	HD 384.4.43 S2:2001
Protection against overcurrent	HD 384.4.43 S2:2001
Protection against overvoltages	HD 384.4.443 S1:2000
Protection against thermal effects	HD 384.4.42 S1 A2:1994
Rooms and cabins containing sauna heaters	HD 384.7.703:2005
Selection and erection of equipment – common rules	prHD 60364-5-51:2003
Solar photovoltaic (PV) power supply systems	HD 60364-7-712:2005
Swimming pools and other basins	HD 384.7.702 S2:2002
Temporary electrical installations for structures, amusement devices and booths at fairgrounds, amusement parks and circuses	prHD 60364-7-740:2006

Listed by Directive

prHD 60364-:2007	Fundamental principles, assessment of general characteristics and definitions
HD 308 S2:2001	Identification of cores in cables and flexible cords
HD 384.4.41 S2/AI:2002	Protection against electric shock
HD 384.4.42 S1 A2:1994	Protection against thermal effects
HD 384.4.482 51:1997	Protection against fire where particular risks or danger exist
HD 384.4.43 S2:2001	Protection against overcurrent

(*continued*)

HD 384.4.473 AI:1980	Application of measures for protection against overcurrent
HD 384.4.443 S1:2000	Protection against overvoltages
prHD 60364-5-51:2003	Selection and erection of equipment – common rules
HD 384.4.43 S2:2001	Protection against overcurrent
prHD 60364-5-54:2004	Earthing arrangements, protective conductors and protective bonding conductors
HD 384.7.714 51:2000	Outdoor lighting installations
HD 60364-7-715:2005	Extra-low-voltage lighting installations
HD 384.6.61 S2:2003	Initial verification
HD 60364-7-701:2007	Locations containing a bath or shower
HD 384.7.702 S2:2002	Swimming pools and other basins
HD 384.7.703:2005	Rooms and cabins containing sauna heaters
HD 60364-7-704:2007	Construction and demolition site installations
HD 60364-7-705:2007	Agricultural and horticultural premises
HD 60364-7-706:2007	Conducting locations with restricted movement
HD 384.7.708:2005	Caravan parks, camping parks and similar locations
prHD 60364-7-709:2007	Marinas and similar locations
HD 384.7.711:2003	Exhibitions, shows and stands
HD 60364-7-712:2005	Solar photovoltaic (PV) power supply systems
HD 60364-7-717:2004	Mobile or transportable units
prHD 60364-7-721:2007	Electrical installations in caravans and motor caravans
prHD 60364-7-740:2006	Temporary electrical installations for structures, amusement devices and booths at fairgrounds, amusement parks and circuses

 BS 7671 will continue to be amended from time to time to take account of the publication of new or amended CENELEC standards.

2

Domestic buildings

"All electrical installations must be accommodated in ways that meet the requirement of the Building Regulations"

Building Regulations
Approved Document P (3.1)

Figure 2.1 Mandatory requirements for domestic buildings

2.1 The Building Act 1984

By Act of Parliament, the Secretary of State is responsible for ensuring that the health, welfare and convenience of persons living in or working in (or nearby) buildings are secured. This Act is called the Building Act 1984 and one of its prime purposes is to assist in the conservation of fuel and power, and to prevent waste, undue consumption, and the misuse and contamination of water.

It imposes on owners and occupiers of buildings a set of requirements concerning the design and construction of buildings and the provision of services, fittings and equipment used in (or in connection with) buildings.

2.2 The Building Regulations

The current legislation in England and Wales is the Building Regulations 2000 (Statutory Instrument No 2531), which is made by the Secretary of State for the Environment under powers delegated by Parliament under the Building Act 1984.

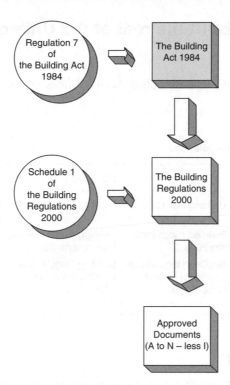

Figure 2.2 Implementing the Building Act

They are a set of minimum requirements and basic performance standards designed to secure the health, safety and welfare of people in and around buildings and to conserve fuel and energy in England and Wales.

2.3 Approved Documents

The Building Regulations are supported by a series of separate documents which correspond to the different areas covered by the Regulations. These are called 'Approved Documents' and they contain practical and technical guidance on ways in which the requirements of Schedule 1 and Regulation 7 of the Building Act 1984 can be met.

Each Approved Document reproduces the actual *requirements* contained in the Building Regulations relevant to the subject area. This is then followed by *practical and technical guidance* (together with examples) showing how the requirements can be met in some of the more common building situations. There may, however, be alternative ways of complying with the requirements to those shown in the Approved Documents and you are, therefore, under no obligation to adopt any particular solution in an Approved Document if you prefer to meet the requirement(s) in some other way.

2.4 What about the rest of the United Kingdom?

As shown in Table 2.1 the Building Act 1984 does not apply to Scotland or Northern Ireland.

Table 2.1 Building Regulations

	Act	Regulations	Implementation
England & Wales	Building Act 1984	Building Regulations 2000	Approved Documents
Scotland	Building (Scotland) Act 2003	Building (Scotland) Regulations 2004	Technical Handbooks
Northern Ireland	Building Regulations (Northern Ireland) Order 1979	Building Regulations (Northern Ireland) 2000	'Deemed to Satisfy' by meeting supporting publications

2.4.1 Scotland

Within Scotland, the requirements for buildings are controlled by the *Building (Scotland) Act 2003*, and the *Building (Scotland) Regulations 2004* then set the functional standards under this Act. The methods for implementing these requirements are similar to those for England & Wales, except that the guidance documents (i.e. for achieving compliance) are contained in two *Technical Handbooks*, one for domestic work, and one for non-domestic. Each handbook has a general section, which is then followed by 6 technical sections.

The main procedural difference between the Scottish system and the others is that a building warrant is **still** required before work can start in Scotland and certain facilities in dwellings are still required in Section 3 of the domestic technical handbook.

2.4.2 Northern Ireland

On the other hand, *Building Regulations (Northern Ireland) Order 1979* (as amended by the *Planning and Building Regulations (Amendment) (NI) Order 1990*) is the main legislation for Northern Ireland and the *Building Regulations (Northern Ireland) 2000* then details the requirements for meeting this legislation.

Supporting publications (such as British Standards, BRE publications and/ or technical booklets published by the Department) are then used to ensure that the requirements are implemented (i.e. deemed to satisfy).

Table 2.2 Legislative cross-reference

England & Wales		Scotland		Northern Ireland	
Part A	Structure	Section 1	Structure	Technical Booklet D	Structure
Part B	Fire safety	Section 2	Fire	Technical Booklet E	Fire safety
Part C	Site preparation and resistance to contaminants and water	Section 3	Environment	Technical Booklet C	Preparation of site and resistance to moisture
Part D	Toxic substances	Section 3	Environment	Technical Booklet B	Materials and workmanship
Part E	Resistance to the passage of sound	Section 5	Noise	Technical Booklet G	Sound insulation of dwellings
Part F	Ventilation	Section 3	Environment	Technical Booklet K	Ventilation
Part G	Hygiene	Section 3	Environment	Technical Booklet P	Sanitary appliances and unvented hot water storage systems
Part H	Drainage and waste disposal	Section 3	Environment	Technical Booklet J	Solid waste in buildings
				Technical Booklet N	Drainage
Part J	Combustion appliances and fuel storage systems	Section 3	Environment	Technical Booklet L	Heat-producing appliances and liquefied petroleum gas installations
		Section 4	Safety		
Part K	Protection from falling, collision and impact	Section 4	Safety	Technical Booklet H	Stairs, ramps and protection form impact
Part L	Conservation of fuel and power	Section 6	Energy	Technical Booklet F	Conservation of fuel and power
Part M	Access and facilities for disabled people	Section 4	Safety	Technical Booklet R	Access for facilities and disabled people
Part N	Glazing	Section 6	Energy	Technical Booklet V	Glazing
Part P	Electrical safety	Section 4	Safety		

2.5 Electrical Safety

For many years, the UK has managed to maintain relatively high electrical safety standards with the support of voluntary controls based on BS 7671, but with a growing number of electrical accidents occurring in the 'home', the government has been forced to consider the legal requirement for safety in electrical installation work in dwellings.

As from 1 January 2005, therefore, all new electrical wiring or electrical components for domestic premises (or small commercial premises linked to domestic accommodation) have had to be designed and installed in accordance with the Building Regulations. Part P, which is based on the fundamental principles set out in Chapter 13 of BS 7671:2008 (i.e. *the IEE Wiring Regulations*). In addition, all fixed electrical installations (i.e. wiring and appliances fixed to the building fabric such as socket outlets, switches, consumer units and ceiling fittings) have to be designed, installed, inspected, tested and certified to BS 7671.

 Part P also introduced new requirements for cable core colours for a.c. power circuits and with effect from 31 March 2006, **all** new installations or alterations to existing installations **must** use the new (harmonised) colour cables. (Further information concerning cable identification colours for extra-low-voltage and d.c. power circuits is available from the IEE website at www.iee.org/cablecolours.)

Table 2.3 Identification of conductors in a.c. power and lighting circuits

Conductor	Colour
Protective conductor	Green-and-yellow
Neutral	Blue
Phase of single-line circuit	Brown
Phase 1 of 3-line circuit	Brown
Phase 2 of 3-line circuit	Black
Phase 3 of 3-line circuit	Grey

 For single-line installations in domestic premises, the new colours are the same as those for flexible cables to appliances (namely green-and-yellow, blue and brown for the protective, neutral and line

 Note: Part P only applies to fixed electrical installations that are intended to operate at low voltage or extra-low voltage which are **not** controlled by the Electricity Safety, Quality and Continuity Regulations 2002 (as amended) or the Electricity at Work Regulations 1989 (as amended).

2.5.1 What is the aim of Approved Document P?

The aim of Part P is to increase the safety of householders by improving the design, installation, inspection and testing of electrical installations in dwellings when they (i.e. the installations) are being newly built, extended or altered.

 The government is currently introducing a scheme whereby domestic installations shall be checked at regular intervals (as well as when they are sold and/or

purchased) to make sure that they comply. This will mean, of course, that if you had an installation which was not correctly certified, then your house insurance might well **not** be valid!

2.5.2 Who is responsible for electrical safety?

Basically, there are three people who are responsible for the electrical safety of (and within) buildings. These are:

- **The owner** – needs to determine whether the work carried out is either minor or notifiable. If the work is notifiable, then the owner needs to make sure that the person(s) carrying out the work is either registered under one of the self-certified schemes (see Figure 2.3) or is able to certify their work under the local authority Building Control Approval route.
- **The designer** – needs to ensure that all electrical work is designed, constructed, inspected and tested in accordance with BS 7671 (current issue) and either falls under a Competent Persons Scheme or the local authority Building Control Approval route.
- **The builder/developer** – needs to ensure that they have electricians who can self-certify their work or who are qualified/experienced enough to enable them to sign off under the Electrical Installation Certification form.

2.5.3 What are the statutory requirements?

All electrical installations need to:

- be designed and installed to protect against mechanical and thermal damage;
- be designed and installed so that they will **not** present an electrical shock and/or fire hazard;
- be tested and inspected to meet relevant equipment/installation standards;
- provide sufficient information so that persons wishing to operate, maintain or alter an electrical installation can do so with reasonable safety;
- comply with such requirements placed by the Building Regulations.

2.5.3.1 What does all this mean?

With a few exceptions, **any** electrical work undertaken in a home which includes the addition of a new electrical circuit, or involves work in the:

- kitchen
- bathroom
- garden area

must be reported to the local authority Building Control for inspection. This includes any work undertaken professionally, or by you or another family member or by a friend.

The **ONLY** exception is when the installer has been approved by a Competent Persons organisation such as ELECSA (see Figure 2.3).

Authorised competent persons self-certification schemes for installers who can do all electrical installation work	Authorised competent persons self-certification schemes for installers who can do electrical work only if it is necessary when they are carrying out other work
	BRE Certification Ltd www. brecertification. co.uk Phone: 01923 664100		**CORGI Competent Persons Scheme** www. corgi-group.. com Phone: 0870 401 2300
BSI	**British Standards Institution** www. bsi-global.com Phone: 01442 230442	ELECSA	**ELECSA Limited** www.elecsa. org.uk Phone: 0870 749 0080
ELECSA	**ELECSA Limited** www.elecsa. org.uk Phone: 0870 749 0080	**NAPIT**	**NAPIT Certification Limited** www.napit. org.uk Phone: 0870 444 1392
NAPIT	**NAPIT Certification Limited** www.napit. org.uk Phone: 0870 444 1392		**NICEIC Certification Services Ltd** www.niceic. org.uk Phone: 01582 531000
	NICEIC Certification Services Ltd www.niceic. org.uk Phone: 01582 531000	**Registration Services**	**OFTEC** www.oftec.org Phone: 0845 658 5080

Figure 2.3 Authorised competent persons self-certification schemes for installers

2.5.4 What types of building does Approved Document P cover?

Part P applies to **all** electrical installations in (**and around**) buildings or parts of buildings comprising:

- dwelling houses and flats;
- dwellings and business premises that have a common supply;
- land associated with domestic buildings;
- fixed lighting and pond pumps in gardens;
- shops and public houses with a flat above;
- common access areas in blocks of flats such as corridors and stairways;
- shared amenities of blocks of flats such as laundries and gymnasiums.

Table 2.4 provides the details of works that are notifiable to local authority and/or must be completed by a company registered as a 'Competent Firm'.

Table 2.4 Notifiable work

Locations where work is being completed	Extensions and modifications to circuits	New circuits
Bathrooms	Yes	Yes
Bedrooms	Yes	Yes
Ceiling (overhead) heating	Yes	Yes
Communal area of flats	Yes	Yes
Computer cabling		
Conservatories		Yes
Dining rooms		Yes
Garages (integral)		Yes
Garages (detached)		Yes*
Garden – lighting	Yes	Yes
Garden – power	Yes	Yes
Greenhouses	Yes	Yes
Halls		Yes
Hot air saunas	Yes	Yes
Kitchen	Yes	Yes
Kitchen diners	Yes	Yes
Landings		Yes
Lounge		Yes
Paddling pools	Yes	Yes
Remote buildings	Yes	Yes
Sheds	Yes	Yes
Shower rooms	Yes	Yes
Small-scale generators	Yes	Yes
Solar power systems	Yes	Yes
Stairways		Yes
Studies		Yes
Swimming pools	Yes	Yes
Telephone cabling		Yes
TV Rooms		Yes
Underfloor heating	Yes	Yes
Workshops (remote)	Yes	Yes

*if the installation requires outdoor wiring

2.5.5 What is a competent firm?

For the purposes of Part P, the government has defined 'Competent Firms' as electrical contractors:

- who work in conformance with the requirements to BS 7671;
- whose standard of electrical work has been assessed by a third party;
- who are registered under the NICEIC Approved Contractor scheme and the Electrotechnical Assessment Scheme.

2.5.6 What is a competent person responsible for?

When a competent person undertakes installation work, that person is responsible for:

- ensuring compliance with BS 7671: 2001 and all relevant Building Regulations;
- providing the person ordering the work with a signed Building Regulations self-certification certificate;
- providing the relevant Building Control Body with an information copy of the certificate;
- providing the person ordering the work with a completed Electrical Installation Certificate.

2.5.7 Who is entitled to self-certify an installation?

Part P affects **every** electrical contractor carrying out fixed installation and/or alteration work in homes. **Only** registered installers are entitled to self-certify the electrical work, however, and they **must** be registered as a competent person under one of the schemes shown in Figure 2.3.

Working with industry and consumer organisations, the government has developed the TrustMark initiative for builders and specialist firms that work on (and in) the home. Schemes that are capable of delivering 'agreed competence and customer care standards' are approved to use the TrustMark brand by a Board consisting of industry and consumer representatives, with government observers. The brand is owned by the DTI, which licences the Board.

www.trustmark.org.uk

Figure 2.4 The TrustMark Initiative (logo courtesy of TrustMark)

The TrustMark replaces the Quality Mark scheme, which closed on 31 December 2004 because too few firms joined. For more information about TrustMark, see their website at www.trustmark.org.uk.

2.5.8 When do I have to inform the local authority Building Control Body?

All proposals to carry out electrical installation work **must** be notified to the local authority's Building Control Body before work begins, **unless** the proposed installation work is undertaken by a person who is a competent person registered under a government-approved Part P Self-Certification Scheme or the work is agreed non-notifiable work, such as:

- connecting an electric gate or garage door to an existing isolator (but be careful; the installation of the circuit up to the isolator is notifiable!);
- fitting and replacing cookers and electric showers (unless a new circuit is required);
- installing equipment (e.g. security lighting, air-conditioning equipment and radon fans) that is attached to the outside wall of a house (unless there are exposed outdoor connections and/or the installation is a new circuit, or an extension of a circuit in a kitchen, or special location, or is associated with a special installation);
- installing fixed equipment where the final connection is via a 13 A plug and socket (unless it involves fixed wiring and the installation of a new circuit or the extension of a circuit in a kitchen or special location);
- installing prefabricated, 'modular' systems such as kitchen lighting systems and armoured garden cabling that are linked by plug and socket connectors (provided that products are CE-marked and that any final connections in kitchens and special locations are made to existing connection units or points, e.g. a 13 A socket outlet);
- installing or upgrading main or supplementary equipotential bonding (provided that the work complies with other applicable legislation, such as the Gas Safety (Installation and Use) Regulations);
- installing mechanical protection to existing fixed installations (provided that the circuit's protective measures and current-carrying capacity of conductors are unaffected by increased thermal insulation);
- re-fixing or replacing the enclosures of existing installation components;
- replacement, repair and maintenance jobs;
- replacing fixed electrical equipment (e.g. socket outlets, control switches and ceiling roses) which do not require the provision of any new fixed cabling;
- replacing the cable of a single circuit cable (where damaged, for example, by fire, rodent or impact – provided that the replacement cable has the same current-carrying capacity, follows the same route and does not serve more than one sub-circuit through a distribution board);

- work that is not in a kitchen or special location, which does not involve a special installation and which only consists of:
  - adding lighting points (light fittings and switches) to an existing circuit
  - adding socket outlets and fused spurs to an existing ring or radial circuit (provided that the existing circuit protective device is suitable and supplies adequate protection for the modified circuit)
- work that is not in a special location and only concerns:
  - adding a telephone, extra-low-voltage wiring and equipment for communications, information technology, signalling, command, control and other similar purposes
  - adding prefabricated equipment sets (and their associated flexible leads) with integral plug and socket connections.

All of this work can be completed by a DIY enthusiast (family member or friend) but still needs to be carried out in accordance with manufacturers' instructions and done in such a way that it does not present a safety hazard. This work does not need to be notified to a local authority Building Control Body (unless it involves installation in an area of high risk such as a kitchen or a bathroom etc.) but all DIY electrical work (unless completed by a qualified professional, who is responsible for issuing a Minor Electrical Installation Certificate) will still need to be checked, certified and tested by a competent electrician.

Any work that involves adding a new circuit to a dwelling will need to be either notified to the Building Control Body (who will then inspect the work) or carried out by a competent person who is registered under a government-approved Part P Self-Certification Scheme.

Work involving any of the following will also have to be notified:

- consumer unit replacements;
- electric floor or ceiling heating systems;
- extra-low-voltage lighting installations (other than pre-assembled, CE-marked lighting sets);
- garden lighting and/or power installations;
- installation of a socket outlet on an external wall;
- installation of outdoor lighting and/or power installations in the garden or that involves crossing the garden;
- installation of new central heating control wiring;
- solar photovoltaic (PV) power supply systems;
- small-scale generators (such as microCHP units).

 Note: Where a person who is **not** registered to self-certify intends to carry out the electrical installation, then a Building Regulation (i.e. a Building Notice or Full Plans) application will need to be submitted together with the appropriate fee, based on the estimated cost of the electrical installation. The Building Control Body will then arrange to have the electrical installation inspected at first-fix stage and tested upon completion.

In any event, the electrical work will still need to be certified under BS 7671 by a suitably competent person who will be responsible for the design, installation, inspection and testing of the system (on completion) and have the confidence of completing a certificate to say that the work is satisfactory and complies with current codes of practice.

The main things to remember are as follows:

- Is the work notifiable or non-notifiable?
- Does the person undertaking the work need to be registered as a competent person?
- What records (if any) need to be kept of the installation?

2.5.9 What if the work is completed by a friend, a relative or me?

You do **not** need to tell your local authority's Building Control Department about non-notifiable work such as:

- repairs, replacements and maintenance work;
- extra power points or lighting points or other alterations to existing circuits (unless they are in a kitchen or bathroom, or are outdoors).

You **do**, however, need to tell them about most other work.

 If you are not sure about this, or you have any questions, ask the local authority's Building Control Department.

2.5.10 What if the work is completed by a contractor or an installer?

If the work is of a notifiable nature then the installer(s) must be registered with one of the schemes shown in Figure 2.3.

Figure 2.5 provides a quick guide to the requirements.

2.6 What inspections and tests will have to be completed and recorded?

As shown in Table 2.5, there are four types of electrical installation certificates and one Building Regulation compliance Certificate that have to be completed.

Copies of these various Certificates and forms are contained in Chapter 6.

2.6.1 What should be included in the records of the installation?

All 'original' Certificates should be retained in a safe place and be shown to any person inspecting or undertaking further work on the electrical installation

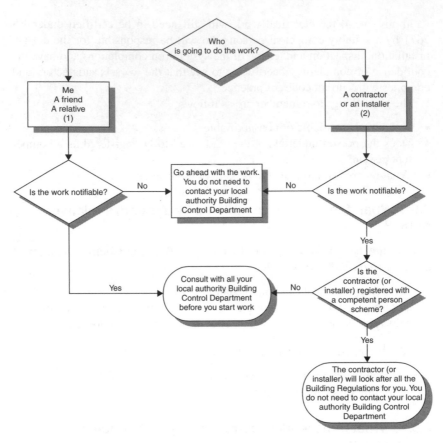

Figure 2.5 How to meet the new rules

in the future. If you later vacate the property, this Certificate will demonstrate to the new owner that the electrical installation complied with the requirements of BS 7671 at the time the Certificate was issued. The Construction (Design and Management) Regulations require that for a project covered by those Regulations, a copy of this Certificate, together with schedules, is included in the project health and safety documentation.

Figure 2.6 indicates how to choose what type of inspection is required.

2.6.2 Where can I get more information about the requirements of Part P?

Further guidance concerning the requirements of Part P (Electrical safety) is available from:

- the IET (Institution of Engineeering Technology) at www.iee.org/Publish
- the NICEIC (National Inspection Council for Electrical Installation Contracting) at www.niceic.org.uk

Table 2.5 Types of installation

Type of inspection	When is it used?	What should it contain?	Remarks
Minor Electrical Installation Works Certificate	For a new electrical installation or for new work associated with an alteration or addition to an existing installation	Relevant provisions of Part 6 of BS 7671	An example of a Minor Electrical Installation could (for example) be the addition of a lighting point to an existing circuit
Electrical Installation Certificate (short form)	For use when one person is responsible for the design, construction, inspection and testing of an installation	A schedule of inspections and a schedule of test results as required by Part 6 (of BS 7671)	For safety reasons, the electrical installation will need to be inspected at appropriate intervals by a competent person
Full Electrical Installation	For the design, construction, inspection and testing of an installation	A schedule of inspections and test results as required by Part 6 (of BS 7671). A Certificate, including guidance for recipients (standard form from Appendix 6 of BS 7671)	An Electrical Installation Certificate is not to be used for a periodic inspection
Periodic Inspection Report	For the inspection of an existing electrical installation	A schedule of inspections and a schedule of test results as required by Part 6 (of BS 7671)	For safety reasons, the electrical installation will need to be inspected at appropriate intervals by a competent person
Building Regulations Compliance Certificate	Confirmation that the work carried out complies with the Building Regulations	The basic details of the installation, the location, the completion date and the name of the installer	A purchaser's solicitor may request this document when you come to sell your property. Looking further ahead, it may be required as one of the documents that will make up your 'Home Information Pack'

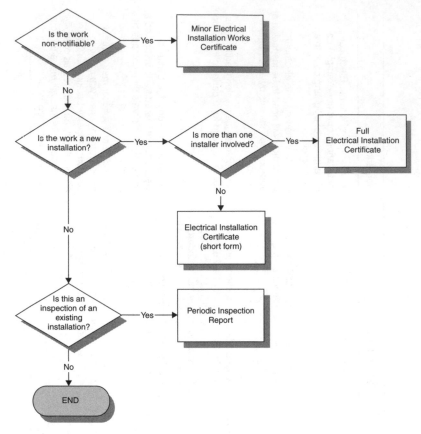

Figure 2.6 Choosing the correct Inspection Certificate

- the ECA (Electrical Contractors' Association) at www.niceic.org.uk or www.eca.co.uk.

For details of fixed wire colour changes, see http://www.niceic.org.uk

2.7 Requirements from the Approved Documents

Although:

- Part E (Resistance to the passage of sound);
- Part J (Combustion appliances and fuel storage systems); and
- Part K (Protection from falling, collision and impact)

have a number of requirements concerning electrical safety and electrical installations (see below for details), the main requirements are contained in:

- Part P (Electrical safety) together with:
  - Part M (Access and use of buildings)
  - Part L (Conservation of fuel and power)
  - Part B (Fire safety).

Figure 2.7 Building Regulations

2.7.1 Part P – Electrical safety

Reasonable provision shall be made in the design, installation, inspection and testing of electrical installations in order to protect persons from fire or injury.

(Approved Document P1)

Sufficient information shall be provided so that persons wishing to operate, maintain or alter an electrical installation can do so with reasonable safety.

(Approved Document P2)

 Note: While Part P makes requirements for the safety of fixed electrical installations, this does not cover system functionality (such as electrically powered fire alarm systems, fans and pumps) which are covered by other Parts of the Building Regulations and other legislation.

2.7.2 Part M – Access and facilities for disabled people

In addition to the requirements of the Disability Discrimination Act 1995 precautions need to be taken to ensure that:

- *new non-domestic buildings and/or dwellings (e.g. houses and flats used for student living accommodation etc.)*
- *extensions to existing non-domestic buildings*
- *non-domestic buildings that have been subject to a material change of use (e.g. so that they become a hotel, boarding house, institution, public building or shop)*

are capable of allowing people, regardless of their disability, age or gender to:

- *gain access to buildings*
- *gain access within buildings*

- *use sanitary conveniences in the principal storey of any new dwelling*
- *be able to use the facilities of the buildings (both as visitors and as people who live or work in them).*

(Approved Document M)

2.7.3 Part L – Conservation of fuel and power

Energy efficiency measures shall be provided which:

- *provide lighting systems that utilise energy-efficient lamps with manual switching controls or, in the case of external lighting fixed to the building, automatic switching, or both manual and automatic switching controls as appropriate, such that the lighting systems can be operated effectively as regards the conservation of fuel and power*
- *provide information, in a suitably concise and understandable form (including results of performance tests carried out during the works) that shows building occupiers how the heating and hot water services can be operated and maintained.*

(Approved Document L1)

 Responsibility for achieving compliance with the requirements of Part L rests with the person carrying out the work. That 'person' may be, for example, a developer, a main (or sub) contractor, or a specialist firm directly engaged by a private client.

 Note: The person responsible for achieving compliance should either themselves provide a Certificate, or obtain a Certificate from the sub-contractor, that commissioning has been successfully carried out. The Certificate should be made available to the client and the building control body.

2.7.4 Part B – Fire safety

The building shall be designed and constructed so that there are appropriate provisions for the early warning of fire, and appropriate means of escape in case of fire from the building to a place of safety outside the building capable of being safely and effectively used at all material times.

(Approved Document B1)

 Requirement B1 does not apply to any prison provided under Section 33 of the Prison Act 1952 (power to provide prisons, etc.).
   To inhibit the spread of fire within the building, the internal linings shall:

- *adequately resist the spread of flame over their surfaces; and*
- *have, if ignited, a rate of heat release or a rate of fire growth which is reasonable in the circumstances.*

(Approved Document B2)

'Internal linings' means the materials or products used in lining any partition, wall, ceiling or other internal structure.

1. *The building shall be designed and constructed so that, in the event of fire, its stability will be maintained for a reasonable period.*
2. *A wall common to two or more buildings shall be designed and constructed so that it adequately resists the spread of fire between those buildings. For the purposes of this sub-paragraph a house in a terrace and a semi-detached house are each to be treated as a separate building.*
3. *Where reasonably necessary to inhibit the spread of fire within the building, measures shall be taken, to an extent appropriate to the size and intended use of the building, comprising either or both of the following:*
   - *sub-division of the building with fire-resisting construction*
   - *installation of suitable automatic fire suppression systems.*
4. *The building shall be designed and constructed so that the unseen spread of fire and smoke within concealed spaces in its structure and fabric is inhibited.*

(Approved Document B3)

Requirement B3(3) does not apply to material alterations to any prison provided under Section 33 of the Prison Act 1952.

2.7.5 Design

Electrical installations must be designed and installed (suitably enclosed and appropriately separated) so that they:

• are safe to use, maintain and alter	P1.7
• comply with the requirements of BS 7671	P1.4
• comply with Part P (and any other relevant parts) of the Building Regulations	P1.7 and 3.1
• comply with the relevant equipment and installation standards	P0.1b
• do not present an electric shock or fire hazard to people	P0.1a
• provide adequate protection against mechanical and thermal damage	P0.1a
• provide adequate protection for persons against the risks of electric shock, burn or fire injuries.	P0.1a

Note: See Appendix A of Part P to the Building Regulations for details of the types of electrical service normally found in dwellings, some of the ways that they can be connected and the complexity of wiring and protective systems that can be used to supply them.

2.7.6 Extensions, material alterations and material changes of use

In accordance with Regulation 4(2) the whole of the existing installation does not have to be upgraded to current standards, but only to the extent necessary for the new work to meet the current standards except where upgrading is required by the energy efficiency requirements of the Building Regulations.

Where any electrical installation work is classified as an extension, a material alteration or a material change of use, the work must consider and include:

• confirmation that the mains supply equipment is suitable (and can) carry the additional loads envisaged	P2.1b–P2.2
• the amount of additions and alterations that will be required to the existing fixed electrical installation in the building	P2.1a
• the earthing and bonding systems required – that they are satisfactory and meet the requirements	P2.1a–P2.2c
• the necessary additions and alterations to the circuits which feed them	P2.1a
• the rating and the condition of existing equipment (belonging to both the consumer and the electricity distributor) – that they are sufficient	P2.2a
• the protective measures required to meet the requirements.	P2.1a – P2.2b

See Figure 2.8 for details of some of the types of electrical services normally found in dwellings, some of the ways they can be connected and the complexity of wiring and protective systems that can be used to supply them.

Note: Appendix C to Part P of the Building Regulations offers guidance on some of the older types of installation that might be encountered during alteration work while Appendix D provides guidance on the application of the now harmonised European cable identification system.

2.7.7 Electricity distributors' responsibilities

The Electricity Distributor is responsible for:

• ensuring that electricity is mechanically protected and can be safely maintained	P1.5

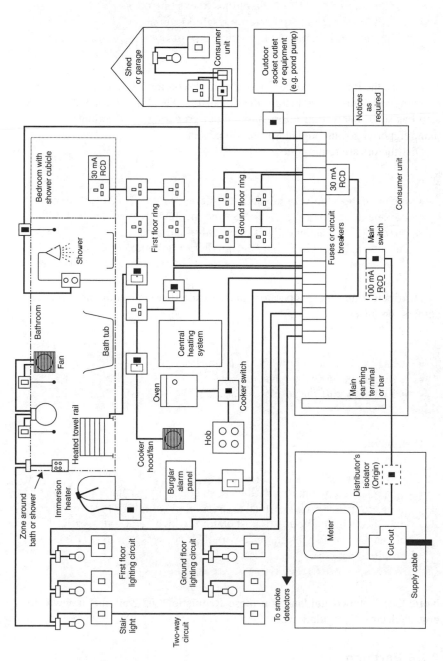

Figure 2.8 Typical fixed installations that might be encountered in new (or upgraded) existing dwellings

Consumer unit

Shed or garage

Outdoor socket outlet or equipment (e.g. pond pump)

Notices as required

Consumer unit

Bedroom with shower cubicle

30 mA RCD

First floor ring

Ground floor ring

30 mA RCD

Shower

Main switch

100 mA RCD

Bathroom

Bath tub

Central heating system

Fuses or circuit breakers

Fan

Oven

Cooker switch

Main earthing terminal or bar

Heated towel rail

Cooker hood/fan

Hob

Zone around bath or shower

Immersion heater

Burglar alarm panel

Distributor's isolator (Origin)

Meter

First floor lighting circuit

Ground floor lighting circuit

To smoke detectors

Cut-out

Stair light

Two-way circuit

Supply cable

- evaluating and agreeing proposals for new   P1.2
  installations or significant alterations to existing ones
- installing the cut-out and meter in a safe location   P1.5
- taking into consideration the possible risk of flooding.   P1.5

Note: See the Planning Portal's publication 'Preparing for flooding' at: www.
planningportal.gov.uk/england/professionals/en. The Environmental Agency
also has some interesting information about preparing for flooding at: www.
planningportal.gov.uk/england/professionals/en
Distributors are required to:

- maintain the supply within defined tolerance limits   P3.8
- provide an earthing facility for new connections   P3.8
- provide certain technical and safety information to   P3.8
  consumers to enable them to design their installations.

Distributors and meter operators must ensure that their equipment on consumers' premises:

- clearly shows the polarity of the conductors   P3.9
- is safe in its particular environment   P3.9
- is suitable for its purpose.   P3.9

Distributors:

- are prevented by the Regulations from connecting install-   P3.12
  ations to their networks which do not comply with BS 7671
- may disconnect consumers' installations which are a source   P3.12
  of danger or cause interference with their networks or other
  installations.

Note: See Detailed guidance on these Regulations is available at www.dti.
gov.uk/electricity-regulations.

2.7.8 Earthing

Note: The most usual type of earthing is an electricity distributor's earthing
terminal, which is provided for this purpose, near the electricity meter.

Distributors are required to provide an earthing facility for all new connections.	P3.8
All electrical installations shall be properly earthed.	P AppC1
All lighting circuits shall include a circuit protective conductor.	P AppC
All socket outlets which have a rating of 32 A or less and which may be used to supply portable equipment for use outdoors, shall be protected by a residual current device (RCD).	P AppC
It is **not** permitted to use a gas, water or other metal service pipe as a means of earthing for an electrical installation (this does not rule out, however, equipotential bonding conductors being connected to these pipes).	P AppC
New or replacement, non-metallic light fittings, switches or other components must not require earthing (e.g. non-metallic varieties) unless new circuit protective (earthing) conductors are provided.	P AppC
Socket outlets that will accept unearthed (2-pin) plugs must **not** use supply equipment that needs to be earthed.	P AppC
Where electrical installation work is classified as an extension, a material alteration or a material change of use, the work must consider and include satisfactory earthing and bonding systems which meet the requirements.	P2.1a–P2.2c

 See Figure 2.9 for details of some earth and bonding conductors that might be part of an electrical installation.

Accessible consumer units should be fitted with a child-proof cover or installed in a lockable cupboard.	P1.6

2.7.9 Electrical installations

All electrical installations shall provide adequate protection for persons against the risks of electric shock, burn or fire injuries, and should be designed and installed (suitably enclosed and appropriately separated) to provide mechanical and thermal protection.

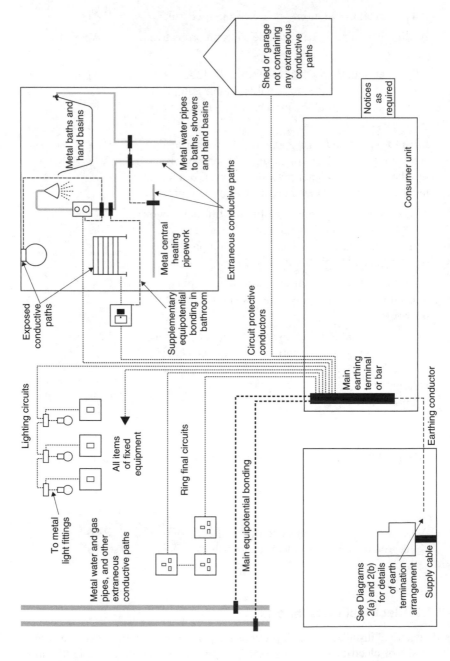

Figure 2.9 Typical earth and bonding conductors that might be part of electrical installation consumer units

Electrical installations must be inspected and tested during installation, at the end of installation and before they are taken into service to verify that they:

are safe to use, maintain and alter	P1.7
comply with Part P (and any other relevant Parts) of the Building Regulations	P1.7
meet the relevant equipment and installation standards	P0.1b
meet the requirements of the Building Regulations.	P3.1

Any proposal for a new mains supply installation (or where significant alterations are going to be made to an existing mains supply) **must** be agreed with the electricity distributor.

2.7.10 Electrical installation work

• is to be carried out professionally	P1.1
• is to comply with the Electricity at Work Regulations 1989 (as amended)	P1.1
• may only be carried out by persons that are competent to prevent danger and injury while doing it, or who are appropriately supervised.	P3.4a

 Note: Persons installing domestic combined heat and power equipment must advise the local distributor of their intentions before (or at the time of) commissioning the source.

2.7.10.1 Types of wiring or wiring system

Cables concealed in floors and (in certain circumstances) walls are required to have an earthed metal covering, be enclosed in steel conduit, or have additional mechanical protection (see BS 7671 for more information).	P AppA 2d
Cables to an outside building (e.g. garage or shed) if run underground, should be routed and positioned so as to give protection against electric shock and fire as a result of mechanical damage to a cable.	P AppA 2d
Heat-resisting flexible cables are required for the final connections to certain equipment (see makers' instructions).	P AppA 2d

PVC insulated and sheathed cables are likely to be suitable for much of the wiring in a typical dwelling.

2.7.10.2 Equipotential bonding conductors

Main equipotential bonding conductors are required for water service pipes, gas installation pipes, oil supply pipes plus and certain other 'earthy' metalwork that may be present on the premises.	P AppC
The installation of supplementary equipotential bonding conductors is required for installations and locations where there is an increased risk of electric shock (e.g. bathrooms and shower rooms).	P AppC
The minimum size of supplementary equipotential bonding conductors (without mechanical protection) is $4\,\text{mm}^2$.	P AppC

2.7.11 Inspection and Test

Electrical installations must be inspected and tested during, at the end of installation and before they are taken into service to verify that they:

• are reasonably safe **and** that they comply with BS 7671: 2008	P1.7
• meet the relevant equipment and installation standards.	P0.1b

All electrical work should be inspected (during installation as well as on completion) to verify that the components have:

• been selected and installed in accordance with BS 7671	P1.11a ii
• been made in compliance with appropriate British Standards or harmonised European Standards	P1.11a i
• been evaluated against external influences (such as the presence of moisture)	P1.11a ii
• not been visibly damaged (or are defective) so as to be unsafe	P1.11a iii

- been tested to check satisfactory performance   P1.1b
  with respect to continuity of conductors, insulation
  resistance, separation of circuits, polarity, earthing
  and bonding arrangements, earth fault loop impedance
  and functionality of all protective devices including
  residual current devices
- been inspected and tested for conformance with the   P1.9
  requirements of BS 7671:2008
- been tested using appropriate and accurate instruments. P1.10

Note: Inspections and testing of DIY work should **also** meet the above requirements.

2.7.12 Additional requirements and facilities for disabled people

During 2002/3, Approved Document M was thoroughly overhauled and restructured in order to meet the changed requirements of the Disability Discrimination Act 1995.

This major rewrite of Part M became effective on 1 May 2004 and was as a result of amendments made to Section 6 of the Disability Discrimination Act 1995 (DDA) which previously stated that 'reasonable adjustments to physical features of premises... shall be made... in certain circumstances' and 'shall apply to all employers with 15 or more employees'. As such, an employer with only a few employees (as well as occupations such as police, fire fighters and prison officers) was not required to alter the physical characteristics of a building that met existing requirements at the time the building works were carried out and **still** continued to meet those particular requirements.

From 1 October 2004, however, under the Disability Discrimination Act 1995 (Amendment) Regulations 2003 (SI 2003/1673), this exemption ceased and 'all those who provide services to the public, irrespective of their size' are required 'to take reasonable steps to remove, alter or provide a reasonable means of avoiding a physical feature of their premises, which makes it unreasonably difficult or impossible for disabled people to make use of their services'.

These amendments have naturally resulted in Part M being completely overhauled and it now covers:

- the conversion of a building for use as a shop being re-defined as a 'material change of use';
- amendments to omit specific references to (and a definition of) disabled people;
- expansion of the terms to include parents with children, elderly people and people with all types of disabilities (e.g. mobility, sight and hearing etc.);

- the use of a building by disabled people as residents, visitors, spectators, customers or employees, or participants in sports events, performances and conferences (which resulted in amendments being made to M1 (accessibility), M2 (sanitary accommodation) and M3 (audience and/or spectator seating).

 The current edition, therefore, no longer primarily concentrates on wheelchair users, but includes people using walking aids, people with impaired sight (and other mobility and sensory problems), and mothers with prams as well as people with luggage, etc.

 Reasonable provision should, therefore, be made to make sure that dwellings (including any purpose-built student living accommodation, other than traditional halls of residence providing mainly bedrooms and not equipped as self-contained accommodation) provide sufficient access for disabled people.

2.7.13 Electrical components and installations

 New or replacement, non-metallic light fittings, switches or other components must **not** require earthing (e.g. non-metallic varieties) **unless** new circuit protective (earthing) conductors are provided.

2.7.13.1 Controls and switches

The aim should be to ensure that all controls and switches should be easy to operate, visible and free from obstruction, and: M (5.4i)

- they should be located between 750 mm and 1200 mm above the floor
- they should not require the simultaneous use of both hands (unless necessary for safety reasons) to operate
- switched socket outlets should indicate whether they are 'ON'
- mains and circuit isolator switches should clearly indicate whether they 'ON' or 'OFF'
- individual switches on panels and on multiple socket outlets should be well separated
- controls that need close vision (e.g. thermostats) should be located between 1200 mm and 1400 mm above the floor
- front plates should contrast visually with their backgrounds.

Controls that need close vision (e.g. thermostats) should   M (4.30f) be located between 1200 mm and 1400 mm above the floor.

The operation of all switches, outlets and controls should not require the simultaneous use of both hands (unless necessary for safety reasons).	M (4.30j)
Where possible, light switches with large push pads should be used in preference to pull cords.	M (5.3)
The colours red and green should not be used in combination as indicators of 'ON' and 'OFF' for switches and controls.	M (5.3)

2.7.13.2 Lighting circuits

All lighting circuits shall include a circuit protective conductor.

2.7.13.3 Fixed lighting

In locations where lighting can be expected to have most use, fixed lighting (e.g. fluorescent tubes and compact fluorescent lamps – but not GLS tungsten lamps with bayonet cap or Edison screw bases) with a luminous efficacy greater than 40 lumens per circuit-watt should be available.	L

Note: The following is an indication of the recommended number of locations (excluding garages, lofts and outhouses) that need to be equipped with efficient lighting.

Number of rooms created (Hall, stairs and landing(s) count as one room as does a conservatory)	Recommended minimum number of locations
1–3	1
4–6	2
7–9	3
10–12	4

Hall, stairs and landing(s) count as one room as does a conservatory.

2.7.13.4 External lighting fixed to the building

External lighting (including lighting in porches, but not lighting in garages and carports) should:

automatically extinguish when there is enough daylight, and when not required at night.	L

have sockets that can only be used with lamps having an   L
efficacy greater than 40 lumens per circuit-watt (such as
fluorescent or compact fluorescent lamp types, and **not** GLS
tungsten lamps with bayonet cap or Edison screw bases).

2.7.13.5 Emergency alarms

Emergency alarm pull cords should: M (4.30e)

- be coloured red
- be located as close to a wall as possible
- have two red 50 mm diameter bangles.

Front plates should contrast visually with their backgrounds.   M (4.30m)

The colours red and green should **not** be used in   M (4.28)
combination as indicators of 'ON' and 'OFF' for switches
and controls.

2.7.13.6 Fire alarms

Fire alarms should emit an audio and visual signal to   M (5.4 g)
warn occupants with hearing or visual impairments.

Emergency assistance alarm systems should have: M (5.4 h)

- visual and audible indicators to confirm that an
  emergency call has been received
- a reset control reachable from a wheelchair, WC, or a
  shower/changing seat
- a signal that is distinguishable visually and audibly
  from the fire alarm.

2.7.13.7 Heat emitters

Heat emitters should either be screened or have their   M (5.4j)
exposed surfaces kept at a temperature below 43°C.

In toilets and bathrooms, heat emitters (if located) should   M (5.10p)
not restrict:

- the minimum clear wheelchair manoeuvring space
- the space beside a WC used to transfer from the
  wheelchair to the WC.

2.7.13.8 Power-operated doors

Doors to accessible entrances shall be provided with a power-operated door opening and closing system if a force greater than 20 N is required to open or shut a door.

 Once open, all doors to accessible entrances should be wide enough to allow unrestricted passage for a variety of users, including wheelchair users, people carrying luggage, people with assistance dogs, and parents with pushchairs and small children.

> The effective clear width through a single leaf door   M (2.13b)
> (or one leaf of a double leaf door) should be in
> accordance with Table 2.6.

Table 2.6 Minimum effective clear widths of doors

Direction and width of approach	New buildings (mm)	Existing buildings (mm)
Straight-on (without a turn or oblique approach)	800	750
At right angles to an access route at least 1500 mm wide	800	750
At right angles to an access route at least 1200 mm wide	825	775
External doors to buildings used by the general public	1000	775

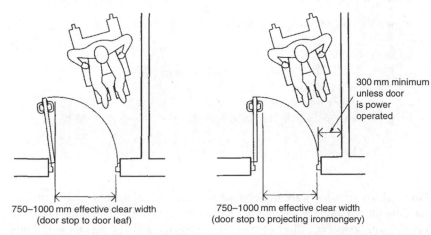

750–1000 mm effective clear width
(door stop to door leaf)

300 mm minimum unless door is power operated

750–1000 mm effective clear width
(door stop to projecting ironmongery)

Figure 2.10 Effective clear width and visibility requirements of doors

Power-operated entrance doors should have a sliding, swinging or folding action controlled manually (by a push pad, card swipe, coded entry, or remote control) or automatically controlled by a motion sensor or proximity sensor such as a contact mat.

Power-operated entrance doors should:

• be provided with a manual or automatic opening device in the event of a power failure where and when necessary for health or safety	K5 (5.2d)
• open towards people approaching the doors	M (2.21a)
• provide visual and audible warnings that they are operating (or about to operate)	M (2.21c)
• incorporate automatic sensors to ensure that they open early enough (and stay open long enough) to permit safe entry and exit	M (2.21c)
• incorporate a safety stop that is activated if the doors begin to close when a person is passing through	M (2.21b)
• have a readily identifiable and accessible stop switch	K5 (5.2d)
• have safety features to prevent injury to people who are struck or trapped (such as a pressure-sensitive door edge which operates the power switch)	K5 (5.2d)
• revert to manual control (or fail safe) in the open position in the event of a power failure	M (2.21d)
• when open, not project into any adjacent access route	M (2.21e)
ensure that its manual controls:	M (2.21f)
• are located between 750 mm and 1000 mm above floor level	
• are operable with a closed fist	
• be set back 1400 mm from the leading edge of the door when fully open	M (2.21g)
• be clearly distinguishable against the background	M (2.21g)
• contrast visually with the background.	M (2.19 and 2.21g)

 Note: Revolving doors are **not** considered 'accessible' as they create particular difficulties (and possible injury) for people who are visually impaired, for people with assistance dogs or mobility problems and for parents with children and/or pushchairs.

2.7.13.9 Switches and socket outlets

Switches and socket outlets for lighting and other equipment should be located so that they are easily reachable.	M2 (8.2)
Switches and socket outlets (for lighting) should be installed between 450 mm and 1200 mm from the finished floor level (See Figure 2.11).	M2 (8.3)

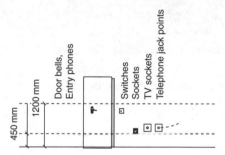

Figure 2.11 Heights of switches and sockets etc.

The aim is to help people with limited reach (e.g. seated in a wheelchair) access a dwelling's wall-mounted switches and socket outlets.

Socket outlets

Older types of socket outlet designed non-fused plugs must not be connected to a ring circuit.	P AppC
Socket outlets that will accept unearthed (2-pin) plugs must not be used to supply equipment needing to be earthed.	P AppC
Sensitive RCD protection is required for all socket outlets which have a rating of 32 A or less and which may be used to supply portable equipment for use outdoors.	P AppC
Socket outlets should comply with the requirements of Part M (see Section 2.13.24).	P1.5

Portable equipment for use outdoors

Sensitive RCD protection is required for all socket outlets which have a rating of 32 A or less and which may be used to supply portable equipment for use outdoors.	P AppC

Switched socket outlets

Switched socket outlets should indicate whether they are 'ON'.	M (4.30k)
Mains and circuit isolator switches should clearly indicate whether they are 'ON' or 'OFF'.	M (4.30l)
Individual switches on panels and on multiple socket outlets should be well separated.	M (4.29)
All socket outlets should be wall-mounted.	M (4.30a and b)
Socket outlets should be located no nearer than 350 mm from room corners.	M (4.30g)
Front plates should contrast visually with their backgrounds.	M (4.30m)
The colours red and green should **not** be used in combination as indicators of 'ON' and 'OFF' for switches and controls.	M (4.28)

Wall sockets

Wall sockets shall meet the following requirements:

Table 2.7 Building Regulations requirements for wall sockets

Type of wall	Requirement	Section
Timber framed	Power points may be set in the linings, provided there is a similar thickness of cladding behind the socket box.	E (p14)
	Power points should not be placed back to back across the wall.	E (p14)
Solid masonry	Deep sockets and chases should not be used in separating walls.	E 2.32
	Stagger the position of sockets on opposite sides of the separating wall.	E 2.32f
Cavity masonry	Stagger the position of sockets on opposite sides of the separating wall.	E 2.65e
	Deep sockets and chases should not be used in a separating wall.	E 2.65d2
	Deep sockets and chases in a separating wall should not be placed back to back.	E 2.65d2
Framed walls with absorbent material	Sockets should: • be positioned on opposite sides of a separating wall;	E 2.146b
	• not be connected back to back;	E 2.146b2
	• be staggered a minimum of 150 mm edge to edge.	E 2.146b2

Light switches

Light switches should:	M (4.30h and I)
• have large push pads	
• align horizontally with door handles	
• be within the 900 to 1100 mm from the entrance door opening.	
Switches and controls should be located between 750 mm and 1200 mm above the floor.	M (4.30c and d)
Where possible, light switches with large push pads should be used in preference to pull cords.	M (5.3)
The colours red and green should not be used in combination as indicators of 'ON' and 'OFF' for switches and controls.	M (5.3)

Telephone points and TV sockets

All telephone points and TV sockets should be located between 400 mm and 1000 mm above the floor (or 400 mm and 1200 mm above the floor for permanently wired appliances).	M (4.30a and b)

2.7.13.10 Other considerations

Lecture/conference facilities

Artificial lighting should be designed to give good colour rendering of all surfaces.	M (4.9)
Artificial lighting should be designed to be compatible with other electronic and radio frequency installations.	M (4.12.1)
Artificial lighting should be designed to be compatible with other electronic and radio frequency installations.	M (4.36f)
Artificial lighting should be designed to give good colour rendering of all surfaces.	M (4.34)

Swimming pools and saunas

Swimming pools and saunas are subject to special requirements specified in Part 7 of BS 7671:2008.	P AppA 1

Cellars or basements

LPG storage vessels and LPG fired appliances fitted with automatic ignition devices or pilot lights must not be installed in cellars or basements.	J (3.5i)

3

Earthing

From an electrical point of view, the world is effectively a huge conductor at zero potential and is used as a reference point which is called 'earth' (in the UK) or 'ground' (in the USA). People and animals are normally in contact with earth and so if another part, which is open to touch, becomes charged at a different voltage from earth, then a shock hazard will exist.

This chapter reminds the reader about the different types of earthing systems and earthing arrangements. It then lists the main requirements for safety protection (direct and indirect contact), protective conductors and protective equipment before briefly touching on the test requirements for earthing.

 Similar to other chapters it should be noted that these lists of requirements are **only** the author's impression of the most important aspects of the Wiring Regulations and electricians should **always** consult BS 7671 to satisfy compliance.

 New requirements for earthing

BS 7671:2008 includes changes necessary to maintain technical alignment with CENELEC harmonisation documents. Earthing requirements now include:

- protection of low-voltage installations against temporary overvoltages due to earth faults in high- and low-voltage systems
- the requirement that a metallic pipe of a water utility supply should **not** be used as an earth electrode is retained and that other metallic water supply pipework (e.g. belonging to a privately owned water supply network) shall likewise **not** be used as an earth electrode unless precautions are taken against its removal and it has been considered for such a use.

New appendices concerning earthing have also been included in the new BS 7671: 2008. These include:

- measurement of earth fault loop impedance (consideration of the increase of the resistance of conductors with increase of temperature)
- methods for measuring the insulation resistance/impedance of floors and walls to earth or to the protective conductor system.

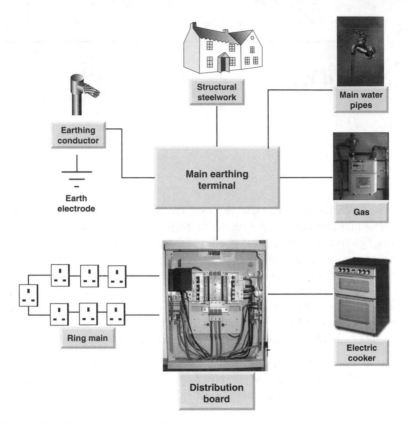

Figure 3.1 Bonding and earthing

3.1 What is earth?

In electrical terms, 'earth' is defined as:

> *The conductive mass of the earth, whose electric potential at any point is conventionally taken as zero.*

From an astrological and geophysical point of view, on the other hand:

> *Earth (also known as the Earth, Terra, and – mostly in the 19th century – Tellus) is the third planet outward from the Sun. It is the largest of the solar system's terrestrial planets and the only planetary body that modern science confirms as harbouring life. The planet formed around 4.57 billion (4.57 × 10⁹) years ago and 'shortly' thereafter (i.e. about 4.533 billion years ago!) acquired its single natural satellite, the moon. Its astronomical symbol consists of a circled cross, representing a meridian and the equator.*

3.2 What is meant by 'earthing' and how is it used?

Earthing: connection of the exposed conductive parts of an installation to the main earthing terminal of that installation.

Earthing is a process that is used to connect all of the parts that could become charged, to the general mass of earth and in so doing, provide a path for fault currents which will hold these parts as close as possible to earth (i.e. zero) potential. In doing so, this will prevent a potential difference happening between earth and earthed parts, as well as letting the flow of fault current to operate the protective systems.

An *earthing system*, on the other hand, defines the electrical potential of the conductors relative to that of the earth's conductive surface. The choice of earthing system has implications for the safety and electromagnetic compatibility of the power supply.

An *earth electrode* is the part of the system that is directly in contact with the earth and this can be just a metal (usually copper) rod or stake driven into the earth or a connection to a buried metal service, pipe or a complex system of buried rods and wires.

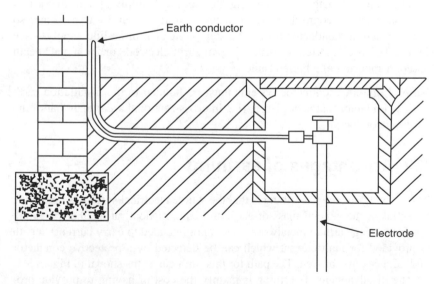

Figure 3.2 Earth conductor and electrode

The resistance of the electrode-to-earth connection will determine its quality and this can be improved by:

- increasing the surface area of the electrode that is in contact with the earth;
- increasing the depth to which the electrode is driven;

- using several connected ground rods;
- increasing the moisture of the soil;
- improving the conductive mineral content of the soil;
- increasing the land area covered by the ground system.

A protective earth (PE) connection ensures that all exposed conductive surfaces are at the same electrical potential as the surface of the earth and thus avoiding the risk of an electrical shock if a person, or an animal, touches equipment (or a device) in which an insulation fault has occurred. PE also ensures that if an insulation fault occurs, then a high fault current will flow which will trigger an overcurrent protection device (e.g. a fuse) that will disconnect the power supply.

A functional earth (FE) connection, as well as providing protection against electrical shock, can also carry a current during the normal operation of a device – a facility that is often required by devices such as surge suppression and electromagnetic-compatibility filters, and some types of antennas as well as a number of measuring instruments.

In a mains (i.e. a.c. power) wiring installation, the 'ground' wire is (directly or indirectly) connected to one or more earth electrodes and carries currents away under fault conditions. These earth electrodes may be located locally or in the supplier's network some distance away, and the ground wire is also usually bonded to pipework so as to keep it at the same potential as the electrical ground during a fault.

The standard method of attaching the electrical supply system to earth is to make a direct connection between the two at the supply transformer so that the neutral conductor (often the star point of a three-line supply – see Figure 3.3) is connected to earth using an earth electrode or the metal sheath and/or armouring of a buried cable.

 Note: Lightning conductor systems must be bonded to the installation earth with a conductor that is not larger (i.e. in cross-sectional area) than that of the earthing conductor.

3.3 Advantages of earthing

The main advantage to earthing is that the whole electrical system is tied to the potential of the general mass of earth and cannot 'float' at another potential. By connecting earth to metalwork (that is not intended to carry current), a path is provided for fault current which can be detected by a protective conductor and, if necessary, broken. The path for this fault current is shown in Figure 3.4.

The disadvantage of earthing is mainly the cost of having to provide protective conductors and earth electrodes, etc.

Earthing arrangements may be used jointly or separately for protective and functional purposes, according to the requirements of the installation. They should, however ensure that:

- they are sufficiently robust (or have additional mechanical protection) to external influences;

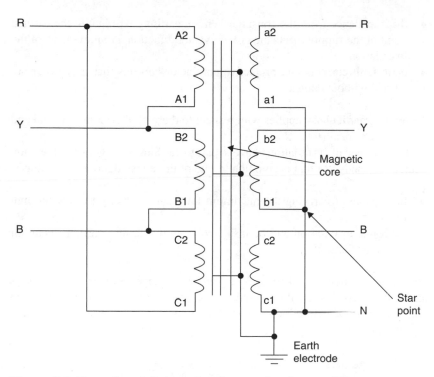

Figure 3.3 Three-line delta/star transformer showing earthing arrangements

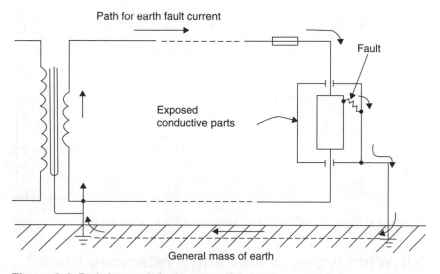

Figure 3.4 Path for earth fault current (shown by arrows)

- the impedance from the consumer's main earthing terminal to the earthed point of the supply meets the protective and functional requirements of the installation;
- earth fault currents and protective conductor currents that may occur are carried without danger.

Note: this particularly applies with respect to thermal, thermomechanical and electromechanical stresses.

If a number of installations have separate earthing arrangements then any protective conductor that is common to one of these installations shall either:

- be capable of carrying the maximum fault current likely to flow through them; or
- earth one installation and be insulated from the earthing arrangements of other installation(s).

Precautions should be taken against possible damage to other metallic parts through electrolysis and if the protective conductor forms part of a cable, then this shall only be earthed in the installation containing the associated protective device.

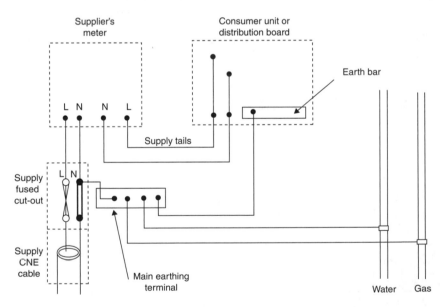

Figure 3.5 Domestic earthing arrangement earth electrodes

3.4 What types of earthing system are there?

In the Regulations, an electrical system is defined as consisting of 'a single source of electrical energy and an installation' and the type of system depends on the link between the source and the exposed conductive parts of the installation, to earth.

Note: In this context, an 'exposed conductive part' means a conductive part of an item of equipment which can be touched and which is not (i.e. currently) a live part, but which 'may' become live under fault conditions.

The three basic systems available are classified as TN (which is then further subdivided into TN-C, TN-C-S and TN-S), TT and IT (as shown in Figure 3.6).

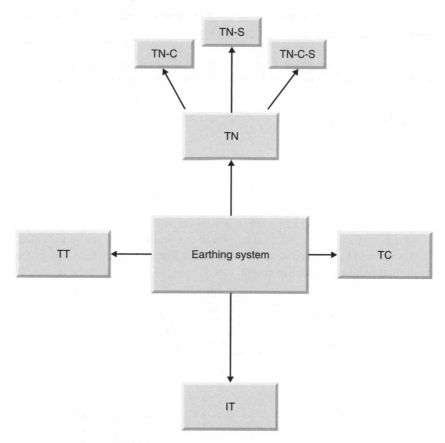

Figure 3.6 Earthing systems

3.4.1 System classification

In order to identify the different systems, a unique four-lettered code is used.
The **first letter** indicates the type of supply earthing, so that:

- **T** indicates that one or more points of the supply are directly earthed (for example, the earthed neutral at the transformer);
- **I** indicates either that the supply system is not earthed (at all) or that the earthing includes a deliberately inserted impedance, the purpose of which is to limit fault current.

The **second letter** indicates the earthing arrangement in the installation, so that:

- **T** indicates that all exposed conductive metalwork is connected directly to earth;
- **N** indicates that all exposed conductive metalwork is connected directly to an earthed supply conductor provided by the electricity supply company.

The **third and fourth letters** indicate the arrangement of the earthed supply conductor system and:

- **S** – ensure that neutral and earth conductor systems are quite separate; and
- **C** – ensure that neutral and earth are combined into a single conductor.

3.4.2 TN systems

In a TN system:

- the integrity of the earthing of the installation depends on a reliable and effective connection of the PEN or PE conductors to earth;
- one or more points in the generator or transformer are connected to earth (usually the star point in a three-line system). The body of the electrical device is then connected with earth via this earth connection at the transformer and exposed conductive parts of the installation are then connected to that point by protective conductors.

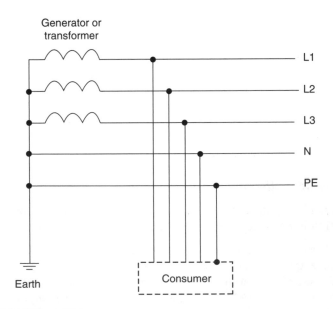

Figure 3.7 TN system

The conductor that connects together the exposed metallic parts of the consumer installation is called the protective earth (PE), while the conductor that connects to the star point in a three-line system (or which carries the return current in a single-line system) is called the neutral (N).

There are three variants of TN systems: TN-C, TN-S and TN-C-S.

3.4.3 TN-C system

A TN-C system is one where the neutral (N) and protective earth (PE) functions are combined in a single conductor (i.e. a PEN or protective multiple earth conductor) throughout the system and this combined neutral and earth wiring is then used both by the supply and from within the installation itself.

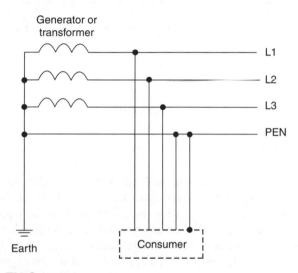

Figure 3.8 TN-C system

There are a number of disadvantages with using a TN-C network. For example:

- RCDs (residual current devices) are far less likely to detect an insulation fault.
- They are vulnerable to unwanted triggering caused by contact between earths of circuits, on different RCDs or with real ground.
- Any connection between the combined neutral and earth core and the body of the earth could end up carrying significant current under normal conditions.
- If there is a contact problem in the PEN conductor, then all parts of the earthing system beyond the break will rise to the potential of the live conductor(s).

 TN-C systems normally use an earthed concentric system, which can only be installed under special conditions.

3.4.4 TN-S system

TN-S systems have separate protective earth (PE) and neutral (N) conductors from the transformer to the consuming device and these conductors remain separated throughout the system (see Figure 3.9).

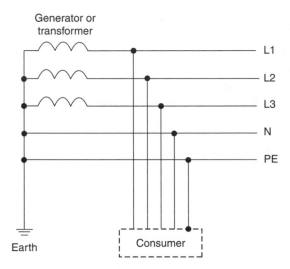

Figure 3.9 TN-S system

The TN-S is the most common earthing system in the UK and one where the electricity supply company provides an earth terminal at the incoming mains position. This earth terminal is then connected by the supply PE back to the star point (neutral) of the secondary winding of the supply transformer, which is also connected, at that point, to an earth electrode. The earth conductor is usually the armour and sheath (if applicable) of the underground supply cable. In TN-S systems an RCD can be used as an additional protection.

Electromagnetic compatibility

One of the advantages of a TN-S system concerns electromagnetic compatibility, whereby the consumer has a low-noise connection to earth and does not suffer from the voltage that appears on the neutral conductor as a result of the return currents and the impedance of that conductor. This is of particular importance with some types of telecommunication and measurement equipment.

 TN-S networks also save costs by having a fairly low-impedance earth connection near each consumer.

3.4.5 TN-C-S system

A TN-C-S system is one that uses a combined PEN conductor between the transformer and the building distribution point substation and the entry point into the building and then splits up into separate PE and N lines within the building (see Figure 3.10) to fixed indoor wiring and flexible power cords.

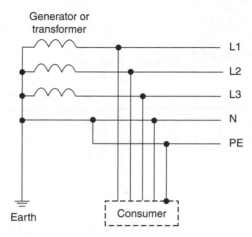

Figure 3.10 TN-C-S system

Note: In the UK, this system is also known as protective multiple earthing as it connects the combined neutral and earth to real earth at many locations to thereby reduce the risk of broken neutrals.

Although TN-C networks save the cost of an additional conductor to separate N and PE connections, special cable types and lots of connections to earth are required.

As any connection between the combined neutral and earth core and the body of the earth could end up carrying significant current under normal conditions, the use of TN-C-S is **not** recommended for locations such as petrol stations etc., where there is a combination of lots of buried metalwork and explosive gases.

In addition, owing to the possibility of a lost neutral, the use of TN-C-S supplies is **banned** for caravans and boats in the UK and it is often recommended to make outdoor wiring TT with a separate rod.

3.4.6 TT system

A TT system is one which has one point of the energy source directly earthed and the exposed conductive parts of the consumer's installation are provided with a local connection to earth, independent of any earth connection at the generator (see Figure 3.11).

This type of installation is usually found in rural locations where the system is not provided with an earth terminal by the electricity supply company and the installation is fed from an overhead supply. Neutral and earth (protective) conductors must be kept quite separate throughout the installation and the final earth terminal must be connected to an earth electrode – via an earthing conductor.

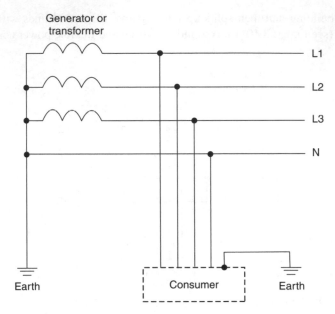

Figure 3.11 TT system

TT systems (similar to TN-S systems) have a low-noise connection to earth, which is particularly important with some types of telecommunication and measurement equipment.

3.4.7 IT system

An IT system is one which has no direct connection between live parts and earth and where the exposed conductive parts of the electrical installation are earthed (see Figure 3.12).

An IT system is similar to a TT system except that the supply earthing in an IT system can either be from an unearthed supply or one which (although not totally earthed) is connected to earth through a current-limiting impedance.

This lack of earth will usually mean that normal protective methods cannot be used and for this reason, IT systems are not normally allowed in the UK public supply system – except for hospitals and other medical locations where they are recommended for use with circuits supplying medical equipment intended for life-support of patients.

3.5 Earth fault loop impendence

As shown in Figure 3.13, the earth fault loop starts at the point of the fault and comprises:

- the circuit protective conductor (CPC);
- the consumer's earthing terminal and earthing conductor;

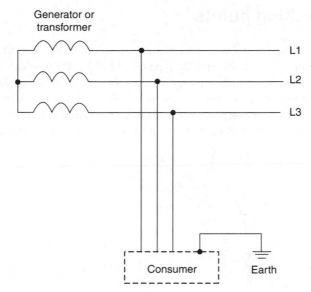

Figure 3.12 IT system

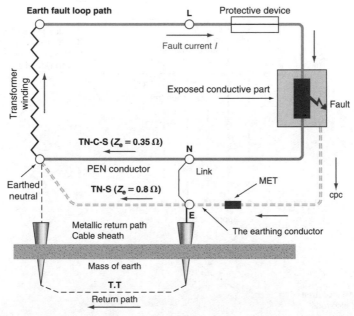

Figure 3.13 Earth fault loop impendence path (courtesy of Brian Scaddan)

- (for TN systems) the metallic return path or (for TT and IT systems) the earth return path;
- the path through the earthed neutral point of the transformer;
- the transformer winding;
- the line conductor from the transformer to the point of fault.

3.6 Earthing points

It has been proved that the resistance area around an earth electrode depends on the size of the electrode and the type of soil and that this electrode resistance is particularly important with regard to the voltage at the surface of the ground. For example, as shown in Figure 3.14, for a 2 m rod with its top at ground level, approximately 80–90% of the voltage appearing at the electrode under fault conditions will be dropped in the first 2.5–3 m from the electrode.

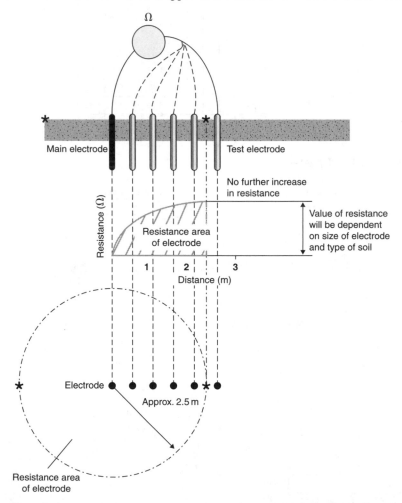

Figure 3.14 The resistance area of an earth electrode (courtesy of Brian Scaddan)

This can be particularly dangerous where livestock is concerned. For example, in some circumstance a grazing cow might have its forelegs inside the resistive area, whilst its hindlegs are outside of the area. Bearing in mind that a potential difference of 25 V can it be lethal, measures have to be taken to reduce this risk. One method is to house the earth electrode in a pit that is below ground level (as shown in Figure 3.15).

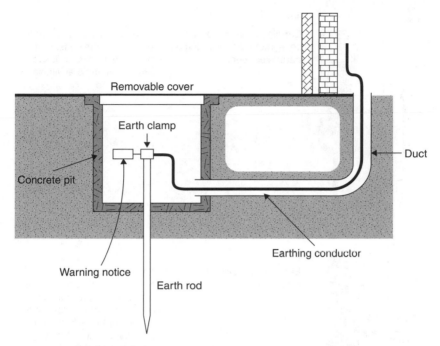

Figure 3.15 An earth electrode protected by a pit below ground level (courtesy of Brian Scaddan)

3.7 Main earthing terminals

The main earthing terminal acts as the single reference point and can comprise a bar, a plate or even a copper internal 'ring' conductor. This is often connected directly to an effective earth electrode and this connection must be of copper because of the corrosion risk if aluminium or copper clad aluminium were used.

The main earthing terminal shall be connected to earth as shown below and in Figure 3.16.

System	Main earthing terminal
TT or IT	Connected via an earthing conductor to an earth electrode
TN-S	Connected to the earthed point of the energy source
TN-C-S	Connected (by the distributor) to the neutral of the energy source

The earth electrode should be positioned as close as possible to the main earthing terminal.

System	Provision of earth	Remarks
TT	Earthing is provided by the consumer's own installation earth electrode.	No earthing facility is made available to the consumer by the distributor or (if such a facility is made available) it is not used.
TN-S	Earthing is provided by the distributor.	This is usually provided either by a direct connection to the supply cable sheath or via a separate protective conductor in the form of a split-concentric cable or overhead conductor.
TN-C-S	Earthing is provided by the distributor.	This is connected to the incoming supply neutral to give a protective multiple earth (PME) supply, where the supply neutral and protective conductors are in the form of a combined neutral and earth (CNE) conductor.

Figure 3.16 Earth provision

3.8 Earth electrodes

An earth electrode is a conductor, or group of conductors, that connects the main earthing terminal of an installation to an earth electrode or to other means of earthing. Earth electrodes shall be designed and constructed so that they can:

- withstand damage;
- take into account a possible increase in resistance due to corrosion.

The following types of earth electrode may be used for electrical installations:

- earth rods or pipes;
- earth tapes or wires;
- earth plates;
- lead sheaths and other metal cable covers;
- other suitable underground metalwork;
- structural metalwork embedded in foundations;
- welded metal reinforced concrete (except pre-stressed concrete) embedded in the earth.

A metallic pipe:

- for gases; or
- flammable liquids; or
- of a water utility supply; or
- other metallic water supply pipework

shall not be used as an earth electrode – unless precautions are taken against its removal and it is suitable for its intended use.

3.9 Earthing conductors

  All earthing conductors **shall** meet the requirements   542-03-01
for protective conductors.

The earthing conductor is an important part of the earth fault loop impedance as it is a protective conductor that connects the main earthing terminal of an installation to an earth electrode, or to some other means of earthing.

3.10 Requirements from the Regulations

3.10.1 Additions and alterations to an installation

No addition or alteration, temporary or permanent, shall be   WR-131.8
made to an existing installation unless it has been

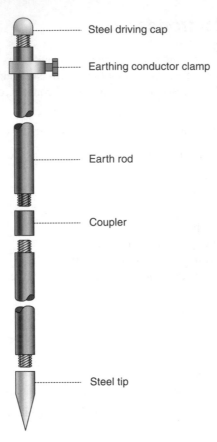

Steel driving cap

Earthing conductor clamp

Earth rod

Coupler

Steel tip

Figure 3.17 An example of an earthing conductor (courtesy of Brian Scaddan)

ascertained that the earthing and bonding arrangements used as a protective measure for the safety of the addition or alteration, are adequate.

3.10.2 Automatic disconnection of supply

Automatic disconnection of supply is a protective measure, in which fault protection is provided by protective earthing, protective equipotential bonding and automatic disconnection in case of a fault.

3.10.3 Autotransformers and step-up transformers

   A step-up autotransformer shall **not** be connected to an IT system.   WR-555.1.2

3.10.4 Cables

A cable passing through a joist within a floor or ceiling construction or through a ceiling support (e.g. under floorboards), shall:

522.6.5

- incorporate an earthed metallic covering; or
- be enclosed in earthed conduit; or
- be enclosed in earthed trunking or ducting; or
- be mechanically protected against damage sufficient to prevent penetration of the cable by nails, screws etc.; or
- be at least 50 mm measured vertically from the top, or bottom as appropriate, of the joist or batten.

A cable concealed in a wall or partition at a depth of less than 50 mm from a surface of the wall or partition shall:

WR-522.6.6

- incorporate an earthed metallic covering; or
- be enclosed in earthed conduit; or
- be enclosed in earthed trunking or ducting complying; or
- be mechanically protected against damage sufficient to prevent penetration of the cable by nails, screws etc.; or
- be installed in a zone within 150 mm from the top of the wall or partition or within 150 mm of an angle formed by two adjoining walls or partitions.

 Where the installation is not intended to be under the supervision of a skilled or instructed person, consideration shall be given to providing additional protection by means of an RCD.

If the cables of an installation not intended to be under the supervision of a skilled or instructed person are concealed in a wall or partition (the internal construction of which includes metallic parts, other than metallic fixings such as nails, screws and the like) then it shall:

WR-522.6.8

- incorporate an earthed metallic covering; or
- be enclosed in earthed conduit; or
- be enclosed in earthed trunking or ducting; or
- be mechanically protected sufficiently to avoid damage to the cable during construction of the wall or partition and during installation of the cable; or
- be provided with additional protection by means of an RCD.

A cable buried in the ground (that is not installed in a conduit or duct) shall incorporate an earthed armour or metal sheath or both, suitable for use as a protective conductor.   WR-522.8.10

3.10.5 Earthing conductors

The connection of an earthing conductor to the earth electrode shall be: WR-542.3.2

- soundly made
- electrically and mechanically satisfactory
- suitably labelled
- suitably protected against corrosion.

Where buried in the ground, the earthing conductor shall have a cross-sectional area not less than that stated in Table 3.1.   WR-542.3.1

Table 3.1 Minimum cross-sectional area of a buried earthing conductor (reproduced with permission of IET)

	Protected against mechanical damage	Not protected against mechanical damage
Protected against corrosion by a sheath	$2.5\,mm^2$ copper $10\,mm^2$ steel	$16\,mm^2$ copper $16\,mm^2$ coated steel
Not protected against corrosion	$25\,mm^2$ copper $50\,mm^2$ steel	

Where PME (protective multiple earthing) exists, the requirements for main equipotential bonding conductors (i.e. for the cross-sectional area of a main equipotential bonding conductor) shall be met.   WR-542.3.1

The thickness of tape or strip conductors shall be capable of withstanding mechanical damage and corrosion (see BS 7430).   WR-542.3.2

3.10.6 Earth electrodes

The following types of earth electrode are recognised as being suitable for the purposes of these Regulations:

- earth rods or pipes;
- earth tapes or wires;
- earth plates;
- underground structural metalwork embedded in foundations;
- welded metal reinforcement of concrete (except pre-stressed concrete) embedded in the earth;
- lead sheaths and other metal coverings of cables;
- other suitable underground metalwork.

 Note: Further information on earth electrodes can be found in BS 7430.

The type and embedded depth of an earth electrode shall be such that soil drying and freezing will not increase its resistance above the required value.	WR-542.2.2
The design used for (and the construction of) an earth electrode shall be such as to withstand damage and to take account of possible increase in resistance due to corrosion.	WR-542.2.3

 A metallic pipe:

- for gases; or
- flammable liquids; or
- of a water utility supply; or
- other metallic water supply pipework

shall **not** be used as an earth electrode unless precautions are taken against its removal and it is suitable for its intended use.

The use, as an earth electrode, of the lead sheath or other metal covering of a cable shall be subject to all of the following conditions:	WR-542.2.5
• Adequate precautions have been taken to prevent excessive deterioration by corrosion • The sheath or covering shall be in effective contact with earth • The consent of the owner of the cable has been obtained	

- Arrangements exist for the owner of the
  electrical installation to be warned of any
  proposed change to the cable which might
  affect its suitability as an earth electrode.

Earth electrodes shall be designed and constructed   WR-542.2.3
so that they can:

- withstand damage
- take into account a possible increase in
  resistance due to corrosion.

3.10.7 Earth fault current

The choice and type of wiring system and the method   WR-132.7
of installation shall include consideration of the
electromechanical stresses likely to occur due to short-
circuit and earth fault currents.

The characteristics of protective equipment shall be   WR-132.8
determined with respect to their function, including
protection against the effects of earth fault current.

3.10.8 Earth faults

This particular Regulation provides requirements for the safety of the low-
voltage installation in the event of:

- a fault between the high-voltage system and earth in the transformer sub-
  station that supplies the low-voltage installation;
- loss of the supply neutral in the low-voltage system;
- short-circuit between a line conductor and neutral in the low-voltage
  installation;
- accidental earthing of a line conductor of a low-voltage IT system.

An earthing arrangement may be considered electrically   WR-442.1.2
independent of another earthing arrangement if a rise of
potential with respect to earth in one earthing
arrangement does not cause an unacceptable rise of
potential with respect to earth in the other earthing
arrangement.

The magnitude and duration of the power frequency stress voltages on the low-voltage equipment in the low-voltage installation due to an earth fault in the high-voltage system shall not exceed the requirements given in Table 3.2.

WR-442.2.2

Table 3.2 Permissible power frequency stress voltage (reproduced with permission of IET)

Duration of the earth fault in the high-voltage system	Permissible power frequency stress voltage on equipment in low-voltage installations U
$>5\,s$	$U_o + 250\,V$
$<5\,s$	$U_o + 1200\,V$

3.10.9 Earthing arrangements

For a TN-S system, means shall be provided for the main earthing terminal of the installation to be connected to the earthed point of the source of energy. Part of the connection may be formed by the distributor's lines and equipment.

WR-542.1.2

For a TN-C-S system, where protective multiple earthing is provided, the main earthing terminal of the installation shall be connected by the distributor to the neutral of the source of energy.

WR-542.1.3

For a TT or IT system, the main earthing terminal shall be connected via an earthing conductor to an earth electrode.

WR-542.1.4

The earthing arrangements may be used jointly or separately for protective and functional purposes, according to the requirements of the installation.

WR-542.1.5

The earthing arrangements shall be such that:

WR-542.1.6

- the value of impedance from the consumer's main earthing terminal to the earthed point of the supply for TN systems, or to earth for TT and IT systems, is in accordance with the protective and functional requirements of the installation, and considered to be continuously effective; and
- earth fault currents and protective conductor currents which may occur are carried without danger, particularly from thermal, thermo-mechanical and electromechanical stresses; and

- they are adequately robust or have additional mechanical protection appropriate to the assessed conditions of external influence.

Precautions shall be taken against the risk of damage to other metallic parts through electrolysis.   WR-542.1.7

Where a number of installations have separate earthing arrangements, any protective conductors common to any of these installations shall either: WR-542.1.8

- be capable of carrying the maximum fault current likely to flow through them; or
- be earthed within one installation only and insulated from the earthing arrangements of any other installation.

Where earthing for combined protective and functional purposes is required, the requirements for protective measures shall take precedence.   WR-543.5.1

Equipment having a protective conductor current exceeding 10 mA shall be connected to the supply: WR-543.7.1.2

- permanently via the wiring of the installation; or
- via a flexible cable with a plug and socket-outlet; or
- via a protective conductor with an earth monitoring system.

The wiring of every final circuit and distribution circuit intended to supply one or more items of equipment, such that the total protective conductor current is likely to exceed 10 mA, shall have a high integrity protective connection complying with one or more of the following: WR-543.7.1.3

- a single protective conductor with a cross-sectional area greater than 10 mm^2
- a single copper protective conductor having a cross-sectional area of not less than 4 mm^2
- two individual protective conductors
- an earth monitoring system which, in the event of a continuity fault occurring in the protective conductor, automatically disconnects the supply to the equipment
- connection of the equipment to the supply by means of a double-wound transformer or equivalent unit, such as a motor-alternator set.

3.10.10 Earthing conductors

Where buried in the ground, the earthing conductor shall have a cross-sectional area not less than that stated in Table 3.3.

WR-542.3.1

Table 3.3 Minimum cross-sectional area of a buried earthing conductor (reproduced with permission of IET)

	Protected against mechanical damage	Not protected against mechanical damage
Protected against corrosion by a sheath	$2.5\,mm^2$ copper $10\,mm^2$ steel	$16\,mm^2$ copper $16\,mm^2$ coated steel
Not protected against corrosion	$25\,mm^2$ copper $50\,mm^2$ steel	

The connection of an earthing conductor to an earth electrode or other means of earthing shall be:

WR-542.3.2

- soundly made; and
- electrically and mechanically satisfactory; and
- labelled; and
- suitably protected against corrosion.

3.10.11 Earthing systems

The type of earthing system to be used for the installation shall be determined, taking into consideration the characteristics of the source of energy and (in particular) any earthing facilities.

WR-312.3.1

Note: The types of earthing system are: TN-C, TN-S, TN-C-S, TT and IT (See Section 3.4).

3.10.12 Electrical separation

Electrical separation is a protective measure in which:

- basic protection is provided by basic insulation of live parts or by barriers or enclosures; and
- fault protection is provided by simple separation of the separated circuit from other circuits and from earth.

Electrical separation may only supply one item of current-using equipment from one unearthed source with simple separation.

WR-413.1.2

3.10.13 Electrical services

A voltage Band I circuit shall not be contained in the same wiring system as a Band II circuit unless (for a multicore cable or cord) the cores of the Band I circuit are separated from the cores of the Band II circuit by an earthed metal screen of equivalent current-carrying capacity to that of the largest core of a Band II circuit.

WR-528.1

3.10.14 Emergency switching

Emergency switching may be emergency switching ON or emergency switching OFF.

Unless the neutral conductor can be regarded as being reliably connected to earth in a TN-S or TN-C-S system, then the neutral conductor need **not** be isolated or switched.

WR-537.4.1.2

3.10.15 Earthing terminals or bars

In every installation a main earthing terminal shall be provided to connect the following to the earthing conductor:

WR-542.4.1

- circuit protective conductors
- protective bonding conductors
- functional earthing conductors (if required)
- lightning protection system bonding conductor, if any.

To enable the resistance of the earthing arrangements to be measured, the earthing conductor needs to be capable of being easily disconnected.

WR-542.4.2

Joints in an earthing conductor shall be capable of being disconnected **only** by means of a tool.

WR-542.4.2

The neutral (star) point of the secondary windings of three-line transformers and generators (or the midpoint of the secondary windings of a single-line transformer or generator) shall be connected to earth. WR 411.8.4.2

Main protective bonding conductors of each installation shall connect extraneous conductive parts to the main earthing terminal including the following: WR-411.3.1.2

- central heating and air conditioning systems
- exposed metallic structural parts of the building
- gas installation pipes
- water installation pipes
- other installation pipework and ducting.

 Where an installation serves more than one building the above requirement shall be applied to each building.

In a TN system, exposed conductive parts of the installation shall be connected by a protective conductor to the main earthing terminal of the installation (which shall, in turn, be connected to the earthed point of the power supply system). WR-411.4.2

In a TT system, exposed conductive parts of the installation shall be protected by a single protective device connected, via the main earthing terminal, to a common earth electrode. WR-411.5.1

For a TN-S system, the main earthing terminal of the installation shall be connected to the earthed point of the source of energy. WR-542.1.2

 Part of the connection may be formed by the distributor's lines and equipment.

For a TN-C-S system, where protective multiple earthing is provided, the main earthing terminal of the installation shall be connected, by the distributor, to the neutral of the source of energy. WR-542.1.3

For a TT or IT system, the main earthing terminal shall be connected via an earthing conductor to an earth electrode. WR-542.1.4

3.10.16 Fault protection

All exposed conductive parts of the reduced low-voltage system shall be connected to earth.	WR-411.8.3
The earth fault loop impedance at every point of utilisation, including socket-outlets, shall be such that the disconnection time does not exceed 5 s.	WR-411.8.3
Where a circuit-breaker is used, the maximum value of earth fault loop impedance (Z_s) shall be determined by the formula $Z_s \times I_a \leq U_o$.	WR-411.8.3

where:

Z_s is the impedance in ohms (Ω) of the fault loop

I_a is the current in amperes (A) causing the automatic operation of the disconnecting device within the time specified in Table 41.1 (p. 46 of BS 7671:2008)

U_o is the nominal a.c. rms or d.c. line voltage to earth in volts (V).

Where a fuse is used, the maximum values of earth fault loop impedance (Z_s) for 5 s disconnection time and U_o of 55 V (single-line) and 63.5 V (three-line) shall be in accordance with the values stated in Table 41.6 of BS 7671:2008, p. 54.

Where fault protection is provided by an RCD, the product of the rated residual operating current (IM) in amperes and the earth fault loop impedance in ohms shall not exceed 50.	WR-411.8.3
Fault protection may be omitted for unearthed street furniture supplied from an overhead line and inaccessible in normal use.	WR-410.3.9
Live parts of the separated circuit shall **not** be connected at any point to another circuit or to earth or to a protective conductor.	WR-413.3.3
No exposed conductive part of the separated circuit shall be connected either to the protective conductor or exposed conductive parts of other circuits, or to earth.	WR-413.3.6

3.10.17 High leakage current

Any insulation or insulating arrangement of extraneous   WR-612.5.2
conductive parts shall not pass a leakage current
exceeding 1 mA in normal conditions of use.

In supply systems, RCMs may be installed to reduce the   WR-538.4.1
risk of operation of the protective device in the event
of excessive leakage current of the installation or the
connected appliances.

Note: A residual current monitor permanently monitors any leakage current
in the downstream installation or part of it.

Where an RCM is used in an a.c. IT system, the use of   WR-538.4
a directionally discriminating RCM is recommended in
order to avoid inopportune signalling of leakage current
where high leakage capacitances are liable to exist
downstream from the point of installation of the RCM.

3.10.18 Isolation

Isolation is intended, for reasons of safety, to make a circuit dead by separating
an installation or section from every source of electric energy.

Every circuit shall be capable of being isolated from   WR-537.2.1.1
each of the live supply conductors.

Note: In a TN-S or TN-C-S system, it is not necessary to isolate or switch
the neutral conductor where it is regarded as being reliably connected to earth
by suitably low impedance.

3.10.19 Protective and neutral (PEN) conductors

PEN conductors may only be used within an installation   WR-543.4.1
where the installation is supplied by a privately owned   and 543.4.2
transformer or converter in such a way that there is no
metallic connection (except for the earthing connection)
with the distributor's network.

The PEN conductor shall be connected to the terminals   WR-543.4.9
or bar intended for the protective earthing conductor
and the neutral conductor.

3.10.20 Protective bonding conductors (PME)

Except where protective multiple earthing (PME)   WR-544.1.1
conditions apply, a main protective bonding conductor
shall have a cross-sectional area not less than half the
cross-sectional area required for the earthing conductor
of the installation and not less than 6 mm^2.

Where supplementary bonding is to be applied to a fixed   WR-544.2.5
appliance (which is supplied via a short length of flexible
cord from an adjacent connection unit or other accessory,
incorporating a flex outlet) the circuit protective conductor
within the flexible cord shall be deemed to provide the
supplementary bonding connection to the exposed
conductive parts of the appliance, from the earthing
terminal in the connection unit or other accessory.

3.10.21 Protective conductors

Where the protective conductor is formed by conduit,   WR-543.2.7
trunking, ducting or the metal sheath and/or armour of
a cable, the earthing terminal of each accessory shall be
connected by a separate protective conductor to an
earthing terminal incorporated in the associated box or
other enclosure.

Except where the circuit protective conductor is formed   WR-543.2.9
by a metal covering or enclosure containing all of the
conductors of the ring, the circuit protective conductor
of every ring final circuit shall also be run in the form
of a ring having both ends connected to the earthing
terminal at the origin of the circuit.

A switching device shall **not** be inserted in a protective   WR-543.3.4
conductor unless:

- the switch has been inserted in the connection between
  the neutral point and the means of earthing; and
- the switch is a linked switch arranged to disconnect
  and connect the earthing conductor
  for the appropriate source, at substantially the
  same time as the related live conductors.

Where electrical monitoring of earthing is used, no   WR-543.3.5
dedicated devices (e.g. operating sensors, coils) shall be
connected in series with the protective conductor.

3.10.22 Protective devices

The rated breaking capacity of a protective device shall **not** be less than the maximum prospective short-circuit or earth fault current at the point at which the device is installed unless back-up protection is provided.	WR-536.1
In a TN-S or TN-C-S system the neutral conductor need not be isolated or switched where it can be regarded as being reliably connected to earth by a suitably low impedance.	WR-537.1.2

 Note: A lower breaking capacity is permitted if another protective device (a back-up protective device) having the necessary breaking capacity is installed on the supply side and the characteristics of the devices are suitably co-ordinated such that the energy let-through of the upstream device does not exceed that which can be withstood without damage by the downstream device.

Where an installation is supplied from more than one source of energy (one of which requires a means of earthing independent of the means of earthing of other sources and there is a need to ensure that not more than one means of earthing is applied at any time), a switch may be inserted in the connection between the neutral point and the means of earthing, **provided** that the switch is a linked switch arranged to disconnect and connect the earthing conductor for the appropriate source, at substantially the same time as the related live conductors.	WR-537.1.5

3.10.23 Protective devices and switches

Switches, circuit-breaker (except where linked) or fuses shall be inserted in an earthed neutral conductor.	WR-132.14.2
Any linked switch or linked circuit-breaker inserted in an earthed neutral conductor shall be capable of breaking all of the related line conductors.	WR-132.14.2

3.10.24 Protective earthing

Exposed conductive parts shall be connected to a protective conductor.	WR-411.3.1.1
Simultaneously accessible exposed conductive parts shall be connected to the same earthing system individually, in groups or collectively.	WR-411.3.1.1
Conductors for protective earthing shall comply with Chapter 54 of BS 7671:2008.	WR-411.3.1.1
A circuit protective conductor shall be run to and terminated at each point in wiring and at each accessory (except a lampholder having no exposed conductive parts and suspended from such a point).	WR-411.3.1.1

3.10.24.1 Protective equipotential bonding – mobile and transportable units

An IT system can be provided by: WR-717.411.6.2

- an isolating transformer or a low-voltage generating set, with an insulation monitoring device installed; or
- a transformer providing simple separation via:
  - automatic disconnection of the supply in case of a first fault between live parts and the frame of the unit, or
  - an RCD; or

The protective measures of: WR-717.417
WR-717.418

- obstacles and placing out of reach, are **not** permitted
- non-conducting location, is **not** permitted
- earth-free local equipotential bonding, is not recommended.

Where an alternative system for the automatic disconnection of supply is available, an IT system shall **not** be used.

Note: Connection of a lightning protection system to the protective equipotential bonding shall be made in accordance with BS EN 62305.

3.10.25 Protective measures

A protective measure shall consist of:

- a combination of basic protection and an independent provision for fault protection; or
- an enhanced protective provision which provides both basic **and** fault protection.

 Note: An example of an enhanced protective measure is reinforced insulation.

Protective measures such as earth-free local equipotential   WR-410.3.6
bonding shall **only** be applied where the installation is
under the supervision of skilled or instructed persons so
that unauthorised changes cannot be made.

Protection by extra-low voltage is a protective measure which consists of either SELV or PELV, and these are considered to be protective measures in all situations.

Protection by extra-low voltage provided by SELV requires   WR-414.1.1
basic insulation between the SELV system and earth.

3.10.26 RCDs

The rated residual operating current of the protective   WR-531.2.3
device shall comply with the requirements appropriate
to the type of system earthing.

In a TN system, where, for certain equipment in a   WR-531.3.1
certain part of the installation, the characteristics of
a protective device do not satisfy the requirements,
that part may be protected by an RCD. In these sorts
of circumstance, the exposed conductive parts of
that part of the installation shall be connected to the
TN earthing system's protective conductor or to a
separate earth electrode which affords an impedance
appropriate to the operating current of the RCD.

For a TN-S system where the neutral is not isolated, any   WR-551.6.2
RCD shall be positioned to avoid incorrect operation
due to the existence of any parallel neutral–earth path.

In an IT system (where protection is provided by an RCD and disconnection following a first fault is not envisaged) the non-operating residual current of the device shall be at least equal to the current which circulates on the first fault to earth of negligible impedance affecting a line conductor.

WR-531.5.1

 In an IT system, an RCD may **not** operate unless one of the earth faults is on a part of the system on the supply side of the device.

3.10.27 Requirements for SELV and PELV circuits

- SELV circuits shall have basic insulation between live parts and earth.   WR-414.4.1
- The PELV circuits and/or exposed conductive parts of equipment supplied by the PELV circuits may be earthed.   WR-414.4.1
- The earthing of PELV circuits may be achieved by a connection to earth or to an earthed protective conductor within the source itself.   WR-414.4.1
- Exposed conductive parts of a SELV circuit shall **not** be connected to earth.   WR-414.4.1

 Note: Basic protection is generally unnecessary in normal dry conditions for PELV circuits where the nominal voltage does not exceed 25 V a.c. or 60 V d.c. and exposed conductive parts and/or the live parts are connected by a protective conductor to the main earthing terminal.

3.10.28 Sources

The neutral (star) point of the secondary windings of three-line transformers and generators (or the midpoint of the secondary windings of single-line transformers and generators) shall be connected to earth.

WR-411.8.4.2

3.10.28.1 Low-voltage generating sets

The prospective short-circuit current and prospective earth fault current shall be assessed for each source of supply or combination of sources which can operate independently of other sources or combinations.

WR-551.2.2

Protection by automatic disconnection of supply shall not rely upon the connection to the earthed point of the system for distribution of electricity to the public when the generator is operating as a switched alternative to a TN system. — WR-551.4.3.2.1

A suitable means of earthing **shall** be provided. — WR-551.4.3.2

 Connection of live parts of the generator with earth may affect the fault protection provided by the equipment.

3.10.29 Warning notices: earthing and bonding connections

A permanent label to BS 951 will be available for all earthing and bonding connections with the words: — WR-514.13.1

Safety Electrical Connection – Do Not Remove

Figure 3.18 Warning notice – earthing and bonding

This shall: — WR-514.13.1

- be permanently fixed in a visible position at or near:
  - the point of connection of every earthing conductor to an earth electrode, and
  - the point of connection of every bonding conductor to an extraneous conductive part, and
  - the main earth terminal, where separate from main switchgear.

Where electrical separation to the supply to more than one item of current-using equipment, is used (Regulations 418.2.5 or 418.3), the warning notice shall read as per that shown in Figure 3.19. — WR-514.13.2

3.10.30 Warning notices: periodic inspection and testing

Where an installation incorporates an RCD a notice shall be fixed in a prominent position at or near the origin of the installation and shall read as per Figure 3.20. — WR-514.12.2

> The protective bonding conductors associated with the electrical installation in this location MUST NOT BE CONNECTED TO
>
> # EARTH
>
> Equipment having exposed-conductive-parts connected to earth must not be brought into this location

Figure 3.19 Warning notice – protective bonding conductors (reproduced with permission of IET)

> This installation, or part of it, is protected by a device which automatically switches off the supply if an Earth fault develops. Test quarterly by pressing the button marked 'T' or 'Test'. The device should switch off the supply and should then be switched on to restore the supply. If the device does not switch off the supply when the button is pressed, seek expert advice.

Figure 3.20 Warning notice – RCD protection (reproduced with permission of IET)

3.10.31 Testing

3.10.31.1 Insulation resistance

The insulation resistance shall be measured between live conductors and between live conductors and the protective conductor connected to the earthing arrangement.	WR-612.3.1
The insulation resistance measured with the test voltages indicated in Table 3.4 shall be considered satisfactory if the main switchboard and each distribution circuit tested separately, with all its final circuits connected but with current-using equipment disconnected, has an insulation resistance not less than the appropriate value given in Table 3.4.	WR-612.3.2

More stringent requirements are applicable for the wiring of fire alarm systems in buildings; see BS 5839-1.

Table 3.4 Minimum values of insulation resistance (reproduced with permission of IET)

Circuit nominal voltage (V)	Test voltage d.c. (V)	Minimum insulation resistance
SELV and PELV	250	>0.5
Up to and including 500V with the exception of the above systems	500	>1.0
Above 500V	1000	>1.0

Where the circuit includes electronic devices which are likely to influence the results or be damaged, only a measurement between the live conductors connected together and the earthing arrangement shall be made.

WR-612.3.3

3.10.31.2 Insulation resistance/impedance of floors and walls

In a non-conducting location (see Regulation 418.1 of BS 7671:2008) at least three measurements shall be made in the same location, one of these measurements being approximately 1 m from any accessible extraneous conductive part in the location. The other two measurements shall be made at greater distances.

These measurements shall be repeated for each relevant surface of the location.

WR-612.5.1

 Note: Further information on measurement of the insulation resistance/impedance of floors and walls can be found in Appendix 13 of BS 7671: 2008.

3.10.31.3 Protection by electrical separation

The separation of the live parts from those of other circuits and from earth shall be confirmed by a measurement of the insulation resistance. The resistance values obtained shall be in accordance with Table 3.4.

WR-612.4.3

3.10.31.4 Polarity

A test of polarity shall be made and it shall be verified that (except for E14 and E27 lampholders to BS EN 60238 and/or circuits with an earthed neutral conductor), centre contact bayonet and Edison screw lampholders have the outer (or screwed contacts) connected to the neutral conductor.	WR-612.6

3.10.31.5 Earth electrode resistance

Where the earthing system incorporates an earth electrode as part of the installation, the electrode resistance to earth shall be measured.	WR-612.7
Where the installation incorporates an earth electrode, the test of Regulation 612.7 shall also be carried out before the installation is energised.	WR-612.1
If any test indicates a failure to comply, that test and any preceding test, the results of which may have been influenced by the fault indicated, shall be repeated after the fault has been rectified.	WR-612.1

3.10.31.6 Earth fault loop impedance

Where protective measures are used which require a knowledge of earth fault loop impedance, the relevant impedances shall be measured.	WR-612.9

 Note: Further information on measurement of earth fault loop impedance can be found in Appendix 14 of BS 7671:2008.

3.10.31.7 Prospective fault current

The prospective short-circuit current and prospective earth fault current shall be measured at the origin and at other relevant points in the installation.	WR-612.11

3.10.32 Special locations and installations

3.10.32.1 Agricultural and horticultural premises

 In agricultural and horticultural premises, a TN-C system shall **not** be used in the installation of a protective device used for the automatic disconnection of supply.

The protective measures of non-conducting location and earth-free local equipotential bonding are **not** permitted.	WR-705.410.3.6
In locations intended for livestock, supplementary bonding shall connect all exposed conductive parts and extraneous conductive parts that can be touched by livestock.	WR-705.415.2.1
In circuits (whatever the type of earthing system used) the following disconnection device shall be provided:	WR-705.411.1

- in final circuits supplying socket-outlets with rated current not exceeding 32 A, an RCD with a rated residual operating current not exceeding 30 mA
- in final circuits supplying socket-outlets with rated current more than 32 A, an RCD with a rated residual operating current not exceeding 100 mA
- in all other circuits, RCDs with a rated residual operating current not exceeding 300 mA.

3.10.32.2 Conducting locations with restricted movement

If a functional earth is required for certain equipment (for example measuring and control equipment), equipotential bonding shall be provided between all exposed conductive parts and extraneous conductive parts inside the conducting location with restricted movement and the functional earth.	WR-706.411.1.2
The unearthed source shall have simple separation and shall be situated outside the conducting location with restricted movement, unless the source is part of the fixed installation within the conducting location with restricted movement.	WR-706.413.1.2

A TN-C-S system shall **not** be used for the supply   WR-704.411.3.1
to a construction site, except for the supply to a
fixed building of the construction site.

3.10.32.3 Electrical installations in caravans, motor caravans and camping parks

 The use of a TN-C-S system for the supply to a caravan **is prohibited** by ESQCR.

The protective measures of non-conducting location and earth-free local equipotential bonding are **not** permitted.	WR-721.410.3.6
Structural metallic parts which are accessible from within the caravan shall be connected through main protective bonding conductors to the main earthing terminal within the caravan.	WR-721.411.3.1.2
Every low-voltage socket-outlet, other than those supplied by an individual winding of an isolating transformer, shall incorporate an earth contact.	WR-721.55.2.1
In a TN system, the final circuits for the supply to caravans or similar shall **not** include a PEN conductor.	WR-740.411.4
The protective measures of non-conducting location and earth-free local equipotential bonding are **not** permitted.	WR-708.410.3.6
If an electrical installation in a caravan, camping park, exhibition, show and/or stand is supplied from a TN system, then **only** a TN-S installation shall be installed.	WR-708.411.4

3.10.32.4 Electrode water heaters and boilers

If an electrode water heater or electrode boiler is connected to a three-line low-voltage supply, the shell of the electrode water heater or electrode boiler shall be connected to the neutral of the supply as well as to the earthing conductor.	WR-554.1.5

If the supply to an electrode water heater or electrode boiler is single-line and one electrode is connected to a neutral conductor earthed by the distributor, the shell of the electrode water heater or electrode boiler shall be connected to the neutral of the supply as well as to the earthing conductor.	WR-554.1.6
If the electrode water heater or electrode boiler is not piped to a water supply or is in physical contact with any earthed metal (and where the electrodes and the water in contact with the electrodes are so shielded in insulating material that they cannot be touched while the electrodes are live), a fuse in the line conductor may be substituted for the circuit-breaker and the shell of the electrode water heater or electrode boiler need not be connected to the neutral of the supply.	WR-554.1.7

3.10.32.5 Exhibitions, shows and stands

The protective measures of non-conducting location and earth-free local equipotential bonding are **not** permitted.	WR-711.410.3.6
Structural metallic parts which are accessible from within the stand, vehicle, wagon, caravan or container shall be connected through the main protective bonding conductors to the main earthing terminal within the unit.	WR-711.411.3.1.2
Where protective equipotential bonding is used for the automatic disconnection of supply (and the type of system earthing is TN) the installation shall be a TN-S system.	WR-711.411.4

3.10.32.6 Floor and ceiling heating systems

The protective measures of non-conducting location and earth-free local equipotential bonding are **not** permitted.	WR-753.410.3.6

3.10.32.7 Locations containing a bath or shower

The protective measures of non-conducting location and earth-free local equipotential bonding are **not** permitted.	WR-701.410.3.6

3.10.32.8 Marinas and similar locations

 The use of a TN-C-S system for the supply to boat or similar construction **is prohibited** by ESQCR.

The protective measures of non-conducting location and earth-free local equipotential bonding are **not** permitted.	WR-709.410.3.6
For marinas, particular attention shall be given to the likelihood of corrosive elements, movement of structures, mechanical damage, presence of flammable fuel and the increased risk of electric shock due to:	WR-709.512.2
• presence of water • reduction in body resistance • contact of the body with earth potential.	
For a TN system in a marina (or similar location) the final circuits for providing the supply to pleasure craft or houseboats shall **not** include a PEN conductor.	WR-709.411.4

3.10.32.9 Mobile or transportable units

Accessible conductive parts of the unit, such as the chassis, shall be connected through the main protective bonding conductors to the main earthing terminal within the unit.	WR-717.411.3.1.2
In the UK a TN-C-S system shall **not** be used to supply a mobile or transportable unit except: • where the installation is continuously under the supervision of a skilled or instructed person; and • the suitability and effectiveness of the means of earthing has been confirmed before the connection is made.	WR-717.411.4

Identification – a permanent notice shall be fixed to   WR-717.514
the unit in a prominent position (preferably adjacent
to the supply inlet connector) which shall state in clear
and unambiguous terms the following:

- the type of supply which may be connected to the
  unit
- the voltage rating of the unit
- the number of lines and their configuration
- the on-board earthing arrangement
- the maximum power requirement of the unit.

In each installation main protective bonding   WR-411.3.1.2
conductors shall connect extraneous conductive parts
to the main earthing terminal including the following:

- central heating and air conditioning systems
- exposed metallic structural parts of the building
- gas installation pipes
- water installation pipes
- other installation pipework and ducting.

Where an installation serves more than one building the above requirement
shall be applied to each building.

Protective equipotential bonding shall be applied   WR-411.3.1.2
to any metallic sheath of any telecommunication
cable.

Earth-free local equipotential bonding is intended to prevent the appearance
of a dangerous touch voltage and shall **only** be used in special circumstances.

 Note: In an IT system, the protective measure of earth-free local equipotential bonding is not recommended.

3.10.32.10 Outdoor lighting

The earthing conductor of a street electrical fixture   WR-559.10.3.4
shall have a minimum copper equivalent cross-
sectional area not less than that of the supply neutral
conductor at that point or not less than 6 mm^2,
whichever is the smaller.

> For an outdoor lighting installation, where the   WR-559.10.4
> protective measure for the whole installation is
> by double or reinforced insulation, no protective
> conductor shall be provided and the conductive parts
> of the lighting column shall not be intentionally
> connected to the earthing system.

For an outdoor lighting installation, a metallic structure (such as a fence or grid) which is in the proximity of but is not part of the outdoor lighting installation need not be connected to the main.

3.10.32.11 Solar photovoltaic (PV) power supply systems

> Earthing of one of the live conductors of the d.c.   WR-712.312.2
> side is permitted, if there is at least simple separation
> between the a.c. side and the d.c. side.

Note: Any connections with earth on the d.c. side should be electrically connected so as to avoid corrosion (see BS 7361-1:1991).

> PV string cables, PV array cables and PV d.c. main   WR-712.522.8.1
> cables shall be selected and erected so as to minimise
> the risk of earth faults and short-circuits.

The protective measures of non-conducting location (Regulation 418.1) and earth-free local equipotential bonding (Regulation 418.2) are **not** permitted on the d.c. side

3.10.32.12 Swimming pools and other basins

Note: The following requirements apply to the basins of swimming pools, the basins of fountains and the basins of paddling pools as well as to the surrounding zones of these basins. In these areas, in normal use, the risk of electric shock is increased by a reduction in body resistance and contact of the body with earth potential.

> The protective measures of non-conducting location   WR-702.410.3.6
> and earth-free local equipotential bonding are **not**
> permitted.

It is permitted to install an electric heating unit embedded in the floor, provided that it:

WR-702.55.1

- is protected by SELV; or
- incorporates an earthed metallic sheath connected to the supplementary equipotential bonding and its supply circuit is additionally protected by an RCD; or
- is covered by an embedded earthed metallic grid connected to the supplementary equipotential bonding and its supply circuit is additionally protected by an RCD.

Special requirements may be necessary for swimming pools for medical purposes.

3.10.32.13 Rooms and cabins containing sauna heaters

The protective measures of non-conducting location earth-free local equipotential bonding are **not** permitted.

WR-703.410.3.6

3.10.32.14 Temporary electrical installations for structures, amusement devices and booths at fairgrounds, amusement parks and circuses

The protective measures of non-conducting location and earth-free local equipotential bonding are **not** permitted.

WR-740.410.3.6

Where the type of system earthing is TN:

WR-740.411.4

- a PEN conductor shall not be used downstream of the origin of the temporary electrical installation
- the final circuits for the supply to caravans or similar shall not include a PEN conductor.

Where a generator supplies a temporary installation, forming part of a TN, TT or IT system, care shall be taken to ensure that the earthing arrangements are in accordance with the Regulations.

WR-740.551.8

3.10.32.15 Water heaters having immersed and uninsulated heating elements

All metal parts of the heater or boiler which are in contact with the water (other than current-carrying parts) shall be solidly and metallically connected to a metal water pipe through which the water supply to the heater or boiler is provided and that water pipe shall be connected to the main earthing terminal by means independent of the circuit protective conductor.	WR-554.3.2

3.10.33 TN System

In a TN installation, all exposed conductive parts **shall** be connected (by a protective conductor) to the main earthing terminal of the installation and that terminal shall be connected to the earthed point of the supply source.	WR-411.4.2 WR-413-02-06
The neutral or midpoint of the power supply system **shall** be earthed.	WR-411.4.2

 Where a fuse is used to satisfy the requirements of Regulation 411.4.2, maximum values of earth fault loop impedance (Z_s) corresponding to a disconnection time of 0.4 s shall meet the requirements of Table 41.2 on p. 48 of BS 7671:2008 for a nominal voltage (U_o) of 230 V.

The maximum disconnection time for final circuits not exceeding 32 A shall be in accordance with Table 41.1 on p. 46 of BS 7671:2008.	WR-411.3.2.2
In a TN system, a disconnection time not exceeding 5 s is permitted for a distribution circuit and/or a circuit not covered by Table 41.1 on p. 46 of BS 7671:2008.	WR-411.3.2.3
This maximum disconnection time of 5 s shall apply to **all** circuits feeding fixed equipment used in highway power supplies.	WR-559.10.3.3

Note: Where a circuit-breaker is used to satisfy the requirements of Regulation 411.3.2.3, the maximum value of earth fault loop impedance (Z_s) shall be determined by the following formula:

$$Z_s \times I_a \le U_o$$

where:

Z_s is the impedance in ohms (Ω) of the fault loop
I_a is the current in amperes (A) causing the automatic operation of the disconnecting device within the time specified in Table 41.1 on p. 46 of BS 7671:2008
U_o is the nominal a.c. rms or d.c. line voltage to earth in volts (V).

Alternatively, for a nominal voltage (U_o) of 230 V and a disconnection time of 5 s (in accordance with Regulation WR-411.3.2.3) the values specified in Table 41.3 on p. 49 of BS 7671:2008, for the types and ratings of overcurrent devices listed may be used instead of the above calculation.	WR-411.4.7
Where a fuse is used for a distribution circuit or a final circuit in accordance with Regulation 411.3.2.3, maximum values of earth fault loop impedance (Z_s) corresponding to a disconnection time of 5 s are stated in Table 41.3 on p. 49 of BS 7671:2008, for a nominal voltage (U_o) of 230 V.	WR-411.4.8

Normally (see Regulation WR-431.1.1) detection of overcurrent shall be provided for all line conductors and shall cause the disconnection of the conductor in which the overcurrent is detected.

In a TN system, however, a circuit supplied between line conductors (and in which the neutral conductor is not distributed) overcurrent detection need not be provided for one of the line conductors, provided that:

• there exists, in the same circuit or on the supply side, differential protection intended to detect unbalanced loads and cause disconnection of all the line conductors; **and** • the neutral conductor is not distributed from an artificial neutral point of the circuits situated on the load side of this differential protective device.	WR-431.1.2

The following types of protective device may be used   WR-411.4.4
for fault protection in a TN system:
- an overcurrent protective device
- an RCD (in which case the circuit should also
  incorporate an overcurrent protective device).

 Note: Compliance with Regulation 411.4 shall be verified by:

- measurement of the earth fault loop impedance;
- verification of the characteristics and/or the effectiveness of the associated
  protective device.

3.10.33.2 PEN Conductors in a TN system

In a fixed installation, a single conductor may serve   WR-411.4.3
both as a protective conductor **and** as a neutral
conductor (i.e. PEN conductor).

For a TN system in a marina (or similar location) the   WR-709.411.4
final circuits for providing the supply to pleasure craft
or houseboats shall **not** include a PEN conductor.

Where the type of system earthing for temporary   WR-740.411.4
electrical installations (e.g. for structures, amusement
devices and booths at fairgrounds, amusement parks
and circuses) is TN, a PEN conductor shall not be
used downstream of the origin of the temporary
electrical installation.

In a TN system, the final circuits for the supply to   WR-740.411.4
caravans or similar shall **not** include a PEN conductor.

3.10.33.3 Neutral conductors in a TN system

In a TN system:

- the neutral conductor shall be protected against   WR-431.2.1
  short-circuit current
- overcurrent protective devices that are used as   WR-531.1.1
  fault protection devices shall be selected and
  erected in compliance with the requirements
  (see Chapter 41 of BS 7671:2008).

Consideration shall be given to the fact that, if the neutral conductor in a three-line TN system is interrupted, basic, double and reinforced insulation as well as components rated for the voltage between line and neutral conductors can be temporarily stressed with the line-to-line voltage.

3.10.33.4 RCDs in a TN system

An RCD shall **not** be used in a TN-C system.

If an RCD is used in a TN-C-S system, a PEN conductor shall **not** be used on the load side.

In a TN system, where a protective device fails to satisfy the requirements, that part of the system may be protected by an RCD. Where this happens, the exposed conductive parts of that part of the installation shall be connected to the TN earthing system protective conductor or to a separate earth electrode which provides an impedance appropriate to the operating current of the RCD.	WR-531.3.1

In this latter case the circuit shall be treated as a TT system.

3.10.33.5 Luminaires in a TN system

In a TN system, the outer contact of every Edison screw or single centre bayonet cap type lampholder (other than E1 4 and E27 lampholders) shall be connected to the neutral conductor.	WR-559.6.1.8

This Regulation also applies to track-mounted systems.

3.10.34 TN-C system

An RCD shall **not** be used in a TN-C system.	WR-411.4.4
In agricultural and horticultural premises, a TN-C system shall **not** be used in the installation of a protective device used for the automatic disconnection of supply.	WR-705.411.4

3.10.35 TN-S system

> For a TN-S system, the main earthing terminal of the   WR-542.1.2
> installation **shall** be capable of being connected to
> the earthed point of the source of energy. Part of the
> connection may be formed by the distributor's lines
> and equipment.

Note: This particularly applies to installations where the generating set provides a supply as a switched alternative to the system for distribution of electricity to the public (standby systems).

3.10.35.1 Neutral conductors in a TN system

> For a TN-S system where the neutral is **not** isolated,   WR-551.6.2
> any RCD shall be positioned to avoid incorrect
> operation due to the existence of any parallel neutral–
> earth path.
>
> In a TN-S system the neutral conductor need not be   WR-537.1.2
> isolated or switched where it can be regarded as being   WR-537.2.1.1
> reliably connected to earth by suitably low impedance.

Note: It may be desirable to disconnect the neutral of the installation from the neutral or PEN of the system for distribution of electricity to the public to avoid disturbances such as induced voltage surges caused by lightning.

3.10.35.2 Special installations and locations

> If an electrical installation in a caravan, camping park,   WR-708.411.4
> exhibition, show and/or stand is supplied from a TN   WR-711.411.4
> system, then **only** a TN-S installation shall be installed.

3.10.36 TN-C-S system

> In a TN-C-S system the neutral conductor need not be   WR-537.1.2
> isolated or switched where it can be regarded as being
> reliably connected to earth by a suitably low impedance.

Where an RCD is used in a TN-C-S system, a PEN conductor **shall not** be used on the load side.

3.10.36.1 Protective multiple earthing in a TN-C-S system

Part of the TN-C-S system uses a combined PEN conductor, which is at some point is split up into separate PE and N lines (see Figure 3.9). The combined PEN conductor typically occurs between the substation and the entry point into the building whereas within the building, separate PE and N conductors are used. This type of system is known as protective multiple earthing (PME), because of the practice of connecting the combined neutral-and-earth conductor to real earth at many locations so as to reduce the risk of broken neutrals.

Where protective multiple earthing is provided, the main earthing terminal of the installation shall be connected by the distributor to the neutral of the source of energy.	WR-542.1.3
Where an installation has more than one source of supply to which PME conditions apply, a main protective bonding conductor **shall** be selected according to the largest neutral conductor of the supply.	WR-544.1.1
Except for highway power supplies and street furniture, where PME conditions apply, the main protective bonding conductor shall be selected in accordance with the neutral conductor of the supply and Table 54.8 on p. 134 of BS 7671:2008.	WR-544.1.1

3.10.36.2 Special installations and locations

A TN-C-S system shall **not** be used for the supply to a construction site, except for the supply to a fixed building of the construction site.	WR-704.411.3.1
Where a metal grid is not laid in the floor of locations intended for livestock, a TN-C-S supply is not recommended.	WR-705.415.2.1

 The use of a TN-C-S system for the supply to a boat or to a caravan **is prohibited** by ESQCR.

A TN-C-S system **shall not** be used to supply a   WR-717.411.4
mobile or transportable unit except:

- where the installation is continuously under the
  supervision of a skilled or instructed person; and
- where the suitability and effectiveness of the
  means of earthing has been confirmed before the
  connection is made.

3.10.37 TT system

In a TT system, all exposed conductive parts which   WR-411.5.1
are protected by a single protective device **shall** be
connected, via the main earthing terminal, to a common
earth electrode.

 It cannot be over-emphasised that consideration should be given to the fact
that, if the neutral conductor in a three-line TT system is interrupted, basic,
double and reinforced insulation as well as components rated for the voltage
between line and neutral conductors can be temporarily stressed with the line-
to-line voltage.

For a TT system, the main earthing terminal shall   WR-542.1.4
be connected via an earthing conductor to an earth
electrode.

The earthing arrangements shall be such that: WR-542.1.6

- the value of impedance from the consumer's
  main earthing terminal to the earthed point of
  the supply for TN systems, or to earth for TT
  systems, is in accordance with the protective and
  functional requirements of the installation (and are
  continuously effective); and
- earth fault currents and protective conductor currents
  which may occur are carried without danger,
  particularly from thermal, thermomechanical and
  electromechanical stresses; and
- they are adequately robust or have additional
  mechanical protection appropriate to the assessed
  conditions of external influence.

Where a generator supplies a temporary installation   WR-740.551.8
that is part of a TT system, care shall be taken
to ensure that the earthing arrangements are in
accordance with the Regulations.

3.10.37.1 Protective devices in a TT system

One or more of the following types of protective device   WR-411.5.2
shall be used:

- an RCD (the preferred option)
- an overcurrent protective device

3.10.37.2 Overcurrent protective devices in a TT system

In a TT system, if overcurrent protective devices are   WR-531.1.2
used for fault protection, they shall be selected and
erected in compliance with the requirements (see
Chapter 41 of BS 7671:2008).

3.10.37.3 RCDs in a TT system

If an installation which is part of a TT system is   WR-531.4.1
protected by a single RCD, it shall be placed at the
origin of the installation (unless that part of the
installation between the origin and the device complies
with the requirements for protection by the use of Class
II equipment or equivalent insulation).

 Note: Where there is more than one origin this requirement applies to each origin.

Where an RCD is used for earth fault protection: WR-411.5.2

- the circuit should also incorporate an overcurrent
  protective device
- the disconnection time shall be: WR-411.5.3

$$R_A \times I_{\Delta n} \leq 50 \text{ V}$$

where

R_A is the sum of the resistances of the earth electrode and the protective conductor connecting it to the exposed conductive-parts (in ohms)

$I_{\Delta n}$ is the rated residual operating current of the RCD.

 Note: The requirements of this Regulation are met if the earth fault loop impedance of the final circuit protected by the RCD meets the requirements of Table 41.5 on p. 50 of BS 7671:2008.

Where RCDs are also used for protection against fire, the conditions for protection by automatic disconnection of the supply shall be verified by: WR-612.8

- measurement of the resistance of the earth electrode for exposed conductive parts of the installation
- verification of the characteristics and/or effectiveness of the associated protective device.

In a TT system where protection by automatic disconnection of an installation and/or generating set is not permanently fixed, an RCD (with a rated residual operating current of not more than 30 mA) shall be installed to protect every circuit.   WR-551.4.4.2

In locations where there is a risk of fire due to the nature of processed or stored materials (and except for mineral insulated cables, busbar trunking systems or powertrack systems) a wiring system shall be protected by an RCD having a rated residual operating current not exceeding 300 mA.   WR-422.3.9

3.10.37.4 Protection of line conductors in a TT system

In a TT system, where a circuit is supplied between line conductors and where the neutral conductor is not distributed, overcurrent detection need not be provided for one of the line conductors, provided that: WR-431.1.2

- there exists, in the same circuit (or on the supply side) differential protection intended to detect unbalanced loads and that will cause disconnection of all the line conductors; **and**
- the neutral conductor is not distributed from an artificial neutral point of the circuits situated on the load side of this differential protective device.

3.10.37.5 Luminaires in a TT system

In a TT system, the outer contact of every Edison   WR-559.6.1.8
screw or single-centre bayonet cap type lampholder
(other than E 14 and E 27 lampholders) shall be
connected to the neutral conductor.

This Regulation also applies to track mounted systems.

3.10.38 IT system

Earthing arrangements may be used jointly or separately   WR-542.1.5
for protective and functional purposes, according to the
requirements of the installation, such that:

- the value of impedance from the consumer's main   WR-542.1.6
  earthing terminal to the earthed point of the supply
  for systems, is in accordance with the protective
  and functional requirements of the installation, and
  considered to be continuously effective; and
- earth fault currents and protective conductor currents
  which may occur are carried without danger,
  particularly from thermal, thermomechanical and
  electromechanical stresses; and
- they are adequately robust or have additional
  mechanical protection appropriate to the assessed
  conditions of external influence
- the earthing arrangement may be considered   WR-442.1.2
  electrically independent of another earthing
  arrangement provided that a rise of potential (with
  respect to earth) in one earthing arrangement does
  not cause an unacceptable rise of potential (with
  respect to earth) in the other earthing arrangement.

Where an IT system has been selected for continuity of service, it is recommended that the IMD (insulation monitoring device) is combined with other
devices to enable the fault to be located while the circuit is operating.

In an IT system: WR-411.6.1

- all live parts **shall** be insulated from earth or
  connected to earth through a sufficiently high
  impedance either at the neutral point or midpoint of
  the system – or at an artificial neutral point

- precautions **shall** be taken to avoid the risk of
  a person being in contact with simultaneously
  accessible exposed conductive parts in the event of
  two faults occurring at the same time
- exposed conductive parts **shall** be earthed   WR-411.6.2
  individually, in groups, or collectively
- the main earthing terminal **shall** be connected via an   WR-542.1.4
  earthing conductor to an earth electrode.

 It is strongly recommended that IT systems with distributed neutrals should **not** be employed and where no neutral point or midpoint exists, a line conductor may be connected to earth through a high impedance.

3.10.38.1 Protective equipotential bonding in an IT system

In each installation main protective bonding   WR-411.3.1.2
conductors shall connect extraneous conductive parts
to the main earthing terminal including the following:

- central heating and air conditioning systems
- exposed metallic structural parts of the building
- gas installation pipes
- water installation pipes
- other installation pipework and ducting.

 Where an installation serves more than one building the above requirement shall be applied to each building.

Protective equipotential bonding shall be applied to the   WR-411.3.1.2
metallic sheath of telecommunication cables.

 Earth-free local equipotential bonding is intended to prevent the appearance of a dangerous touch voltage and shall **only** be used in special circumstances.

3.10.38.2 Protection of the neutral conductor in an IT system

In an IT system, the neutral conductor shall **not** be   WR-431.2.2
distributed unless:
- overcurrent detection is provided for the neutral
  conductor of every circuit; or

- the neutral conductor is effectively protected against short-circuit by a protective device installed on the supply side; or
- the circuit is protected by an RCD with a rated residual operating current not exceeding 0.2 times the current-carrying capacity of the corresponding neutral conductor.

If an IT system does not have a neutral conductor, it is permitted to omit the overload protective device in one of the line conductors **provided** that an RCD is installed in each circuit.

3.10.38.3 Overload protection in an IT system

The omission of devices for protection against overload is permitted for circuits supplying current-using equipment where unexpected disconnection of the circuit could cause danger or damage.

Examples of such circuits are:

- the exciter circuit of a rotating machine;
- the supply circuit of a lifting magnet;
- the secondary circuit of a current transformer;
- a circuit supplying a fire extinguishing device;
- a circuit supplying a safety service, such as a fire alarm or a gas alarm;
- a circuit supplying medical equipment used for life support in specific medical locations where an IT system is incorporated.

3.10.38.4 Supplies to an IT system

In an IT system: WR-512.1.1

- equipment **shall** be suitable for the nominal voltage (U_0)
- equipment **shall** be insulated for the nominal voltage between lines.

Where a generator supplies a temporary installation,   WR-740.551.8
forming part of an IT system:

- care shall be taken to ensure that the earthing arrangements are in accordance with the Regulations
- the neutral conductor of the star-point of the generator shall not be connected to the exposed conductive parts of the generator.

3.10.38.5 Monitoring devices in an IT system

Where an IT system is used as a means to ensure continuity of supply, insulation monitoring devices **shall** be provided which give an audible and visual indication of a first fault from a live part to an exposed conductive part or to earth.	WR-560.5.3 WR-411.6.3.1

 Note: This device shall initiate an audible and/or visual signal which shall continue as long as the fault persists.

The following monitoring devices and protective devices   WR-411.6.3
may be used:

- residual current devices (RCDs)
- residual current monitoring (RCM) devices
- insulation monitoring devices (IMDs)
- insulation fault location systems
- overcurrent protective devices.

Consideration should be given to the possibility that, if   WR-442.4
a line conductor of an IT system is earthed accidentally,
the insulation (or components) rated for the voltage
between line and neutral conductors can be temporarily
stressed with the line-to-line voltage.

 The object of this particular Regulation is to provide requirements for the safety of a low-voltage installation in the event of:

- a fault between the high-voltage system and earth in the transformer sub-station that supplies the low-voltage installation;
- loss of the supply neutral in the low-voltage system;
- short-circuit between a line conductor and neutral in the low-voltage installation;
- accidental earthing of a line conductor of a low-voltage IT system.

In locations where there is a strong risk of fire due to the   WR-422.3.9
nature of processed or stored materials, the wiring system
of an IT system (except for mineral insulated cables,
busbar trunking systems or powertrack systems) **shall**
be protected against insulation faults, by an IMD with
audible and visual signals.

Residual current devices (RCDs) in IT systems

In an IT system an RCD with a rated residual operating current of not more than 30 mA **shall** be installed to protect every circuit.	WR-551.4.4.2
Each item of equipment used outside the unit **shall** be protected by a separate RCD.	WR-717.411.6.2
Additional protection by an RCD having the characteristics **shall** be provided for every socket-outlet intended to supply current-using equipment outside the unit, with the exception of socket-outlets which are supplied from circuits with protection by: • SELV; or • PELV; or • electrical separation.	WR-717.415
Where protection is provided by an RCD (and disconnection following a first fault is not envisaged) the non-operating residual current of the device **shall** be at least equal to the current which circulates on the first fault to earth of negligible impedance affecting a line conductor.	WR-531.5.1

 In an IT system, an RCD may **not** operate unless one of the earth faults is on a part of the system on the supply side of the device.

Residual current monitoring (RCM) devices in IT systems

Except where a protective device is installed to interrupt the supply in the event of the first earth fault, an RCM device **shall** be provided to indicate the occurrence of a first fault from a live part to an exposed conductive part or to earth.	WR-411.6.3.2

 Note: This device shall initiate an audible and/or visual signal which shall continue as long as the fault persists.

After the occurrence of a first fault and in the event of a second fault occurring on a different live conductor, automatic disconnection of supply shall occur.	WR-411.6.4

Where an RCM is used in an a.c. IT system, it is recommended that a directionally discriminating RCM is used so as to avoid undue signalling of leakage current where high leakage capacitances are liable to exist downstream from the point of installation of the RCM.	WR-538.4
In an IT system where interruption of the supply in case of a first insulation fault to earth is **not** required or not permitted, an RCM may be installed to assist in the location of a fault.	WR-538.4.2

 An RCM is **not** intended to provide protection against electric shock.

Insulation monitoring devices (IMDs) for IT systems

An IMD is designed to indicate any important reduction of the insulation level of the system in order to find the cause before a second insulation fault occurs, thus avoiding any power supply interruption. An IMD is intended to be permanently connected to an IT system and to continuously monitor the insulation resistance of the complete system (secondary side of the power supply and the complete installation supplied by this power supply) to which it is connected.

The 'earth' or 'functional earth' terminal of the IMD shall be connected to the main earth terminal of the installation.	WR-538.1.2
The IMD shall be connected between earth and a live conductor of the monitored equipment.	WR-538.3
The IMD may be used for this purpose in all types of system earthing, except in TN-C systems.	WR-538.3
An IMD **shall** indicate when an fault earth fault is detected and the fault shall be located and eliminated, as soon as possible, in order to restore normal operating conditions.	WR-538.1
An IMD **shall** be used in a circuit comprising safety equipment which is normally de-energised by a switching device that disconnects all live poles and which is only energised in the event of an emergency. In these circumstances: • the reduction of the insulation level shall be indicated locally by a visual or an audible signal with the choice of remote indication • the IMD shall be connected between earth and a live conductor of the monitored equipment	WR-538.3

- the measuring circuit shall be automatically disconnected when the equipment is energised
- the IMD may be used for this purpose in all types of system earthing, except in TN-C systems.

IMDs, installed in locations where persons other than instructed persons or skilled persons have access to their use, shall be designed or installed in such a way that it shall be impossible to modify the settings, except by the use of a key, a tool or a password.   WR-538.1.3

The earth or functional earth terminal of the IMD shall be connected to the main earth terminal of the installation.

The supply circuit of the IMD shall be connected either to the installation on the same circuit of the connecting point of the line terminal and as close as possible to the origin of the system, or to an auxiliary supply.   WR-538.1.2

Where the installation is supplied from more than one power supply, connected in parallel, one IMD per supply shall be used, provided they are interlocked in such a way that only one IMD remains connected to the system.   WR-538.1.2

For d.c. installations, the line terminal(s) of the IMD shall be connected either directly to the midpoint, if any, or to one or all of the supply conductors.   WR-538.1.2

In some d.c. IT two-conductor installations, a passive IMD that does not inject current into the system may be used, provided that:

- the insulation of all live distributed conductors is monitored; and
- all exposed conductive parts of the installation are interconnected; and
- circuit conductors are selected and installed so as to reduce the risk of an earth fault to a minimum.

In some particular d.c. IT two-conductor installations, a passive IMD that does not inject current into the system may be used, **provided** that: WR-538.1.4

- the insulation of all live distributed conductors is monitored; and
- all exposed conductive parts of the installation are interconnected; and

- circuit conductors are selected and installed so as to reduce the risk of an earth fault to a minimum.

The connecting point to the installation shall be selected so that the IMD is able to monitor the insulation of the installation under all operating conditions.   WR-538.1.2

The line terminal(s) of the IMD shall be connected as close as practicable to the origin of the system to either: WR-538.1.2

- the neutral point of the power supply; or
- an artificial neutral point with impedances connected to the line conductors; or
- a line conductor or two or more line conductors.

Where the IMD is connected between one line and earth, it shall be suitable to withstand at least the line-to-line voltage between its line terminal and its earth terminal.   WR-538.1.2

The IMD shall be set to a lower value corresponding to the normal insulation of the system when operating normally with the maximum of loads connected.   WR-538.1.3

 An IMD is **not** intended to provide protection against electric shock.

Overcurrent protective devices

Overcurrent protective devices that are used in IT systems for fault protection, in the event of a second fault shall: WR-531.1.3

- comply with the relevant requirements (see Chapter 41 of BS 7671:2008)
- be suitable for line-to-line voltage applications for operation in the case of a second insulation fault
- disconnect all corresponding live conductors, including the neutral conductor, if any.

4

Safety protection

Around 1000 electrical accidents at work are reported to HSE every year and about 30 people die of their injuries. Many of these deaths and injuries arise from:

- use of poorly maintained electrical equipment;
- working near overhead power lines;
- contact with underground power cables during excavation work;
- work on or near 230 V domestic electricity supplies;
- use of unsuitable electrical equipment in explosive areas such as car paint spraying booths.

Fires started by poor electrical installations and faulty electrical appliances cause many additional deaths and injuries.

For this reason, protection against electric shock and safety protection methods are an essential part of the Regulations and the following mandatory requirement, therefore, needs to be observed:

 All installations shall comply with the requirements for safety protection in respect of:

- electric shock
- thermal effects
- overcurrent
- undervoltage
- isolation and switching.

 Note: 'Installation' in this context is taken to mean either as a whole or in its several parts.

In electrical installations, risk of injury may result from:

- arcing or burning, likely to cause blinding effects, excessive pressure and/ or toxic gases;
- excessive temperatures likely to cause burns, fires and other injurious effects;

- ignition of a potentially explosive atmosphere;
- mechanical movement of electrically actuated equipment;
- power supply interruptions and/or interruption of safety services;
- shock currents;
- undervoltages, overvoltages and electromagnetic influences likely to cause or result in injury or damage.

4.1 Basic safety requirements

The fundamental safety requirements of the IEE Wiring Regulations are as follows:

4.1.1 Mandatory requirements

- Protective safety measures shall be applied in every installation, part installation and/or item of equipment.
- Installations shall comply with the requirements for safety protection in respect of:
  - electric shock
  - thermal effects
  - overcurrent
  - fault current
  - undervoltage
  - isolation and switching.
- There shall be no detrimental influence between various protective measures used in the same installation, part installation or equipment.

4.1.2 Fundamental safety requirements

The following are précised details of the most important elements of the IEE Wiring Regulations that meet these fundamental design requirements.

4.1.2.1 Design

Electrical installations shall be designed for:

- the protection of persons, livestock and property;
- the proper functioning of the electrical installation;
- protection against mechanical and thermal damage;
- protection of people from an electric shock or fire hazard.

4.1.2.2 Characteristics of available supply or supplies

Detailed design characteristics shall be available for all supplies. These shall include:

- nature of current (a.c. and/or d.c.);
- purpose and number of conductors;

For a.c.	For d.c.
• line conductor(s)	• outer conductor
• neutral conductor	• middle conductor
• protective conductor	• earthed conductor
• PEN conductor	• live conductor
	• protective conductor
	• PEN conductor

- Values and tolerances of:
  - earth fault loop impedance
  - nominal voltage and voltage tolerances
  - nominal frequency and frequency tolerances
  - maximum current allowable
  - particular requirements of the distributor
  - prospective short-circuit current
  - protective measures inherent in the supply (e.g. earth, neutral or mid-wire).

4.1.2.3 Electricity distributors responsibilities

The electricity distributor is responsible for:

- evaluating and agreeing proposals for new installations or significant alterations to existing ones;
- ensuring that their equipment on consumers' premises:
  - is suitable for its purpose
  - is safe in its particular environment
  - clearly shows the polarity of the conductors;
- installing the cut-out and meter in a safe location;
- ensuring that the cut-out and meter is mechanically protected and can be safely maintained;
- providing an earthing facility for all new connections;
- maintaining the supply within defined tolerance limits;
- providing certain technical and safety information to the consumer to enable them to design their installations.

4.1.2.4 Installation and erection

- All electrical joints and connections shall meet stipulated requirements concerning conductance, insulation, mechanical strength and protection.
- Conductors shall be identified by colour, lettering and/or numbering.
- Connections and joints shall be accessible for inspection, testing and maintenance, unless:
  - they are in a compound-filled or encapsulated joint
  - the connection is between a cold tail and a heating element
  - the joint is made by welding, soldering, brazing or compression tool.
- Design temperatures shall not be exceeded by the installation of electrical equipment.

- Electrical equipment shall be arranged so that it is fully accessible (i.e. for operation, inspection, testing, maintenance and repair) and that there is sufficient space for later replacement.
- Equipment used for the supply of safety services shall be arranged to allow easy access for periodic inspection, testing and maintenance.
- Exposed parts of electrical equipment shall be located (or guarded) so as to prevent accidental contact and/or injury to persons or livestock.
- Good workmanship and proper materials shall be used.
- Installed electrical equipment shall minimise the risk of igniting flammable materials.
- Installed equipment must be accessible for operational, inspection and maintenance purposes.
- Installations shall be divided into circuits in order to:
  - avoid danger and minimise inconvenience in the event of a fault
  - facilitate safe operation, inspection, testing and maintenance.
- The process of erection shall not impair the characteristics of electrical equipment.

4.1.2.5 Identification and notices

Wiring shall be marked and/or arranged so that it can be quickly identified for inspection, testing, repair or alteration of the installation.

4.1.2.6 Inspection and testing

- Every electrical installation must be inspected and tested during erection and on completion **before** being put into service.
- Details of the general design characteristics of the electrical installation must be made available. These shall include the result of the assessment of general characteristics.
- Information (e.g. diagrams, charts, tables and/or schedules) must be made available to the person carrying out the inspection and testing and these (as a minimum) shall indicate:
  - the type and composition of each circuit (points of utilisation served, number and size of conductors, type of wiring)
  - the method used
  - the identification (and location) of all protection, isolation and switching devices
  - circuits or equipments that are susceptible to a particular test.
- If the inspection and tests are satisfactory, a signed Electrical Installation Certificate together with a Schedule of Inspections and a Schedule of Test Results (see Chapter 9) are to be given to the person responsible for ordering the work.
- Precautions shall be taken to avoid danger to persons and to avoid damage to property and installed equipment during inspection and testing.

	Qty		Sales Order	
ISBN	1		F 9676596 1	
9780750689731				
Customer P/O No			Cust P/O List	
PO40500/502/0004			21.99 GBP	

Title: Wiring regulations in brief : a complete guide to the requirements of

Format: P (Paperback)
Author: Tricker, Ray
Publisher: Butterworth – Heinemann
Fund:
Location:
Loan Type:
Coutts CN: 8210947

Order Specific Instructions

4.1.2.7 Maintenance

An assessment shall be made of the frequency and type of maintenance (e.g. periodic inspection, testing, maintenance and repair) that an installation can reasonably be expected to receive during its intended life.

4.1.3 Building Regulations requirements

The following are précised details of the most important elements of the Building Regulations Approved Documents and Standards concerning safety protection. They include:

- design, installation, inspection and testing of electrical installations;
- conservation of fuel and power;
- access and facilities for disabled people;
- extensions, material alterations and material changes of use.

4.1.4 Design, installation, inspection and testing of electrical installations

- All proposals to carry out electrical installation work **must** be notified to the local authority's Building Control Body before work begins, **unless** the proposed installation work:
  - is undertaken by a person who is a competent person registered with an electrical self-certification scheme; and
  - does not include the provision of a new circuit.
- Any work that involves adding a new circuit to a dwelling needs to be either notified to the Building Control Body (who will then inspect the work) or needs to be carried out by a competent person who is registered under a Government Approved Part P Self Certification Scheme.

 Note: Where a person who is **not** registered to self-certify, intends to carry out the electrical installation, then a Building Regulation (i.e. a Building Notice or Full Plans) application will need to be submitted together with the appropriate fee, based on the estimated cost of the electrical installation. The Building Control Body will then arrange to have the electrical installation inspected at first fix stage and tested upon completion.

- Reasonable provision shall be made in the design, installation, inspection and testing of electrical installations in order to protect persons from fire or injury.
- Sufficient information shall be provided so that persons wishing to operate, maintain or alter an electrical installation can do so with reasonable safety.

Work involving any of the following will also have to be notified to the Building Control Body:

- locations containing a bath tub or shower basin;
- swimming pools or paddling pools;
- hot air saunas;
- electric floor or ceiling heating systems;
- garden lighting or power installations;
- solar photovoltaic (PV) power supply systems;
- small-scale generators such as microCHP units;
- extra-low-voltage lighting installations, other than pre-assembled, CE-marked lighting sets.

Note: Whilst Part P of the Building Regulations makes requirements for the safety of fixed electrical installations, this does not cover system functionality (such as electrically powered fire alarm systems, fans and pumps), which are covered in other Parts of the Building Regulations and other legislation.

4.1.4.1 *Conservation of fuel and power*

Energy efficiency measures shall be provided which:

- provide lighting systems that utilise energy-efficient lamps with manual switching controls or, in the case of external lighting fixed to the building, automatic switching, or both manual and automatic switching controls as appropriate, such that the lighting systems can be operated effectively with regard to the conservation of fuel and power;
- provide information, in a suitably concise and understandable form (including results of performance tests carried out during the works) that shows building occupiers how the heating and hot water services can be operated and maintained.

The person responsible for achieving compliance should either themselves provide a certificate, or obtain a certificate from the sub-contractor, that commissioning has been successfully carried out. The certificate should be made available to the client and the building control body.

Responsibility for achieving compliance with these requirements rests with the person carrying out the work. That 'person' may be, for example, a developer, a main (or sub)contractor, or a specialist firm directly engaged by a private client.

4.1.4.2 *Access and facilities for disabled people*

In addition to the requirements of the Disability Discrimination Act 1995, precautions need to be taken to ensure that:

- new non-domestic buildings and/or dwellings (e.g. houses and flats used for student living accommodation); and

- extensions to existing non-domestic buildings; and
- non-domestic buildings that have been subject to a material change of use (e.g. so that they become a hotel, boarding house, institution, public building or shop)

are capable of allowing people, regardless of their disability, age or gender to:

- be able safely to use the facilities of the buildings (both as visitors and as people who live or work in them).

4.1.4.3 Extensions, material alterations and material changes of use

Where any electrical installation work is classified as an extension, a material alteration or a material change of use, the work must consider and include:

- confirmation that the mains supply equipment is suitable and can carry the additional loads envisaged;
- the amount of additions and alterations that will be required to the existing fixed electrical installation in the building;
- the earthing and bonding systems are satisfactory and meet the requirements;
- the necessary additions and alterations to the circuits which feed them;
- the protective measures required to meet the requirements;
- the rating and the condition of existing equipment (belonging to both the consumer and the electricity distributor) is sufficient.

 Note: Appendix C to Part P of the Building Regulations offers guidance on some of the older types of installation that might be encountered during alteration work and Appendix D provides guidance on the application of the now harmonised European cable identification system.

4.1.5 Protection from electric shock

- Protection against electric shock shall be provided.
- Protection against both basic (i.e. direct contact) and fault (i.e. indirect contact) shall be provided.
- Persons and livestock shall be protected against dangers that may arise from contact with exposed conductive parts during a fault.
- Persons and livestock shall be protected against dangers that may arise from contact with live parts of the installation.
- Live parts shall be completely covered with insulation which:
  - can only be removed by destruction
  - is capable of durably withstanding electrical, mechanical, thermal and chemical stresses normally encountered during service.
- Live parts shall be inside enclosures (or behind barriers) protected to at least IP2X or IPXXB.

- Bare (or insulated) overhead lines being used for distribution between buildings and structures shall be installed in accordance with the Electricity Safety, Quality and Continuity Regulations 2002.
- Bare live parts (other than overhead lines) shall not be within arm's reach.
- Bare live parts (other than an overhead line) shall not be within 2.5 m of:
  - an exposed conductive part
  - an extraneous conductive part
  - a bare live part of any other circuit.
- Simultaneously accessible exposed conductive parts shall be connected to the same earthing system either individually, in groups or collectively.
- Exposed parts of electrical equipment shall be located (or guarded) so as to prevent accidental contact and/or injury to persons or livestock.

The following methods are used for protection against direct contact (i.e. basic protection) and for protection against indirect contact (fault protection).

4.2 Basic protection against electric shock

A person may perform work involving direct contact with electrical parts only if the electrical part:

- is isolated from all sources of electricity;
- is tested to ensure its isolation from all sources of electricity; and
- is earthed if it is of high voltage.

Work may be performed by a person, operating plant or vehicle coming within the exclusion zone for an electrical part, only if the electrical part:

- is isolated from all sources of electricity;
- is tested to ensure its isolation from all sources of electricity; and
- is earthed if it is of high voltage.

To meet these requirements, the Regulations state that one of the following, basic, measures shall be used for protection against indirect contact:

- insulating live parts
- using a barrier or an enclosure
- using obstacles
- placing equipment out of reach
- using an RCD.

4.2.1 Protection by insulation of live parts

As the title suggests, this is a basic form of insulation protection and is intended to prevent contact with a live part of an electrical installation from direct contact. Paint, lacquers and varnishes do **not** provide adequate protection.

1–2 mA	Barely perceptible, no harmful effects
5–10 mA	Throw off, painful sensation
10–15 mA	Muscular contraction, can't let go!
20–30 mA	Impaired breathing
50 mA and above	Ventricular fibrillation and death

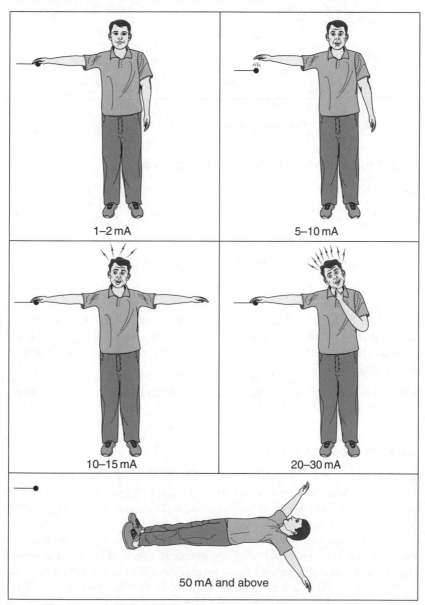

Figure 4.1 The effects of electric shock (courtesy of Brian Scaddan)

```
                        ┌──────────────┐
                        │  Obstacles   │
                        └──────────────┘
  ┌──────────────┐            │              ┌──────────────┐
  │ Barriers or  │            │              │ Placing out  │
  │ enclosures   │            │              │  of reach    │
  └──────────────┘            ▼              └──────────────┘
          ╲          ┌──────────────┐          ╱
           ╲         │              │         ╱
  ┌──────────────┐   │    Basic     │   ┌──────────────┐
  │ Insulation of│──▶│  Protection  │◀──│  Residual    │
  │  live parts  │   │              │   │  Current     │
  └──────────────┘   └──────────────┘   │  Devices     │
                                        └──────────────┘
```

Figure 4.2 Basic protection against electric shock

4.2.2 Protection by barriers or enclosures

The intention of this form of protection is to prevent or deter any contact with a live part. Whilst, generally speaking, this method is for protection against direct contact, it also provides a degree of protection against indirect contact.

4.2.3 Protection by obstacles and placing out of reach

Obstacles and placing out of reach will only provide basic protection and the intention of this form of protection is to prevent unintentional contact with a live part, but **not** an intentional contact caused by deliberately circumnavigating the obstacle. Protection by obstacles and placing out of reach is primarily intended for installations that are controlled or supervised by skilled persons.

Note: Protection by placing out of reach is intended only to prevent unintentional contact with live parts.

4.2.4 Protection by RCDs

In electrical installations, an RCD or an RCB (residual current circuit-breaker) is a circuit breaker that operates to disconnect a particular circuit whenever it detects current leaking out of that circuit (such as current leaking to earth through a ground fault) and it exceeds safety limits.

Figure 4.3 illustrates the construction of an RCD and works on the principle that in a normal (i.e. healthy) circuit, the magnetic effects of the line and neutral currents will cancel out because the same current will pass through the line coil and the load and then back through the neutral coil. In a faulty circuit where the line or the neutral are to earth, the currents will no longer be equal and the out of balance current will produce some residual magnetism in the core. As the magnetism will be alternating, it will link with the turns of the search coil and induce an EMF in it which will drive a current through the trip coil and cause the tripping mechanism to operate.

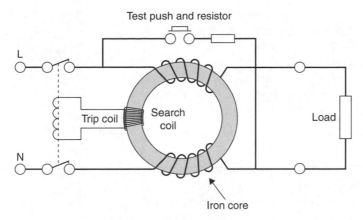

Figure 4.3 RCD (courtesy of Brian Scaddan)

 Although RCDs reduce the risk of electric shock, they should **not** be used as the sole means of protection against direct contact.

The use of RCDs with a rated residual operating current not exceeding 30 mA and an operating time not exceeding 40 ms is recognised in a.c. systems as providing additional protection in the event of failure of:

- one of the other methods of basic protection against electric shock; and/or
- the provision for fault protection; or
- carelessness by users.

 Note: The use of RCDs is not recognised as a sole means of protection and does not obviate the need to apply one of the other protective measures (such as automatic disconnection of supply, double or reinforced insulation, SELV or PELV).

4.2.5 Requirements from the Regulations – basic protection against electric shock

All electrical equipment shall comply with one of the provisions for basic protection (e.g. basic insulation; barriers, enclosures, obstacles or placing out of reach) where appropriate.	WR-411.2 WR-411.7.2
Basic protection is deemed to be provided where:	WR-414.2
• the nominal voltage cannot exceed the upper limit of voltage Band I; and • the supply is from a recognised source such as an safety isolating transformer, battery, diesel-driven generator, insulation testing equipment, monitoring device; motor-generator (see Section 4.1.3 of BS 7671:2008); and	

- exposed conductive parts of a SELV circuit
  shall not be connected:
  - to earth; or
  - to protective conductors; or
  - to exposed conductive parts of another
    circuit.

Basic protection is generally unnecessary in normal   WR-414.4.5
dry conditions for:

- SELV circuits where the nominal voltage does
  not exceed 25 V a.c. or 60 V d.c.
- PELV circuits where the nominal voltage does
  not exceed 25 V a.c. or 60 V d.c. and exposed-
  conductive parts and/or the live parts are
  connected by a protective conductor to the main
  earthing terminal.

In **all** other cases, basic protection is not required if
the nominal voltage of the SELV or PELV system
does not exceed 12 V a.c. or 30 V d.c.

4.2.5.1 Protection by insulation of live parts

 Live parts **shall** be completely covered with   WR 416.1
insulation which can only be removed by
destruction.

4.2.5.2 Protection by barriers or enclosures

The protective measures of placing out of reach   WR-559.10.1
and obstacles shall **not** be used except where:

- the maintenance of equipment is restricted to
  skilled persons who are specially trained
- items of street furniture are within 1.5 m
  of a low-voltage overhead line.

The protective measures of obstacles and placing   WR-705.410.3.5
out of reach are **not** permitted in the following:

- construction and demolition site installations
- conducting locations with restricted movement
- electrical installations in caravan/camping
  parks and similar locations

- electrical installations in caravans and motor caravans
- exhibitions, shows and stands
- floor and ceiling heating systems
- locations containing a bath or shower
- mobile or transportable units WR-717.417
- rooms and cabins containing sauna heaters WR-703.410.3.5
- swimming pools and other basins. WR-702.410.3.5

Live parts shall be inside enclosures or behind WR- 416.2.1
barriers providing at least the degree of protection
IPXXB or IP2X.

A barrier or enclosure shall be: WR-416.2.3

- firmly secured in place
- have sufficient stability and durability to maintain the required degree of protection and appropriate separation from live parts.

If it is necessary to remove a barrier or open an WR-416.2.4
enclosure or remove parts of enclosures, then this
shall **only** be possible:

- by the use of a key or tool; or
- after disconnection of the supply to live parts; or
- where an intermediate barrier (with a degree of protection of at least IPXXB or IP2X) prevents contact with live parts.

A horizontal top surface of a barrier or enclosure WR-416.2.2
which is readily accessible shall provide a degree of
protection of at least IPXXD or IP4X.

This Regulation does not apply to:

- a ceiling rose complying with BS 67
- a cord operated switch complying with BS 3676
- a bayonet lampholder complying with BS EN 61184
- an Edison screw lampholder complying with BS EN 60238.

If an item of equipment (such as a capacitor) is WR-416.2.5
installed behind a barrier or in an enclosure and
that equipment could retain a dangerous electrical
charge after it has been switched off, a warning
label shall be provided.

Where the protective measure automatic
disconnection of supply is used:

WR-559.10.3.1

- all live parts of electrical equipment shall be
  protected: by insulation; or
- by barriers or enclosures providing basic
  protection.

All conductive parts of operational electrical
equipment that are only separated from live parts by
basic insulation shall be contained in an insulating
enclosure affording at least the degree of protection
IPXXB or IP2X.

WR-412.2.2.1

The insulating enclosure shall not:

- be traversed by conductive parts likely to
  transmit a potential; or
- contain any screws or other fixing means
  which might need to be removed (e.g. during
  installation and maintenance) and which
  'could' be replaced by metallic screws or
  some other type of fixing that could affect the
  enclosure's insulation.

If the insulating enclosure must be traversed by
mechanical joints or connections (e.g. for operating
handles of built-in equipment), then these should
be arranged so that protection against shock in the
case of a fault is not impaired.

WR-412.2.2.2

Where a lid or door in an insulating enclosure
can be opened without the use of a tool or key,
all conductive parts which are accessible if the
lid or door is open shall be behind an insulating
barrier (providing a degree of protection not less
than IPXXB or IP2X) preventing persons from
coming unintentionally into contact with those
conductive parts.

WR-412.2.2.3

 This insulating barrier shall be removable only by the use of a tool or key.

No conductive part enclosed in the insulating
enclosure shall be connected to a protective
conductor.

WR-412.2.2.4

No exposed conductive part or intermediate part shall be connected to any protective conductor unless specific provision for this is made in the specification for the equipment concerned.	WR-412.2.2.4
The enclosure shall not affect the operation of the equipment that it is protecting.	WR-412.2.2.5

4.2.5.3 Protection by obstacles and placing out of reach

Obstacles shall prevent: • unintentional bodily approach to live parts; and • unintentional contact with live parts during the operation of live equipment in normal service.	WR-417.2.1
Obstacles shall be secured to prevent unintentional removal.	WR-417.2.2
Obstacles may be removable without using a key or tool.	WR-417.2.2

Protection by placing out of reach is intended only to prevent unintentional contact with live parts.

Simultaneously accessible parts at different potentials shall **not** be within arm's reach.	WR-417.3.1
A bare live part (other than an overhead line) shall **not** be within arm's reach or within 2.5 m of: • an exposed conductive part • an extraneous conductive part • a bare live part of any other circuit.	WR-417.3.1

Note: Bare (or insulated) overhead lines used for distribution between buildings and structures shall be installed in accordance with the Electricity Safety, Quality and Continuity Regulations 2002.

The protective measures of placing out of reach and obstacles shall **not** be used except where: • the maintenance of equipment is restricted to skilled persons who are specially trained	WR-559.10.1

- items of street furniture are within 1.5 m of a
  low-voltage overhead line.

4.2.5.4 Protection by RCDs

Every installation shall be divided into circuits, as necessary, to reduce the possibility of unwanted tripping of RCDs due to excessive protective conductor currents produced by equipment.	WR-314.1
In a.c. systems, additional protection by means of an RCD shall be provided for:	WR-411.3.3

- socket-outlets with a rated current not
  exceeding 20 A that are used by ordinary
  persons unless:
  - they are used under the supervision of
    skilled or instructed persons (e.g. in some
    commercial or industrial locations), or
  - a suitably labelled and/or identified socket-
    outlet is provided for connection of a
    particular item of equipment
  - it is mobile equipment with a current rating
    not exceeding 32 A for use outdoors.

If a generating set is connected, protection by RCDs shall remain effective for every intended combination of sources of supply.	WR-551.4.2

4.3 Fault protection (protection against indirect contact)

Persons and livestock shall be protected against dangers that may arise from contact with exposed conductive parts during a fault.	WR-131.2.2

Indirect contact (i.e. when part of the body touches or is in dangerous proximity to any object that is in contact with energised electrical equipment or
exposed conductive parts which might become live under fault conditions)
has always been a potential problem to the unwary when installing, maintaining or inspecting electrical installations. It should also be remembered that as

voltages increase, the potential for arcing increases and through arcing, injuries and/or fatalities will often occur **even** if actual bodily contact with high-voltage lines and/or equipment is not made!

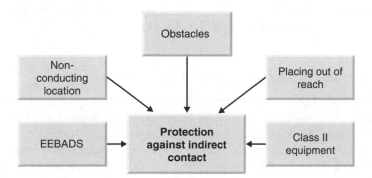

Figure 4.4 Protection against indirect contact

To meet these requirements, the Regulations state that one of the following basic measures shall be used for protection against indirect contact:

- earthed equipotential bonding and automatic disconnection of supply (EEBADS);
- non-conducting location;
- protection by obstacles and placing out of reach;
- Class II equipment or equivalent insulation.

4.3.1 Protection by earthed equipotential bonding and automatic disconnection of supplies (EEBADS)

An earthed equipotential zone is a zone within which exposed conductive parts and extraneous conductive parts are maintained at substantially the same potential by bonding, such that, under fault conditions, the differences in potential between simultaneously accessible exposed and extraneous conductive parts will not cause electric shock.

Earthed equipotential bonding provides a very good form of protection against indirect contact by joining together (i.e. bonding) all of the metallic parts together and then connecting them to earth. This ensures that all metalwork is at (or near) zero volts and so, under fault conditions, all metalwork will rise to a similar potential and simultaneous contact with two metal parts will not result an electric shock as there is no significant PD between them.

For installations and locations with an increased risk of shock (such as bathrooms and saunas etc.) additional measures may be required, such as:

- automatic disconnection of supply by means of an RCD with a rated residual operating current (Isn) not exceeding 30 mA;

- supplementary equipotential bonding;
- reduction of maximum fault clearance time.

Note: The application of protection by earthed equipotential bonding (and automatic disconnection of supply) will depend on the requirements of the type of system earthing in use (e.g. TN, TT, IT).

4.3.2 Protection by non-conducting location

This method of protection is **not** recognised for general application and may not be used in installations such as agricultural and horticultural buildings (saunas, caravans etc.) that are subject to an increased risk of shock.

This method of protection is intended to prevent simultaneous contact with parts which may be at different potentials (i.e. through the failure of the basic insulation of live parts) and a 'non-conducting location' is a location where there is no earthing or protective system because:

- there is nothing which needs to be earthed;
- exposed conductive parts are arranged so that it is impossible to touch two of them (or an exposed conducting part and an extraneous conductive part) at the same time.

4.3.3 Protection by obstacles and placing out of reach

Obstacles and placing out of reach will only provide basic protection and the intention of this form of protection is to prevent unintentional contact with a live part, but **not** an intentional contact caused by deliberately circumnavigating the obstacle. Protection by obstacles and placing out of reach is primarily intended for installations that are controlled or supervised by skilled persons.

4.3.4 Protection by Class II equipment or equivalent insulation

Class II equipment is unique in that as well as providing the basic insulation for live parts, it has a second layer of insulation, which can be used either to prevent contact with exposed conductive parts or to make sure that there can never be any contact between such exposed conductive parts and live parts.

Class II protection is provided by one or more of the following:

- electrical equipment having double or reinforced insulation;
- low-voltage switchgear;
- low-voltage controlgear assemblies;
- supplementary insulation;
- reinforced insulation applied to uninsulated live parts.

4.3.5 Requirements from the Regulations – fault protection against electric shock

4.3.5.1 Protection by earthed equipotential bonding and automatic disconnection of supplies (EEBADS)

In each installation main protective bonding   WR-411.3.1.2
conductors shall connect extraneous conductive
parts to the main earthing terminal including the
following:

- central heating and air conditioning systems
- exposed metallic structural parts of the
  building
- gas installation pipes
- water installation pipes
- other installation pipework and ducting.

Where an installation serves more than one building the above requirement
shall be applied to each building.

Protective equipotential bonding shall be applied to   WR-411.3.1.2
any metallic sheath of a telecommunication cable.

Connection of a lightning protection system to the protective equipotential
bonding shall be made in accordance with BS EN 62305.

4.3.5.2 Protection by obstacles and placing out of reach

Obstacles shall prevent: WR-417.2.1

- unintentional bodily approach to live parts; and
- unintentional contact with live parts during the
  operation of live equipment in normal service.

Obstacles shall be secured to prevent unintentional   WR-417.2.2
removal.

Obstacles may be removable without using a key   WR-417.2.2
or tool.

Protection by placing out of reach is intended only to prevent unintentional
contact with live parts.

Simultaneously accessible parts at different potentials shall **not** be within arm's reach.   WR-417.3.1

A bare live part (other than an overhead line) shall **not** be within arm's reach or within 2.5 m of: WR-417.3.1

- an exposed conductive part
- an extraneous conductive part
- a bare live part of any other circuit.

 Note: Bare (or insulated) overhead lines used for distribution between buildings and structures shall be installed in accordance with the Electricity Safety, Quality and Continuity Regulations 2002.

The protective measures of placing out of reach and obstacles shall **not** be used except where: WR-559.10.1

- the maintenance of equipment is restricted to skilled persons who are specially trained
- items of street furniture are within 1.5 m of a low-voltage overhead line.

4.4 Protection against both direct and indirect contact

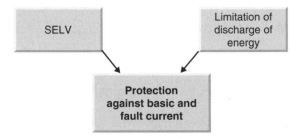

Figure 4.5 Protection against basic and fault contact

The Regulations state that one of the following basic measures shall be used for protection against both direct contact and indirect contact:

- SELV; or
- limitation of discharge of energy.

4.4.1 Protection by SELV

SELV (separated extra-low voltage) is an extra-low-voltage system that is electrically separated from earth and from other systems so that a single fault cannot give rise to the risk of electric shock. PELV (protective extra-low voltage) is an extra-low-voltage system which is not electrically separated from earth, but which otherwise satisfies all the requirements for SELV.

SELV is a term used to describe the highest voltage level that can be contacted by a person without causing injury. It is usually defined as 60 V d.c.

4.4.2 Limitation of discharge of energy

This type of protection may also be used for electric fences supplied from electric fence controllers complying with BS EN 61011 or BS EN 61011-1.

Requirements from the regulations for protection by SELV

The separation of the live parts from those of other circuits and from clarity shall be confirmed by a measurement of the insulation resistance. The resistance values obtained shall be in accordance with Table 4.1 (see below).	WR-612.4.1

More stringent requirements are applicable for the wiring of fire alarm systems in buildings (see BS 5839-1).

Table 4.1 Minimum values of insulation resistance (reproduced with permission of IET)

Circuit nominal voltage (V)	Test voltage d.c. (V)	Minimum insulation resistance
SELV and PELV	250	>0.5
Up to and including 500 V with the exception of the	500	>1.0
clarity Above 500 V	1000	>1.0

4.5 Additional requirements

The following are additional requirements for installations and locations where the risk of electric shock is increased by a reduction in body resistance and/or by contact with earth potential.

4.5.1 Protective bonding conductors

Equipotential bonding ensures that protective devices will operate and remove dangerous potential differences, before a hazardous shock can be delivered.

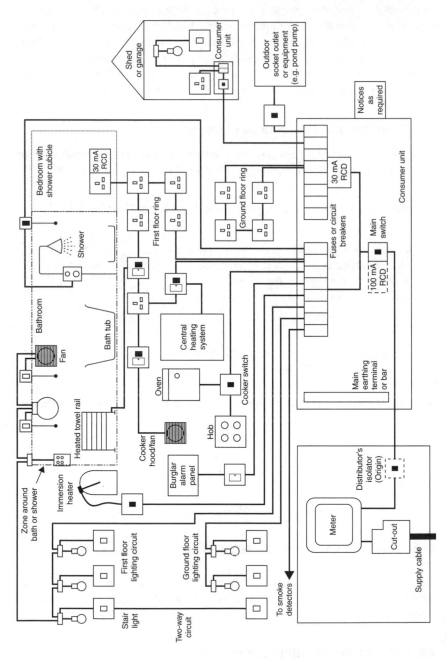

Figure 4.6 Typical fixed installations that might be encountered in new (or upgraded) existing dwellings

This is achieved by making sure that all of the installation's earthed metalwork (i.e. exposed conductive parts) is connected to other metalwork (i.e. extraneous conductive parts) via the earth conductor to provide an earth fault current path that ensures dangerous potential differences cannot occur.

Main equipotential bonding conductors connect together the installation earthing system and the metalwork of other services such as gas, electricity and water as close as possible to their point of entry to the building.

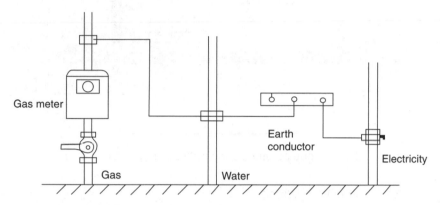

Figure 4.7 Main equipotential bonding

Supplementary bonding conductors connect together extraneous conductive parts – that is, metalwork which is not associated with the electrical installation but which may provide a conducting path that could give rise to shock.

4.5.2 Main equipotential bonding conductors

The Regulations require that the main equipotential bonding conductors for every electrical installation is connected to the main earthing terminal of that particular installation and that these shall include the following:

- water service pipes (but see requirements for domestic buildings in Chapter 2);
- gas installation pipes;
- other service pipes and ducting;
- central heating and air conditioning systems;
- exposed metallic structural parts of the building;
- the lightning protective system.

Note: Where an installation serves more than one building, the above requirement shall be applied to each building.

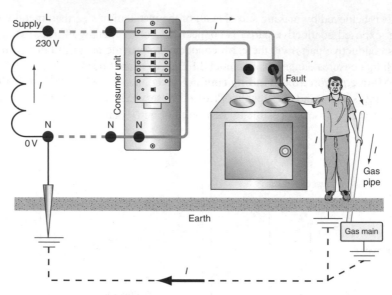

Figure 4.8 Earthed equipotential bonding (courtesy of Brian Scaddan)

4.5.2.1 Protective earthing

Automatic disconnection of supply is a protective measure in which fault protection is provided by protective earthing.

Safety Electrical Connection – Do Not Remove

Figure 4.9 Earthing and bonding notice

4.5.3 Supplementary bonding conductors

For installations and locations where there is an increased risk of shock (such as agricultural and horticultural premises and building sites) additional measures may be required, such as reduction of maximum fault clearance time and supplementary equipotential bonding.

Locations which contain a bath or shower and where body resistance is lowered as a result of water, are potentially very hazardous environments and it is important to ensure that no dangerous potentials exist between exposed and extraneous conductive parts. For this reason, local supplementary equipotential bonding needs to be provided to connect together the terminals of the protective conductors of each circuit supplying Class I and Class II equipment with extraneous conductive parts in those zones, such as:

- metallic pipes supplying services and metallic waste pipes (e.g. water, gas);
- metallic central heating pipes;

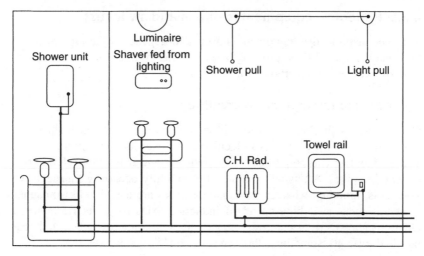

Figure 4.10 Supplementary equipotential bonding (courtesy of Brian Scaddan)

- air conditioning systems;
- accessible metallic structural parts of the building;
- metallic baths and shower basins.

4.5.4 Protective conductors

A protective conductor is a conductor that provides a measure of protection against electric shock and is used to connect together any of the following parts:

- exposed conductive parts;
- extraneous conductive parts;
- the main earthing terminal;
- earth electrode(s);
- the earthed point of the source.

A circuit protective conductor, on the other hand, is an arrangement of conductors that join all of the exposed conductive parts together and connect them to the main earthing terminal. There are many types of circuit protective conductor such as:

- a separate conductor;
- a conductor included in a sheathed cable with other conductors;
- the metal sheath and/or armouring of a cable;
- a conducting cable enclosure (such as conduit or trunking);
- exposed conductive parts (such as the conducting cases of equipment).

4.5.5 Protective equipment (devices and switches)

The type of protective equipment chosen will depend on the type of protection that is required to be provided (e.g. whether overcurrent, earth fault current, overvoltage or undervoltage).

4.5.6 Protection against overvoltage

Overvoltage is the hazardous condition that occurs when the voltage in a circuit (or part of a circuit) is suddenly raised over its upper limit. An overvoltage incident can be permanent or transient and it is often referred to as a 'voltage spike'. A typical example of a naturally occurring transient overvoltage is lightning, whereas man-made sources are usually electromagnetic induction when switching on or off inductive loads (e.g. electric motors or electromagnets). Transient overvoltage might last microseconds and reach hundreds of volts, sometimes thousands of volts, in amplitude.

Figure 4.11 Lightning

In accordance with the Regulations, additional protection against overvoltages of atmospheric origin is not necessary for:

- installations that are supplied by low-voltage systems which do not contain overhead lines;
- installations that are supplied by low-voltage networks which contain overhead lines and with a location subject to less than 25 thunderstorm days per year;
- installations that contain overhead lines with a location subject to less than 25 thunderstorm days per year;

provided that the installations meet the required minimum equipment impulse to withstand voltages shown in Table 4.2.

Table 4.2 Required minimum impulse to withstand voltage (kV)

Nominal voltage of the installation (V)	Category IV (equipment with very high impulse voltage)	Category III (equipment with high impulse voltage)	Category II (equipment with normal impulse voltage)	Category I (equipment with reduced impulse voltage)
230/240 277/480	6	4	2.5	1.5
400/690 1000	8	6	4	2.5

Values to be determined by system engineer or, in the absence of information, the values for 400/690 can be chosen.

 Suspended cables with insulated conductors that have earthed metallic coverings are considered to be an 'underground cable'.

4.5.7 Requirements from the Regulations

4.5.7.1 Accessibility of electrical equipment

Electrical equipment shall be arranged so that: WR-132.12

- there is sufficient space for the initial installation and later replacement of individual items of electrical equipment
- the equipment is accessible for operation, inspection, testing, fault detection, maintenance and repair.

4.5.7.2 Additional protection

For an addition or alteration to an existing installation, it WR-610.4
shall be verified that the addition or alteration complies with the Regulations and does not impair the safety of the existing installation.

4.5.7.3 Additions and alterations to an installation

No addition or alteration, temporary or permanent, shall WR-131.8
be made to an existing installation:

- unless, in particular, the earthing and bonding arrangements used as a protective measure for the safety of the addition or alteration are adequate

- unless it has been ascertained that the rating and
  the condition of any existing equipment (including
  that of the distributor) will be adequate for the
  altered circumstances
- unless the earthing and bonding arrangements
  used as a protective measure for the safety of the
  addition or alteration are adequate.

4.5.7.4 Automatic supply

Safety services may be required to operate at all material times where people
or livestock are at risk – including during mains and local supply failure and
through fire conditions. To meet this requirement, specific sources, equipment, circuits and wiring are necessary.

For safety services required to operate in fire conditions: WR-560.5.2

- a safety source of supply shall be selected which
  will maintain a supply of adequate duration
- equipment shall be provided, either by construction
  or by erection, with protection ensuring fire-
  resistance of adequate duration.

Note: The safety source is generally additional to the normal source. A normal source is, for example, the public supply network.

4.5.7.5 Circuits

Circuits of safety services: WR-560.7.1

- shall be independent of other circuits
- shall not pass through zones exposed to explosion   WR-560.7.2
  risk (BE3)
- shall not pass through locations exposed to fire risk   WR-560.7.3
  (BE2) unless they are fire-resistant.

Protection against overload may be omitted where the loss of supply may cause
a greater hazard provided that the occurrence of an overload is indicated.

Overcurrent protective devices shall be selected and   WR-560.7.4
erected so as to avoid an overcurrent in one circuit
impairing the correct operation of other circuits of
safety services.

4.5.7.6 Combined protective and neutral (PEN) conductors

PEN conductors may only be used within an installation:	WR-543.4.1 and 543.4.2
• where any necessary authorisation for use of a PEN conductor has been obtained and where the installation complies with the conditions for that authorisation; or • where the installation is supplied by a privately owned transformer or converter in such a way that there is no metallic connection (except for the earthing connection) with the distributor's network; or • where the supply is obtained from a private generating plant.	
For a fixed installation, a conductor of a cable not subject to flexing and having a cross-sectional area not less than $10\,mm^2$ for copper or $16\,mm^2$ for aluminium may serve as a PEN conductor provided that the part of the installation concerned is not supplied through an RCD.	WR-543.4.3
The outer conductor of a concentric cable shall not be common to more than one circuit.	WR-543.4.4
The conductance of the outer conductor of a concentric cable (measured at a temperature of 20°C) shall:	WR-543.4.5
• for a single-core cable, be not less than that of the internal conductor • for a multicore cable serving a number of points contained within one final circuit or having the internal conductors connected in parallel, be not less than that of the internal conductors connected in parallel.	
At every joint in the outer conductor of a concentric cable and at a termination, the continuity of that joint shall be supplemented by a conductor additional to any means used for sealing and clamping the outer conductor.	WR-543.4.6
No means of isolation or switching shall be inserted in the outer conductor of a concentric cable.	WR-543.4.7

Other than a cable conforming to BS EN 60702-1, each PEN conductor of every cable shall be insulated or have an insulating covering suitable for the highest voltage to which it may be subjected.	WR-543.4.8
If, from any point of the installation, the neutral and protective functions are provided by separate conductors, those conductors shall not then be reconnected together beyond that point.	WR-543.4.9
At the point of separation, separate terminals or bars shall be provided for the protective and neutral conductors.	WR-543.4.9
The PEN conductor shall be connected to the terminals or bar intended for the protective earthing conductor and the neutral conductor.	WR-543.4.9

4.5.7.7 Conditions of installation

Electrical equipment shall be selected so as to withstand safely the stresses, the environmental conditions and the characteristics of its location.	WR-133.3

4.5.7.8 Cross-sectional area of conductors

The cross-sectional area of conductors shall be determined both for normal operating conditions and, where appropriate, for fault conditions according to: • the admissible maximum temperature • the voltage drop limit • the electromechanical stresses likely to occur due to short-circuit and earth fault currents • other mechanical stresses to which the conductors are likely to be exposed • the maximum impedance for operation of short-circuit and earth fault protection • the method of installation • harmonics • thermal insulation.	WR-132.6

4.5.7.9 Design

The electrical installation shall be designed to provide for: • the protection of persons, livestock and property • the proper functioning of the electrical installation for the intended use.	WR-132.1

4.5.7.10 Disconnecting devices

Disconnecting devices shall be provided so as to allow electrical installations, circuits or individual items of equipment to be switched off or isolated for the purposes of operation, inspection, fault detection, testing, maintenance and repair.	WR-132.10

4.5.7.11 Earthing arrangements and protective conductors

Where protective bonding conductors are installed (especially in photovoltaic power supply systems) they shall be parallel to and in as close contact as possible with d.c. cables and a.c. cables and accessories.	WR-712.54

4.5.7.12 Earthing arrangements for protective purposes

Where overcurrent protective devices are used for fault protection, the protective conductor shall be incorporated in the same wiring system as the live conductors or in their immediate proximity.	WR-543.6.1

4.5.7.13 Earthing requirements for the installation of equipment having high protective conductor currents

Equipment having a protective conductor current exceeding 3.5 mA but not exceeding 10 mA, shall be either permanently connected to the fixed wiring of the installation without the use of a plug and socket-outlet or connected by means of a plug and socket-outlet complying with BS EN 60309-2.	WR-543.7.1.1

Equipment having a protective conductor current exceeding 10 mA shall be connected to the supply:	WR-543.7.1.2

- permanently via the wiring of the installation; or
- via a flexible cable with a plug and socket-outlet; or
- via an protective conductor with an earth monitoring system.

The wiring of every final circuit and distribution circuit intended to supply one or more items of equipment (such that the total protective conductor current is likely to exceed 10 mA) shall have a high integrity protective connection complying with one or more of the following:	WR-543.7.1.3

- a single protective conductor with a cross-sectional area greater than $10 \, mm^2$
- a single copper protective conductor having a cross-sectional area of not less than $4 \, mm^2$
- two individual protective conductors
- an earth monitoring system which, in the event of a continuity fault occurring in the protective conductor, automatically disconnects the supply to the equipment
- connection of the equipment to the supply by means of a double-wound transformer or equivalent unit, such as a motor-alternator set.

Where two protective conductors are used, the ends of the protective conductors shall be terminated independently of each other at all connection points throughout the circuit (such as the distribution board, junction boxes and socket-outlets).	WR-543.7.1.4
At the distribution board, information shall be provided indicating those circuits having a high protective conductor current.	WR-543.7.1.5

 Note: This information shall be positioned so as to be visible to a person who is modifying or extending the circuit.

Protective bonding conductors shall be protected against mechanical damage and corrosion, and shall be selected to avoid electrolytic effects.	WR-705.544.2

Socket-outlet protective conductors shall not be connected to any PEN conductor of the electricity supply	WR-708.553A 14

4.5.7.14 Electrical safety service supply

An electrical safety service supply is either:

- a non-automatic supply, the starting of which is initiated by an operator; or
- an automatic supply, the starting of which is independent of an operator.

4.5.7.15 Emergency control

Where in case of danger there is the necessity for immediate interruption of supply, an interrupting device shall be installed in such a way that it can be easily recognised and effectively and rapidly operated.	WR-132.9

4.5.7.16 Environmental conditions

Equipment likely to be exposed to weather, corrosive atmospheres or other adverse conditions shall be so constructed or protected as may be necessary to prevent danger arising from such exposure.	WR-132.5.1
Equipment in surroundings susceptible to risk of fire or explosion shall be so constructed or protected, and such other special precautions shall be taken, as to prevent danger.	WR-132.5.2

4.5.7.17 Erection of electrical installations

Electrical equipment shall be installed in accordance with the instructions provided by the manufacturer of the equipment.

The characteristics of the electrical equipment shall not be impaired by the process of erection.	WR-134.1.2

Electrical joints and connections shall be properly constructed with regard to conductance, insulation, mechanical strength and protection.	WR-134.1.4
Electrical equipment shall be installed so that design temperatures are not exceeded.	WR-134.1.5
Electrical equipment that is likely to cause high temperatures or electric arcs shall be placed (or guarded) so as to minimise the risk of ignition of flammable materials.	WR-134.1.6
Where the temperature of an exposed part of electrical equipment is likely to cause injury to persons or livestock, that part shall be so located or guarded in order to prevent accidental contact.	WR-134.1.6
Where necessary, suitable safety warning signs and/or notices shall be provided.	WR-134.1.7

4.5.7.18 Erection and initial verification of electrical installations

Where necessary, suitable safety warning signs and/or notices shall be provided.	WR-134.1.7

4.5.7.19 External influences

The selection of equipment according to external influences is necessary for proper functioning and to ensure the reliability of the measures of protection for safety.	WR-5122.4

4.5.7.20 Initial verification

The person or persons responsible for the design, construction, inspection and testing of the installation shall provide the person ordering the work with a Certificate which takes account of their respective responsibilities for the safety of that installation, together with the schedules described in Regulation 632.1 of BS 7671:2008.	WR-632.3

4.5.7.21 Inspection

Note: Inspection shall precede testing and shall normally be done with that part of the installation under inspection disconnected from the supply.

The inspection shall be made to verify that the installed electrical equipment is: • in compliance; and • correctly selected and erected; and • not visibly damaged or defective so as to impair safety.	WR-611.2

4.5.7.22 Installation of equipment for insulation fault location in an IT system

An IMD shall be used on a circuit comprising safety equipment which is normally de-energised by a switching means disconnecting all live poles and which is only energised in the event of an emergency (provided that the IMD is automatically deactivated whenever the safety equipment is activated).	WR-538.3

4.5.7.23 Isolation and switching

Effective means shall be provided so that all voltage may be cut off from every installation, circuit and item of equipment, so as to prevent or remove danger.	WR-132.15.1
Fixed electric motors shall be provided with an efficient means of switching off, readily accessible, easily operated and located so as to prevent danger.	WR-132.15.2

4.5.7.24 Luminaires

A luminaire with a lamp that could eject flammable materials in the case of failure should be: • equipped with a safety protective shield • be constructed with a safety protective shield.	WR-422.3.1 WR-422.4.2

> Any cable or cord between the fixing means and
> the luminaire shall be installed so that any expected
> stresses in the conductors, terminals and terminations
> do not interfere with the safety of the installation.
>
> WR-559.6.1.5

4.5.7.25 New materials and inventions

> Where the use of a new material or invention leads to
> departures from the Regulations, the resulting degree
> of safety of the installation shall **not** be less than that
> obtained by compliance with the Regulations.
>
> WR-120.4

4.5.7.26 Omission of devices for protection against overload for safety reasons

The omission of devices for protection against overload is permitted for circuits supplying current-using equipment where unexpected disconnection of the circuit could cause danger or damage.

Examples of such circuits are:

- the exciter circuit of a rotating machine;
- the supply circuit of a lifting magnet;
- the secondary circuit of a current transformer;
- a circuit supplying a fire extinguishing device;
- a circuit supplying a safety service, such as a fire alarm or a gas alarm;
- a circuit supplying medical equipment used for life support in specific medical locations where an IT system is incorporated.

In such situations consideration should be given to the provision of an overload alarm.

4.5.7.27 Periodic inspection and testing

> Periodic inspection consisting of a detailed
> examination of an installation shall be carried out
> by appropriate tests to show that the requirements
> for disconnection times for protective devices, are
> complied with, to provide for:
>
> - safety of persons and livestock against the
>   effects of electric shock and burns
>
> WR-621.2

- protection against damage to property by fire and heat arising from an installation defect
- confirmation that the installation is not damaged or deteriorated so as to impair safety
- the identification of installation defects and departures from the requirements of these Regulations that may give rise to danger.

4.5.7.28 Precautions within a fire-segregated compartment

The risk of spread of fire shall be minimised by the selection of appropriate materials and erection.	WR-527.1.1
A wiring system shall be installed so that the general building structural performance and fire safety are not reduced.	WR-527.1.2
Where safety depends on the direction of rotation of a motor, provision shall be made for the prevention of reverse operation due to, for example, a phase reversal.	WR-537.5.4.3

4.5.7.29 Preservation of electrical continuity of protective conductors

A protective conductor shall be suitably protected against mechanical and chemical deterioration and electrodynamic effects.   WR-543.3.1

A protective conductor with a cross-sectional area up to and including $6\,mm^2$ shall be protected throughout by a covering at least equivalent to that provided by the insulation of a single-core non-sheathed cable of appropriate size having a voltage rating of at least 450/750 V unless it is: WR-543.3.2

- a protective conductor forming part of a multicore cable
- cable trunking or conduit used as a protective conductor.

Where the sheath of a cable incorporating an uninsulated protective conductor of cross-sectional area up to and including $6\,mm^2$ is removed adjacent to   WR-543.3.2

joints and terminations, the protective conductor shall be protected by insulating sleeving complying with BS EN 60684 series.

Every connection and joint shall be accessible for inspection, testing and maintenance.   WR-543.3.3

A switching device shall **not** be inserted in a protective conductor unless: WR-543.3.4

- the switch has been inserted in the connection between the neutral point and the earthing point and the switch is a linked switch arranged to disconnect and connect the earthing conductor at the same time as the related live conductors
- it is a multipole linked switch or plug-in device in which the protective conductor circuit has not been interrupted before the live conductors and is re-established not later than when the live conductors are reconnected.

Joints intended to be disconnected for test purposes are permitted in a protective conductor circuit.   WR-543.3.4

Where electrical monitoring of earthing is used, no dedicated devices (e.g. operating sensors, coils) shall be connected in series with the protective conductor.   WR-543.3.5

Every joint in metallic conduit shall be mechanically and electrically continuous.   WR-543.3.6

4.5.7.30 Prevention of harmful effects

Electrical equipment shall not cause harmful effects on other equipment or interfere with the supply during normal service, including switching operations.   WR-133.4

4.5.7.31 Prevention of mutual detrimental influence

An electrical installation shall be arranged in such a way that no mutual detrimental influence will occur between electrical installations and non-electrical installations.   WR-132.11

Electromagnetic interference shall be taken into account.

4.5.7.32 Protection against fault current

Conductors other than live conductors, and any other   WR-131.5
parts intended to carry a fault current, shall be capable
of carrying that current without attaining an excessive
temperature. Electrical equipment, including conductors,
shall be provided with mechanical protection against
electromechanical stresses of fault currents as necessary
to prevent injury or damage to persons, livestock or
property.

4.5.7.33 Protection against overcurrent

Persons and livestock shall be protected against injury,   WR-131.4
and property shall be protected against damage, due to
excessive temperatures or electromechanical stresses
caused by any overcurrents likely to arise in live
conductors.

4.5.7.34 Protection against power supply interruption

Where danger or damage is expected to arise due to an   WR-131.7
interruption of supply, suitable provisions shall be made
in the installation or installed equipment.

4.5.7.35 Protection against thermal effects

Electrical installation shall be so arranged that: WR-131.3.1

- the risk of ignition of flammable materials due to
  high temperature or electric arc is minimised
- during normal operation of the electrical
  equipment, there shall be minimal risk of burns to
  persons or livestock.

Persons, fixed equipment and fixed materials adjacent to   WR-131.3.2
electrical equipment shall be protected against harmful
effects of heat or thermal radiation emitted by electrical
equipment.

4.5.7.36 Protection against voltage disturbances and measures against electromagnetic influences

Persons and livestock shall be protected against injury, and property shall be protected against any harmful effects, as a consequence of a fault between live parts of circuits supplied at different voltages.	WR-131.6.1
Persons and livestock shall be protected against injury, and property shall be protected against damage: • as a consequence of overvoltages such as those originating from atmospheric events or from switching	WR-131.6.2
• as a consequence of undervoltage and any subsequent voltage recovery.	WR-131.6.3
The installation shall have an adequate level of immunity against electromagnetic disturbances so as to function correctly in the specified environment.	WR-131.6.4
The installation design shall take into consideration the anticipated electromagnetic emissions generated by the installation or the installed equipment.	WR-131.6.4

4.5.7.37 Protection of low-voltage installations against temporary overvoltages due to earth faults in the high-voltage system and due to faults in the low-voltage system

This particular Regulation provides requirements for the safety of the low-voltage installation in the event of:

- a fault between the high-voltage system and earth in the transformer sub-station that supplies the low-voltage installation;
- loss of the supply neutral in the low-voltage system;
- short-circuit between a line conductor and neutral in the low-voltage installation;
- accidental earthing of a line conductor of a low-voltage IT system.

4.5.7.38 Protective conductors

Where a number of installations have separate earthing arrangements, any protective conductors common to any of these installations shall either: • be capable of carrying the maximum fault current likely to flow through them; or	WR-542.1.8

- be earthed within one installation only and insulated from the earthing arrangements of any other installation.

If the protective conductor: WR-543.1

- is not an integral part of a cable; or
- is not formed by conduit, ducting or trunking; or
- is not contained in an enclosure formed by a wiring system

the cross-sectional area shall be not less than $2.5\,mm^2$ copper equivalent if protection against mechanical damage is provided, and $4\,mm^2$ copper equivalent if mechanical protection is not provided.

Where a protective conductor is common to two or   WR-543.1.2
more circuits, its cross-sectional area shall be:

- calculated for the most onerous of the values of fault current and operating time encountered in each of the various circuits; or
- selected so as to correspond to the cross-sectional area of the largest line conductor of the circuits.

4.5.7.39 Protective bonding conductors

In each installation main protective bonding   WR-411.3.1.2
conductors shall connect extraneous conductive parts
to the main earthing terminal including the following:

- central heating and air conditioning systems
- exposed metallic structural parts of the building
- gas installation pipes
- water installation pipes
- other installation pipework and ducting.

Protective bonding conductors shall interconnect   WR-418.2.2
every simultaneously accessible exposed conductive-
part and extraneous conductive part.

Unless protection by automatic disconnection of   WR-418.2.3
supply can be applied, local protective bonding
conductors shall not be in electrical contact with earth:

- directly; or
- through exposed conductive parts; or
- through extraneous conductive parts.

> The exposed conductive parts of the separated circuit   WR-418.3.4
> shall be connected together by insulated, non-earthed
> protective bonding conductors.

 These conductors shall **not** be connected to the protective conductor or exposed conductive parts of any other circuit or to any extraneous conductive parts.

> Where electrical separation to the supply to more   WR-514.13.2
> than one item of current-using equipment is used
> (Regulations 418.2.5 or 418.3), the warning notice
> shall read as follows:

The protective bonding conductors associated with the electrical installation in this location MUST NOT BE CONNECTED TO

EARTH

Equipment having exposed-conductive-parts connected to earth must not be brought into this location

---

4.5.7.40 Main equipotential bonding conductors

> Except where protective multiple earthing (PME)   WR-544.1.1
> conditions apply, a main protective bonding conductor
> shall have a cross-sectional area not less than half
> the cross-sectional area required for the earthing
> conductor of the installation and not less than $6\,mm^2$.

> Where an installation has more than one source   WR-544.1.1
> of supply to which PME conditions apply, a main
> protective bonding conductor shall be selected
> according to the largest neutral conductor of the
> supply.

> The main equipotential bonding connection to any   WR-544.1.2
> gas, water or other service shall be made as near as
> practicable to the point of entry of that service into
> the premises.

> Where there is an insulating section or insert at that   WR-544.1.2
> point, or there is a meter, the connection shall be
> made to the consumer's hard metal pipework and
> before any branch pipework.

Where practicable the connection shall be made within 600 mm of the meter outlet union or at the point of entry to the building if the meter is external.   WR-544.1.2

Protective bonding conductors shall be protected against mechanical damage and corrosion, and shall be selected to avoid electrolytic effects.   WR-705.544.2

Accessible conductive parts of the unit, such as the chassis, shall be connected through the main protective bonding conductors to the main earthing terminal within the unit.   WR-717.411.3.1.2

The main protective bonding conductors shall be finely stranded.   WR-717.411.3.1.2

4.5.7.41 Types of protective conductor

A gas pipe, an oil pipe, a flexible or pliable conduit, support wires or other flexible metallic parts, or constructional parts subject to mechanical stress in normal service, shall **not** be selected as a protective conductor.   WR-543.2.1

A protective conductor may consist of one or more of the following: WR-543.2.2

- a single-core cable
- a conductor in a cable
- an insulated or bare conductor in a common enclosure with insulated live conductors
- a fixed bare or insulated conductor
- a metal covering, for example, the sheath, screen or armouring of a cable
- a metal conduit, metallic cable management system or other enclosure or electrically continuous support system for conductors
- an extraneous conductive part.

Where a metal enclosure or frame of a low-voltage switchgear or controlgear assembly or busbar trunking system is used as a protective conductor: WR-543.2.4

- its electrical continuity shall be assured, either by construction or by suitable connection, in such a way as to be protected against mechanical, chemical or electrochemical deterioration

- its cross-sectional area shall be in accordance with BS EN 60439-1
- it shall permit the connection of other protective conductors at every predetermined tap-off point.

The metal covering (including the sheath – bare or insulated) of a cable, trunking, ducting and metal conduit, may be used as a protective conductor for the associated circuit.

WR-543.2.5

An extraneous conductive part may be used as a protective conductor if:

WR-543.2.6

- electrical continuity can be assured and the part is either constructed or connected so that it is protected against mechanical, chemical or electrochemical deterioration
- precautions have been taken against its removal
- it has been considered for such a use and, if necessary, suitably adapted.

Where the protective conductor is formed by a conduit, trunking, ducting or the metal sheath and/or armour of a cable, the earthing terminal of each accessory shall be connected by a separate protective conductor to an earthing terminal that is part of the associated box or other enclosure.

WR-543.2.7

An exposed conductive part of equipment shall not be used to form a protective conductor for other equipment.

WR-543.2.8

Except where the circuit protective conductor is formed by a metal covering or enclosure containing all of the conductors of the ring, the circuit protective conductor of every ring final circuit shall also be run in the form of a ring having both ends connected to the earthing terminal at the origin of the circuit.

WR-543.2.9

A separate metal enclosure for cables shall not be used as a PEN conductor.

WR-543.2.10

4.5.7.42 Protective devices and switches

Single-pole fuses, switches or circuit-breakers shall only be inserted in the line conductor.

WR-132.14.1

Switches, circuit-breaker (except where linked) or fuses shall be inserted in an earthed neutral conductor.   WR-132.14.2

Any linked switch or linked circuit-breaker inserted in an earthed neutral conductor shall be capable of breaking all of the related line conductors.   WR-132.14.2

4.5.7.43 Protective earthing

The type of earthing system to be used for the installation shall be determined, taking into consideration the characteristics of the source of energy and (in particular) any earthing facilities.   WR-312.3.1

If an overcurrent protective device is used for protection against electric shock, the protective conductor shall be incorporated in the same wiring system as the live conductors or be located nearby.   WR-543.6.1

Where earthing is required for protective as well as functional purposes, then the requirements for protective measures shall take precedence.   WR-546-01-01

Exposed conductive parts shall be connected to a protective conductor.   WR-411.3.1.1

Simultaneously accessible exposed conductive parts shall be connected to the same earthing system individually, in groups or collectively.   WR-411.3.1.1

A circuit protective conductor shall be run to and terminated at each point in re-wiring and at each accessory (except a lampholder having no exposed conductive parts and suspended from such a point).   WR-411.3.1.1

The characteristics of protective equipment shall be determined with respect to its function, including protection against the effects of earth fault current.   WR-132.8

Switches, circuit-breakers (except where linked) or fuses shall be inserted in an earthed neutral conductor.   WR-132.14.2

Any linked switch or linked circuit-breaker inserted in an earthed neutral conductor shall be capable of breaking all of the related line conductors.   WR-132.14.2

All exposed conductive parts of the reduced low-voltage system shall be connected to earth.   WR-411.8.3

The earth fault loop impedance at every point of utilisation, including socket-outlets, shall be such that the disconnection time does not exceed 5 s.	WR-411.8.3
Live parts of the separated circuit shall not be connected at any point to another circuit or to earth or to a protective conductor.	WR-413.3.3
No exposed conductive part of the separated circuit shall be connected either to the protective conductor or exposed conductive parts of other circuits, or to earth.	WR-413.3.6

4.5.7.44 Protective equipment (devices and switches)

The characteristics of protective equipment shall be determined with respect to their function, including protection against the effects of: • overcurrent (overload, short-circuit) • earth fault current • overvoltage • undervoltage and no-voltage.	WR-132.8
The protective devices shall operate at values of current, voltage and time which are suitably related to the characteristics of the circuits and to the possibilities of danger.	WR-132.8

4.5.7.45 Protective measures

If a protective measure does not satisfy the requirements, then supplementary provisions shall be applied so that together the protective provisions achieve the same degree of safety.	WR-410.3.7

4.5.7.46 Protective measure: Extra-low voltage provided by SELV

Where SELV is used, whatever the nominal voltage, basic protection shall be provided by: • basic insulation; or • barriers or enclosures and affording a degree of protection of at least IPXXB or IP2X.	WR-702.414.4.5

4.5.7.47 Safety isolating transformers and electronic convertors

Safety isolating transformers shall comply with BS EN 61558-2-6 or provide an equivalent degree of safety.	WR-740.55.5
A manually reset protective device shall protect the secondary circuit of each transformer or electronic convertor.	WR-740.55.5
Safety isolating transformers shall be mounted out of arm's reach or be mounted in a location that provides equal protection, and shall have adequate ventilation.	WR-740.55.5
Access by competent persons for testing or by a skilled person competent in such work for protective device maintenance shall be provided.	WR-740.55.5
Electronic convertors shall conform to BS EN 61347-2-2.	WR-740.55.5
Enclosures containing rectifiers and transformers shall be adequately ventilated and the vents shall not be obstructed when in use.	WR-740.55.5

4.5.7.48 Safety services

Safety services need to regulated.	WR-351

 Note: Examples of safety services include:

- emergency lighting
- fire pumps
- fire rescue service lifts
- fire detection and alarm systems
- CO detection and alarm systems
- fire evacuation systems
- smoke ventilation systems
- fire services communication systems
- essential medical systems
- industrial safety systems.

4.5.7.49 Safety sources

Electrical sources for safety services shall be installed as fixed equipment, in such a manner that they cannot be adversely affected by failure of the normal source.	WR-560.6.1

Safety sources for safety services shall be placed in a suitable location and be accessible only to skilled persons or instructed persons.	WR-560.6.2
The location of a safety source shall be properly and adequately ventilated so that exhaust gases, smoke or fumes from the safety source cannot penetrate areas occupied by persons.	WR-560.6.3
Separated independent feeders from a supply network shall not serve as electrical safety sources unless assurance can be obtained that the two supplies are unlikely to fail concurrently.	WR-560.6.4
A safety source shall be selected to accommodate all the safety services in a given premises.	WR-560.6.5
A safety source may, in addition, be used for purposes other than safety services, **provided** that a fault occurring in a circuit for purposes other than safety services does not cause the interruption of any circuit for safety services.	WR-560.6.6
Protection against fault current and against electric shock in the case of a fault shall be ensured whether the installation is supplied separately by either of the two sources or by both in parallel.	WR-560.6.8.1

 Note: The following sources for safety services are recognised:

- storage batteries
- primary cells
- generator sets independent of the normal supply
- a separate feeder of the supply network effectively independent of the normal feeder.

4.5.7.50 Seismic effects

Wiring systems shall be selected and erected with due regard to the seismic hazards of the physical location of the installation.	WR-522.12.1
Where the seismic hazards experienced are low severity (AP2) or higher, particular attention shall be paid to: - the fixing of wiring systems to the building structure	WR-522.12.2

- the connections between the fixed wiring and all items of essential equipment (e.g. safety services) these shall be selected for their flexible quality.

4.5.7.51 Sources for SELV and PELV

The following sources may be used for SELV and PELV systems:

WR-414.3

- an electrochemical source (such as a battery) or another source independent of a higher-voltage circuit (e.g. a diesel-driven generator)
- an electronic device such as insulation testing equipment or monitoring device
- a motor-generator with windings providing isolation equivalent to that of a safety isolating transformer
- a safety isolating transformer.

4.5.7.52 Supplementary bonding conductor

A supplementary bonding conductor connecting two exposed conductive parts shall have a conductance (if sheathed or otherwise provided with mechanical protection) not less than that of the smaller protective conductor connected to the exposed conductive parts.

WR-544.2.1

If mechanical protection is not provided, its cross-sectional area shall be not less than 4 mm^2.

WR-544.2.1

A supplementary bonding conductor connecting an exposed conductive part to an extraneous conductive part shall have a conductance (if sheathed, or otherwise provided with mechanical protection) not less than half that of the protective conductor connected to the exposed conductive part.

WR-544.2.2

A supplementary bonding conductor connecting two extraneous conductive parts shall have a cross-sectional area not less than 2.5 mm^2 if sheathed or otherwise

WR-544.2.2 and WR-544.2.3

provided with mechanical protection or $4\,mm^2$ if mechanical protection is not provided.

Supplementary bonding shall be provided by a supplementary conductor, a conductive part of a permanent and reliable nature, or by a combination of these requirements.   WR-544.2.4

Where supplementary bonding is to be applied to a fixed appliance (which is supplied via a short length of flexible cord from an adjacent connection unit or other accessory, incorporating a flex outlet) the circuit protective conductor within the flexible cord shall be deemed to provide the supplementary bonding connection to the exposed conductive parts of the appliance.   WR-544.2.5

Supplementary bonding conductors shall be protected against mechanical damage and corrosion, and shall be selected to avoid electrolytic effects.   WR-705.544.2

4.5.7.53 Testing

When undertaking testing in a potentially explosive atmosphere, appropriate safety precautions in accordance with BS EN 60079-17 and BS EN 61241-17 are necessary.   WR-612.1

4.5.7.54 Transformers and converters

A safety isolating transformer for an extra-low-voltage lighting installation shall comply with BS EN 61558-2-6 and: WR-559.11.3.1

- either the transformer shall be protected on the primary side by a protective device complying with the requirements of Regulation 559.11.4.2; or
- the transformer shall be short-circuit proof (both inherently and non-inherently) and shall be marked with the symbol shown here.

4.5.7.55 Type of wiring and method of installation

The choice of the type of wiring system and the method of installation shall include consideration of the following:

WR-132.7

- the nature of the location
- the nature of the structure supporting the wiring
- accessibility of wiring to persons and livestock
- voltage
- the electromechanical stresses likely to occur due to short-circuit and earth fault currents
- electromagnetic interference
- other external influences (e.g. mechanical, thermal and those associated with fire) to which the wiring is likely to be exposed during the erection of the electrical installation or in service.

4.5.7.56 Uninterruptible power supply sources (UPS)

An uninterruptible power supply source of the static type shall be able to:

WR-560.6.11

- operate distribution circuit protective devices; and
- start the safety devices

when operating in the emergency condition from the convertor supplied by the battery.

4.5.7.57 Agricultural and horticultural premises

For high density livestock rearing systems operating for the life support of livestock the following applies:

WR-705.560.6

- Where the supply of food, water, air and/or lighting to livestock is not ensured in the event of power supply failure, a secure source of supply shall be provided (such as an alternative or back-up supply) and separate final circuits for ventilation and lighting units shall also be provided.

- Where electrically powered ventilation is necessary in an installation one of the following shall be provided:
  - a standby electrical source ensuring sufficient supply for ventilation equipment; or
  - temperature and supply voltage monitoring.

Note: A notice should be placed adjacent to the standby electrical source, indicating that it should be tested periodically according to the manufacturer's instructions.

4.5.7.58 Construction and demolition site installations

Safety and standby supplies shall be connected by means of devices arranged to prevent interconnection of the different supplies.	WR-704.537.2.2

4.5.7.59 Electrical installations in caravan/camping parks and similar locations

The protective measures of obstacles and placing out of reach are **not** permitted.	WR-708.410.3.5
The protective measures of non-conducting location and earth-free local equipotential bonding are **not** permitted.	WR-708.410.3.6

In the UK the ESQCR (Electricity Safety, Quality and Continuity Regulations 2002) prohibit the use of a TN-C-S system for the supply to a caravan or similar construction.

4.5.7.60 Locations containing a bath or shower

The protective measures of obstacles and placing out of reach are **not** permitted.	WR-701.410.3.5
The protective measures of non-conducting location and earth-free local equipotential bonding are **not** permitted.	WR-701.410.3.6

4.5.7.61 Swimming pools and other basins

Zones 0 and 1

Equipment for use in the interior of basins which is only intended to be in operation when people are not inside zone 0 shall be supplied by a circuit protected by: • SELV; and • automatic disconnection of the supply using an RCD; or • electrical separation.	WR-702.410.3.4.1
The socket-outlet of a circuit supplying such equipment and the control device of such equipment shall have a notice in order to warn the user that this equipment shall be used only when the swimming pool is not occupied by persons.	WR-702.410.3.4.1

Zones 0 and 1 of fountains

In zones 0 and 1, one or more of the following protective measures shall be employed: • SELV • automatic disconnection of supply using an RCD • electrical separation.	WR-702.410.3.4.2

Zone 2 (swimming pools and other basins)

One or more of the following protective measures shall be employed: • SELV • automatic disconnection of supply • electrical separation.	WR-702.410.3.4.3
The protective measures of obstacles and placing out of reach are **not** permitted.	WR-702.410.3.5
The protective measures of non-conducting location and earth-free local equipotential bonding are **not** permitted.	WR-702.410.3.6

Supplementary equipotential bonding

All extraneous conductive parts in zones 0, 1 and   WR-702.411.3.3
2 shall be connected by supplementary protective
bonding conductors to the protective conductors of
exposed conductive parts of equipment situated in
these zones.

5

Electrical equipment, components, accessories and supplies

The amount and number of different types of equipment, components, accessories and supplies for electrical installations currently available is enormous and any attempt to cover every type, model and/or manufacture would prove an impossible task for a book such as this. The intention of this chapter, therefore, is to provide a catalogue of all the different types identified and referred to in the Wiring Regulations (e.g. luminaires, RCDs, plugs and sockets) and then make a list of the specific requirements that are sprinkled throughout the Regulations. For your (hopeful!?) convenience this catalogue has been compiled in alphabetical order.

 Similarly to other chapters, please remember that these lists of requirements are **only** the author's impression of the most important aspects of the Wiring Regulations and electricians should **always** consult BS 7671 to satisfy compliance!

5.1 Installation

Except where specifically designed for direct connection to flexible wiring, equipment shall be fixed so that connections between wiring and equipment shall **not** be subject to undue stress or strain resulting from the normal use of the equipment.	WR-530.4.1
Unenclosed equipment shall be mounted in a suitable mounting box and or fixed to the fabric of the building.	WR-530.4.2
Socket outlets, connection units, plate switches and similar accessories shall be fitted to a mounting box.	WR-530.4.2

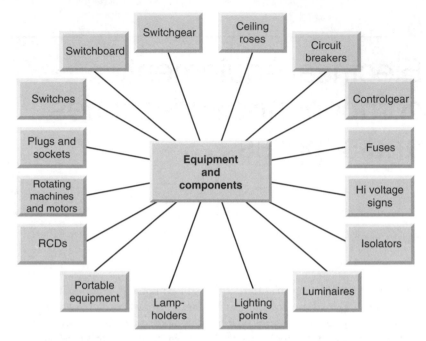

Figure 5.1 Electrical equipment and components

Wherever equipment is fixed on or in cable trunking, skirting trunking or mouldings it shall not be fixed on covers which can be removed inadvertently.	WR-530.4.3

5.2 Ceiling roses

A ceiling rose or lampholder for a filament lamp shall **not** be installed in any circuit operating at a voltage normally exceeding 250 volts.	WR-559.6.1.2
A ceiling rose shall n**ot** be used for the attachment of more than one outgoing flexible cord unless it is specially designed for multiple pendants.	WR-559.6.1.3
At each fixed lighting point one of the following shall be used:	WR-559.6.1.1

• a ceiling rose to BS 67
• a luminaire supporting coupler to BS 6972 or BS 7001
• a batten lampholder or a pendant set to BS EN 60598

- a luminaire to BS EN 60598
- a suitable socket outlet to BS 1363-2, BS 546 or BS EN 60309-2
- a plug-in lighting distribution unit to BS 5733
- a connection unit to BS 1363-4
- appropriate terminals enclosed in a box complying with the relevant part of BS EN 60670 series or BS 4662
- a device for connecting a luminaire (DCL) outlet according to IEC 61995-1.

5.3 Circuit-breakers

Every circuit shall be provided with a means of isolation from all live supply conductors by a linked switch or a linked circuit-breaker.	WR-422.3.13
A circuit-breaker providing protection against both overload and fault current shall be capable of 'making' any overcurrent up to and including the maximum prospective fault current at the point where the device is installed.	WR-432.1
Where a circuit-breaker is used, the maximum value of earth fault loop impedance (Z_s) shall be determined by the formula $Z_s \times I_a \leq U_o$.	WR-411.8.3
Circuit-breakers shall be inserted in the line conductor.	WR-132.14.1
Circuit-breakers (except where linked) shall be inserted in an earthed neutral conductor.	WR-132.14.2

 Note: Any linked circuit-breaker inserted in an earthed neutral conductor shall be capable of breaking all of the related line conductors.

5.3.1 Locations with risks of fire due to the nature of processed or stored materials

Every circuit shall be provided with a means of isolation from all live supply conductors by a linked switch or a linked circuit-breaker.	WR-422.3.13

5.3.2 Protection against fault current only

A circuit-breaker with a short-circuit release providing protection against fault current shall only be installed where overload protection is achieved by other means.	WR-432.3
A device shall be capable of breaking (and for a circuit-breaker, making) the fault current up to and including the prospective fault current.	WR-432.3
Every circuit-breaker shall be provided with an indication of its intended rated current, that is appropriate to the circuit it protects.	WR-533.1
Where a circuit-breaker may be operated by a person other than a skilled person or instructed person, it shall be designed or installed so that it is not possible to modify the setting or the calibration of its overcurrent release without a deliberate act involving the use of either a key or a tool and resulting in a visible indication of its setting or calibration.	WR-533.1.2
A main linked switch or linked circuit-breaker shall be provided as near as practicable to the origin of every installation as a means of switching the supply on load and as a means of isolation.	WR-537.1.4
The supply to the electrode water heater or electrode boiler shall be controlled by a linked circuit-breaker.	WR-554.1.2

 Note: If the electrode water heater or electrode boiler is not piped to a water supply or in physical contact with any earthed metal (and where the electrodes and
the water in contact with the electrodes are so shielded in insulating material that
they cannot be touched while the electrodes are live) a fuse in the line conductor
may be substituted for the circuit-breaker, and the shell of the electrode water
heater or electrode boiler need not be connected to the neutral of the supply.

Single-line water heaters and boilers with an uninsulated heating element immersed in the water shall not have a non-linked circuit-breaker fitted in the neutral conductor, in any part of the circuit between the heater or boiler or the origin of the installation.	WR-554.3.1 and 4

5.4 Electric motors

An electric motor which is automatically or remotely controlled or which is not continuously supervised shall be protected against excessive temperature by a protective device with manual reset.	WR-422.3.7
A motor with star-delta starting shall be protected against excessive temperature in both the star and delta configurations.	WR-422.3.7
All equipment, including cable, of every circuit carrying the starting, accelerating and load currents of a motor shall be suitable for a current at least equal to the full-load current rating of the motor when rated in accordance with the appropriate British Standard.	WR-552.1.1
Every electric motor having a rating exceeding 0.37 kW shall be provided with control equipment incorporating means of protection against overload of the motor.	WR-552.1.2
Except where failure to start after a brief interruption would be likely to cause greater danger, every motor shall be provided with means to prevent automatic restarting after a stoppage due to a drop in voltage or failure of supply, where unexpected restarting of the motor might cause danger.	WR-552.1.3
Fixed electric motors shall be provided with an efficient means of switching off, that is readily accessible, easily operated and so placed as to prevent danger.	WR-132.15.2
Motor control circuits shall be designed so as to prevent any motor from restarting automatically after a stoppage due to a fall in or loss of voltage, if such starting is liable to cause danger.	WR-537.5.4.1
Where reverse-current braking of a motor is provided, provision shall be made for the avoidance of reversal of the direction of rotation at the end of braking if such reversal may cause danger.	WR-537.5.4.2
Where the motor is intended for intermittent duty and for frequent starting and stopping, account shall be taken of any cumulative effects of the starting or braking currents upon the temperature rise of the equipment of the circuit.	WR-552.1.1

Where safety depends on the direction of rotation of a motor, provision shall be made for the prevention of reverse operation due to, for example, a reversal of lines.	WR-537.5.4.3

5.5 Fuses

Single-pole fuses shall only be inserted in the line conductor.	WR-132.14.1
Fuses shall be inserted in an earthed neutral conductor.	WR-132.14.2
For every fuse and circuit-breaker there shall be provided on, or adjacent to it, an indication of its intended rated current as appropriate to the circuit it protects.	WR-533.1

5.5.1 Protection against fault current only

Fuses that provide protection against fault current shall only be installed where overload protection is achieved by other means.	WR-432.3

5.5.2 Devices for protection against overcurrent

A fuse shall preferably be of the cartridge type.	WR-533.1.1.3
Every fuse shall be provided with an indication of its intended rated current as appropriate to the circuit it protects.	WR-533.1
A fuse base shall be arranged so as to exclude the possibility of the fuse carrier making contact between conductive parts belonging to two adjacent fuse bases.	WR-533.1.1.1
A fuse base using screw-in fuses shall be connected so that the centre contact is connected to the conductor from the supply and the shell contact is connected to the conductor to the load.	WR-533.1.1.1

Fuses with fuse-links that are likely to be removed or replaced by persons other than instructed persons or skilled persons shall either:	WR-533.1.1.2

- have marked on or adjacent to it an indication of the type of fuse link that should be used; or
- be of a type such that there is no possibility of inadvertently replacing the fuse with one that has a higher rated current but a higher fusing factor than that intended.

Where a semi-enclosed fuse is selected, it shall be fitted with an element in accordance with the manufacturer's instructions or (in the absence of such instructions) it shall be fitted with a single element of tinned copper wire of the appropriate diameter specified in Table 53.1 on p.114 of BS 7671:2008.	WR-533.1.1.3

5.5.3 Functional switching devices

Functional switching devices:	WR-537.5.2.1 and WR-537.5.2.2

- shall be suitable for the most onerous duty they are intended to perform
- may control the current without necessarily opening the corresponding poles.

Fuses and links shall **not** be used for functional switching.	WR-537.5.2.3

5.5.4 Electrode water heaters and boilers

If the electrode water heater or electrode boiler is not piped to a water supply or in physical contact with any earthed metal (and where the electrodes and the water in contact with the electrodes are so shielded in insulating material that they cannot be touched while the electrodes are live), a fuse in the line conductor may be substituted for the circuit-breaker and the shell of the electrode water heater or electrode boiler need not be connected to the neutral of the supply.	WR-554.1.7

Single-line water heaters and boilers with an   WR-554.3.1
uninsulated heating element immersed in the water   and 4
shall not have a fuse fitted in the neutral conductor,
in any part of the circuit between the heater or
boiler or the origin of the installation.

5.5.5 Devices for isolation and switching

Where it is intended that isolation and switching   WR-559.10.6.1
is carried out **only** by instructed persons, for TN
systems, the means of switching the supply on
load and the means of isolation may be provided
by a suitably rated fuse carrier.

Where a fuse is used to disconnect a circuit supplying socket outlets and other final circuits which supply portable equipment that is intended for manual movement during use, or hand-held Class I equipment:

 Note: Table 5.1 is an indication of the maximum values of earth fault loop impedance corresponding to a disconnection time of 0.4 s for a nominal voltage to earth (U_o) of 230 V.

Table 5.1 Maximum earth fault loop impedance (Z_s) for circuit-breakers U_O of 230 V (courtesy BSI)

General purpose fuses to BS 88-2.1 and BS 88-6							
Rating (amperes)	6	10	16	20	25	32	40	50
Z_s (ohms)	8.89	5.33	2.82	1.85	1.50	1.09	0.86	0.63

Where the device is a general purpose type fuse to BS 88-2.1, a fuse to BS 88-6, a fuse to BS 1361 or a semi-enclosed fuse to BS 3036, the conditions shown in Table 5.2 shall apply.

 Note: The circuit loop impedances given in the table above should not be exceeded when the conductors are at their normal operating temperature. If the conductors are at a different temperature when tested, then the reading should be adjusted accordingly.

All low-voltage fused plug and socket outlets shall conform with the applicable British Standard listed in Table 5.3.

Table 5.2 Coordination between conductor and protective device (courtesy StingRay)

	Design current (I_B) of the circuit	Lowest current-carrying capacity (I_z) of any conductor in a circuit	Operating current of any protective device (I_z)
		Requirements	
	Shall be greater than the nominal current or current setting (I_n) of a protective device.	Shall be greater than the nominal current or current setting (I_n) of a protective device.	Shall not exceed 1.45 times the lowest of the current-carrying capacities (I_z) of any of the conductors of the circuit.
BS 88-2.1 fuses		Yes	Yes
BS 88-6 fuses		Yes	Yes
BS 1361 fuses		Yes	Yes
BS 3036 (semi-enclosed) fuse			Yes provided (I_n) does not exceed 0.725 (I_z)

Table 5.3 Plug and socket outlets for low-voltage circuits (courtesy BSI)

Type of plug and socket outlet	Rating (amperes)	Applicable British Standard
Fused plug and shuttered socket outlets, 2-pole and earth, for a.c.	13	BS 1363 (fuses to BS 1362)
Plug, fused or non-fused, and socket outlets, 2-pole and earth	2, 5, 15, 30	BS 546 (fuses, if any, to BS 646)
Plug, fused or non-fused, and socket outlets, protected-type, 2-pole with earthing contact	5, 15, 30	BS 196
Plug and socket outlets (industrial type)	16, 32, 63, 125	BS EN 60309-2

5.6 Heaters

Measures shall be taken to prevent an enclosure of electrical equipment such as a heater from exceeding the following temperatures: WR-422.3.2

- 90°C under normal conditions; and
- 115°C under fault conditions.

5.6.1 Electrode water heaters and boilers

An electrode water heater and electrode boiler shall **only** be connected to an a.c. system.	WR-554.1.1
The supply to the electrode water heater or electrode boiler shall be controlled by a linked circuit-breaker.	WR-554.1.2
The shell of the electrode water heater or electrode boiler shall be bonded to the metallic sheath and armour, if any, of the incoming supply cable.	WR-554.1.3
If an electrode water heater or electrode boiler is directly connected to a supply at a voltage exceeding low voltage, the installation shall include an RCD.	WR-554.1.4
If an electrode water heater or electrode boiler is connected to a three-line low-voltage supply, the shell of the electrode water heater or electrode boiler shall be connected to the neutral of the supply as well as to the earthing conductor.	WR-554.1.5
If the supply to an electrode water heater or electrode boiler is single-line and one electrode is connected to a neutral conductor earthed by the distributor, the shell of the electrode water heater or electrode boiler shall be connected to the neutral of the supply as well as to the earthing conductor.	WR-554.1.6
If the electrode water heater or electrode boiler is not piped to a water supply or in physical contact with any earthed metal (and where the electrodes and the water in contact with the electrodes are so shielded in insulating material that they cannot be touched while the electrodes are live) a fuse in the line conductor may be substituted for the circuit-breaker and the shell of the electrode water heater or electrode boiler need not be connected to the neutral of the supply.	WR-554.1.7

5.6.2 Heaters for liquids or other substances having immersed heating elements

Heaters for liquid and/or other substances shall have an automatic device to prevent a dangerous rise in temperature.	WR-554.2.1

5.6.3 Water heaters having immersed and uninsulated heating elements

All metal parts of the heater or boiler which are in contact with the water (other than current-carrying parts) shall be solidly and metallically connected to a metal water pipe through which the water supply to the heater or boiler is provided and that water pipe shall be connected to the main earthing terminal by means independent of the circuit protective conductor.

WR-554.3.2

Water heaters and boilers shall be permanently connected to the electricity supply via a double-pole linked switch which is either:

WR-554.3.3

- separate from and within easy reach of the heater/ boiler; or
- part of the boiler/heater (provided that the wiring from the heater or boiler is directly connected to the switch without use of a plug and socket outlet).

The wiring from the heater or boiler shall be connected directly to that switch without the use of a plug and socket outlet.

WR-554.3.3

If the heater or boiler is installed in a room containing a fixed bath, the switch shall comply with Section 701 of BS 7671:2008.

Single-line water heaters and boilers with an uninsulated heating element immersed in the water shall not have a single-pole switch, non-linked circuit-breaker or fuse fitted in the neutral conductor, in any part of the circuit between the heater or boiler or the origin of the installation.

WR-554.3.1 and 4

5.6.4 Electric surface heating systems

The equipment, system design, installation and testing of an electric surface heating system shall be in accordance with BS 6351.

WR-554.5.1

5.6.5 Electric floor heating systems

For electric floor heating systems, only heating cables or thin sheet flexible heating elements shall be erected **provided** that they have either a metal sheath or a metal enclosure or a fine mesh metallic grid.	WR-701.753
The fine mesh metallic grid, metal sheath or metal enclosure shall be connected to the protective conductor of the supply circuit.	WR-701.753

 Compliance with the latter requirement is not required if the protective measure SELV is provided for the floor heating system.

For electric floor heating systems the protective measure 'protection by electrical separation' is **not** permitted.	WR-701.753

5.6.6 Floor and ceiling heating systems

This section applies to the installation of electric floor and ceiling heating systems which are erected as either thermal storage heating systems or direct heating systems. It does not apply to the installation of wall heating systems.

5.6.6.1 Protection against electric shock

The protective measures of obstacles and placing out of reach are **not** permitted.	WR-753.410.3.5
The protective measures of non-conducting location and earth-free local equipotential bonding are **not** permitted.	WR-753.410.3.6

5.6.6.2 Automatic disconnection of supply

RCDs with a rated residual operating current not exceeding 30 mA shall be used as disconnecting devices.	WR-753.411.3.2

5.6.6.3 Electrical separation

The protective measure of electrical separation is **not** permitted.	WR-753.413.1.2

5.6.6.4 Additional protection – RCDs

A circuit supplying heating equipment of Class II construction or equivalent insulation shall be provided with additional protection by the use of an RCD.	WR-753.415.1

5.6.6.5 Protection against burns

In floor areas where contact with skin or footwear is possible, the surface temperature of the floor shall be limited (for example, 35°C).	WR-753.423

5.6.6.6 Heating units

To avoid the overheating of floor or ceiling heating systems in buildings, one or more of the following measures shall be applied within the zone where heating units are installed to limit the temperature to a maximum of 80°C: WR-753.424.1.1

- appropriate design of the heating system
- appropriate installation of the heating system in accordance with the manufacturer's instructions
- use of protective devices.

Heating units shall be: WR-753.424.1.1

- connected to the electrical installation via cold tails or suitable terminals
- inseparably connected to cold tails, for example, by a crimped connection.

As the heating unit may cause higher temperatures or arcs under fault conditions, special measures should be taken when the heating unit is installed close to easily ignitable building structures, such as placing on a metal sheet, in metal conduit or at a distance of at least 10 mm in air from the ignitable structure.

WR-753.424.1.2

5.6.6.7 Compliance with standards

Flexible sheet heating elements shall comply with the requirements of BS EN 60335-2-96.

WR-753.511

Heating cables shall comply with BS 6351 series.

WR-753.511

5.6.6.8 Operational conditions

Precautions shall be taken not to stress the heating unit mechanically; for example, the material by which it is to be protected in the finished installation shall cover the heating unit as soon as possible.

WR-753.512.1.6

5.6.6.9 External influences

Heating units for installation in ceilings shall have a degree of protection of not less than IPX1.

WR-753.512.2.5

Heating units for installation in a floor of concrete or similar material shall have a degree of ingress protection not less than IPX7 and shall have the appropriate mechanical properties.

WR-753.512.2.5

5.6.6.10 Identification and notices

The designer of the installation/heating system or installer shall provide a plan for each heating system, containing the following details:

WR-753.514

- manufacturer and type of heating units
- number of heating units installed

- length/area of heating units
- rated power
- surface power density
- layout of the heating units in the form of a sketch, a drawing, or a picture
- position/depth of heating units
- position of junction boxes
- conductors, shields and the like
- heated area
- rated voltage
- rated resistance (cold) of heating units
- rated current of overcurrent protective device
- rated residual operating current of RCD
- the insulation resistance of the heating installation and the test voltage used
- the leakage capacitance.

This plan shall be fixed to, or adjacent to, the distribution board of the heating system.   WR-753.514

5.6.6.11 Prevention of mutual detrimental influences

Heating units shall not cross expansion joints of the building or structure.   WR-753.515.4

5.6.6.12 Heating-free areas

For the necessary attachment of room fittings, heating-free areas shall be provided in such a way that heat emission is not prevented by such fittings.   WR-753.520.4

5.6.6.13 External influences

For cold tails (circuit wiring) and control leads installed in the zone of heated surfaces, the increase of ambient temperature shall be taken into account.   WR-753.522.1.3

5.6.6.14 Presence of solid foreign bodies

Where heating units are installed there shall be heating-free areas where drilling and fixing by screws, nails and the like are permitted.	WR-753.522.4.3
The installer shall inform other contractors that no penetrating means (such as screws for door stoppers) shall be used in the area where floor or ceiling heating units are installed.	WR-753.522.4.3

5.7 Isolators

The location of each disconnector (isolator) shall be indicated unless there is no possibility of confusion.	WR-514.11.1
Off-load isolators (disconnectors) shall **not** be used for functional switching.	WR-537.5.2.3

5.8 Lampholders

A lampholder for a filament lamp shall **not** be installed in any circuit operating at a voltage normally exceeding 250 volts.	WR-559.6.1.2
Bayonet lampholders B15 and B22 shall comply with BS EN 61184 and shall have the temperature rating T2 described in that standard.	WR-559.6.1.7
In circuits of a TN or TT system (except for E14 and E27 lampholders complying with BS EN 60238) the outer contact of every Edison screw or single centre bayonet cap type lampholder shall be connected to the neutral conductor. This Regulation also applies to track-mounted systems.	WR-559.6.1.8
Insulation piercing lampholders shall **not** be used unless the cables and lampholders are compatible, and provided the lampholders are non-removable once fitted to the cable.	WR-711.559.4.3
Lighting circuits incorporating B15, B22, E14, E27 or E40 lampholders shall be protected by an overcurrent protective device of maximum rating 16A.	WR-559.6.1.6

Lampholders with an ignitability characteristic 'P' as specified in BS 476 Part 5 (or where separate overcurrent protection is provided) shall not be connected to any circuit where the rated current of the overcurrent protective device exceeds the appropriate value (stated in Table 5.4 below).

Table 5.4 Overcurrent protection of lampholders (courtesy BSI)

Type of lampholder			Maximum rating (amperes) of overcurrent protective device protecting the circuit
Bayonet (BS EN 61184)	B15	SBC	6
	B22	BC	16
Edison screw	E14	SES	6
(BS EN 60238)	E27	ES	16
	E40	GES	16

Unless the wiring is enclosed in earthed metal or insulating material.

5.8.1 Polarity

A test of polarity shall be made and it shall be verified that   WR-612.6
(except for E14 and E27 lampholders to BS EN 60238) in
circuits having an earthed neutral conductor, centre contact
bayonet and Edison screw lampholders have:

- the outer or screwed contacts connected to the
  neutral conductor; and
- wiring has been correctly connected to socket
  outlets and similar accessories.

5.9 Luminaires

Every plug, socket outlet, luminaire supporting coupler   WR-411.7.5
(LSC), device for connecting a luminaire and cable
coupler in a FELV system:

- shall have a protective conductor contact; and
- shall **not** be dimensionally compatible with those
  used for any other system in use in the same
  premises.

Every plug, socket outlet, luminaire supporting coupler   WR-411.8.5
(LSC), device for connecting a luminaire (DCL) and
cable coupler of a reduced low-voltage system:

- shall have a protective conductor contact; and

- shall **not** be dimensionally compatible with those used for any other system in use in the same premises.

Note: This protective measure shall **not** be applied to any circuit that includes a socket outlet or cable coupler, or where a user may change items of equipment without authorisation.

Every socket outlet and luminaire supporting coupler in a SELV or PELV system shall require the use of a plug which is incompatible dimensionally with those used for any other system in use in the same premises.	WR-414.4.3
Luminaires shall be kept at an adequate distance from combustible materials.	WR-422.3.1 and WR-422.4.2
A luminaire with a lamp that could eject flammable materials in the case of failure should be equipped with a safety protective shield.	WR-422.3.1 WR-422.4.2
Every electric discharge lighting installation having an open circuit voltage exceeding low voltage and every circuit supplying luminaires at a voltage exceeding low voltage, shall be provided with: • an interlock on a self-contained luminaire • an effective local means for the isolation of the circuit from the supply • a switch with a lock or removable handle, or a distribution board which can be locked.	WR-537.2.1.6
Every luminaire shall: • be appropriate for the location; and • be provided with an enclosure providing a degree of protection of at least IP5X; and • have a limited surface temperature in accordance with BS EN 60598-2-24; and • be of a type that prevents lamp components from falling from the luminaire.	WR-422.3.8

Note: Luminaires marked $\underline{\nabla}\!\!\!\!D$ are designed to provide limited surface temperature.

| Parts of a cable or flexible cord within a luminaire shall be suitable for the temperatures likely to be encountered, or shall be provided with additional insulation suitable for those temperatures. | WR-522.2.2 |
| The connection of suspended current-using equipment (such as a luminaire for a fixed installation) shall be made by cable with flexible cores. | WR-522.7.2 |

5.9.1 Mobile equipment

| Where mobile equipment is likely to be used, provision shall be made so that the equipment can be fed from an adjacent and conveniently accessible socket outlet, taking account of the length of flexible cord normally fitted to portable appliances and luminaires. | WR-553.1.7 |

5.9.2 Outdoor lighting installation

An outdoor lighting installation comprises one or more luminaires, a wiring system and accessories and includes lighting installations for:

- roads, parks, car parks, gardens, places open to the public, sporting areas, illumination of monuments and floodlighting;
- places such as telephone kiosks, bus shelters, advertising panels and town plans;
- road signs and road traffic signal systems;
- temporary festoon lighting.

5.9.3 General requirements for installations

| A track system for luminaires shall comply with the requirements of BS EN 60570. | WR-559.4.4 |
| Every luminaire shall comply with the relevant standard for manufacture and test of that luminaire and shall be selected and erected in accordance with the manufacturer's instructions. | WR-559.4.1 |

| Luminaires without transformers or converters (but which are fitted with extra-low-voltage lamps connected in series) shall be considered as low-voltage equipment not extra-low-voltage equipment. | WR-559.4.2 |
| Where a luminaire is installed in a pelmet, there shall be no adverse effects due to the presence or operation of curtains or blinds. | WR-559.4.3 |

5.9.4 Protection against fire

When selecting and erecting a luminaire, the thermal effects of radiant and convected energy on the surroundings shall be taken into account, including:

- the maximum permissible power dissipated by the lamps;
- the fire-resistance of adjacent material:
  ○ at the point of installation, and
  ○ in the thermally affected areas;
- the minimum distance to combustible materials, including material in the path of a spotlight beam.

5.9.5 Wiring systems

| At each fixed lighting point one of the following shall be used: | WR-559.6.1.1 |

- a ceiling rose to BS 67
- a luminaire supporting coupler to BS 6972 or BS 7001
- a batten lampholder or a pendant set to BS EN 60598
- a luminaire to BS EN 60598
- a suitable socket outlet to BS 1363-2, BS 546 or BS EN 60309-2
- a plug-in lighting distribution unit to BS 5733
- a connection unit to BS 1363-4
- appropriate terminals enclosed in a box complying with the relevant part of BS EN 60670 series or BS 4662
- a device for connecting a luminaire (DCL) outlet according to IEC 61995-1.

In suspended ceilings one plug-in lighting distribution unit may be used for a number of luminaires.

A ceiling rose or lampholder for a filament lamp shall **not** be installed in any circuit operating at a voltage normally exceeding 250 volts.	WR-559.6.1.2
A ceiling rose shall **not** be used for the attachment of more than one outgoing flexible cord unless it is specially designed for multiple pendants.	WR-559.6.1.3
Luminaire-supporting couplers are designed specifically for the mechanical support and electrical connection of luminaires and shall **not** be used for the connection of any other equipment.	WR-559.6.1.4

5.9.5.1 Fixing of the luminaire

A lighting installation shall be appropriately controlled by a switch, a combination of switches or by an automatic control system, which is suitable for discharge lighting circuits.	WR-559.6.1.9
Adequate means to fix the luminaire shall be provided such as mechanical accessories (e.g. hooks or screws), boxes or enclosures which are able to support luminaires or associated supporting devices for connecting a luminaire.	WR-559.6.1.5
Any cable or cord between the fixing means and the luminaire shall be installed so that any expected stresses in the conductors, terminals and terminations will not impair the safety of the installation.	WR-559.6.1.5
Bayonet lampholders B15 and B22 shall comply with BS EN 61184 and shall have the temperature rating T2 described in that Standard.	WR-559.6.1.7
In places where the fixing means is intended to support a pendant luminaire, the fixing means shall be capable of carrying a mass of not less than 5 kg.	WR-559.6.1.5

Note: If the mass of the luminaire is greater than 5 kg, the installer shall ensure that the fixing means is capable of supporting the mass of the pendant luminaire.

The weight of luminaires and their eventual accessories shall be compatible with the mechanical capability of the ceiling or suspended ceiling or supporting structure where installed.	WR-559.6.1.5
Lighting circuits incorporating B15, B22, E14, E27 or E40 lampholders shall be protected by an overcurrent protective device of maximum rating 16 A.	WR-559.6.1.6
In circuits of a TN or TT system, except for E 14 and E27 lampholders complying with BS EN 60238, the outer contact of every Edison screw or single centre bayonet cap type lampholder shall be connected to the neutral conductor. This Regulation also applies to track-mounted systems.	WR-559.6.1.8

5.9.5.2 Through wiring

The installation of through wiring in a luminaire is **only** permitted if the luminaire is designed for such wiring.	WR-559.6.2.1
A cable for through wiring shall be selected in accordance with the temperature information on the luminaire or on the manufacturer's instruction sheet.	WR-559.6.2.2
Groups of luminaires divided between the three line conductors of a three-line system with only one common neutral conductor shall be provided with at least one device that simultaneously disconnects all line conductors.	WR-559.6.2.3

5.9.6 Independent lamp controlgear, e.g. ballasts

Only independent lamp controlgear marked as suitable for independent use (according to the relevant standard) shall be used external to a luminaire.	WR-559.7
Only the following are permitted to be mounted on flammable surfaces:	WR-559.8

A 'class P' thermally protected ballast(s)/ transformer(s), marked with the symbol:

A temperature declared thermally protected ballast(s)/transformer(s), marked with the symbol:

with a marked value equal to or below 130°C. NOTE: The generally recognised symbol is of an independent ballast of EN 60417:

Compensation capacitors having a total capacitance exceeding $0.5\,\mu\text{F}$ shall only be used in conjunction with discharge resistors.

WR-559.8

Capacitors and their marking shall be in accordance with BS EN 61048.

WR-559.8

5.9.7 Stroboscopic effect

Lighting for premises where machines with moving parts are in operation should consider the stroboscopic effects which can give a misleading impression of moving parts being stationary. Such effects may be avoided by selecting luminaires with suitable lamp controlgear, such as high frequency controlgear, or by distributing lighting loads across all the phases of a three-line supply.

WR-559.9

5.9.8 Requirements for outdoor lighting installations, highway power supplies and street furniture

The protective measures of placing out of reach and obstacles shall **not** be used except where:

WR-559.10.1

- the maintenance of equipment is restricted to skilled persons who are specially trained
- items of street furniture are within 1.5 m of a low-voltage overhead line.

The protective measures of non-conducting location and earth-free local equipotential bonding shall **not** be used.

WR-559.10.2

Where the protective measure automatic   WR-559.10.3.1
disconnection of supply is used:

- all live parts of electrical equipment shall
  be protected by insulation or by barriers or
  enclosures providing basic protection.

 A door in street furniture, used for access to electrical lighting equipment, shall not be used as a barrier or an enclosure.

- Enclosures of live parts shall only be   WR-559.10.3.1
  accessible with a key or a tool (unless the
  enclosure is in a location where only skilled
  or instructed persons have access).
- A door giving access to electrical equipment
  and located less than 2.50 m above ground
  level shall:
  - be locked with a key or shall require the
    use of a tool for access
  - be provided with basic protection when
    the door is open either by the use of
    equipment having at least a degree
    of protection of IP2X or IPXXB by
    construction or by installation, or by
    installing a barrier or an enclosure giving
    the same degree of protection.
- For a luminaire at a height of less than 2.80 m
  above ground level, access to the light source
  shall only be possible after removing a barrier
  or an enclosure requiring the use of a tool.

 For an outdoor lighting installation, a metallic structure (such as a fence or grid) which is in the proximity of but is not part of the outdoor lighting installation need not be connected to the main earthing terminal.

It is recommended that equipment such as lighting   WR-559.10.3.2
arrangements in places such as telephone kiosks,
bus shelters and town plans is provided with
additional protection by an RCD.

A maximum disconnection time of 5 s shall apply   WR-559.10.3.3
to all circuits feeding fixed equipment used in
highway power.

The earthing conductor of a street electrical fixture shall have a minimum copper equivalent cross-sectional area not less than that of the supply neutral conductor at that point or not less than $6\,mm^2$, whichever is the smaller.

WR-559.10.3.4

5.9.9 Double or reinforced insulation

For an outdoor lighting installation, where the protective measure for the whole installation is by double or reinforced insulation, no protective conductor shall be provided and the conductive parts of the lighting column shall not be intentionally connected to the earthing system.

WR-559.10.4

A device providing protection against the risk of fire shall meet the following requirements:

WR-559.11.4.2

- The device shall continuously monitor the power demand of the luminaires.
- The device shall automatically disconnect the supply circuit within 0.3 s in the case of a short-circuit or failure which causes a power increase of more than 60 W.
- The device shall provide automatic disconnection while the supply circuit is operating with reduced power or if there is a failure which causes a power increase of more than 60 W.
- The device shall provide automatic disconnection upon connection of the supply circuit if there is a failure which causes a power increase of more than 60 W.
- The device shall be fail-safe.

Suspension devices for extra-low-voltage luminaires, including supporting conductors, shall be capable of carrying five times the mass of the luminaires (including their lamps) intended to be supported, but not less than 5 kg.

WR-559.11.6

5.9.10 Underwater luminaires for swimming pools

A luminaire for use in the water or in contact with the water shall be fixed and shall comply with BS EN 60598-2-18.

WR-702.55.2

Underwater lighting located behind watertight   WR-702.55.2
portholes, and serviced from behind, shall comply
with the appropriate part of BS EN 60598 and
be installed in such a way that no intentional or
unintentional conductive connection between any
exposed conductive part of the underwater luminaires
and any conductive parts of the portholes can occur.

5.9.11 Luminaires in fountains

A luminaire installed in zone 0 or 1 shall be fixed   WR-702.55.3
and shall comply with BS EN 60598-2-18.

Electrical equipment in zone 0 or 1 shall be   WR-702.55.3
provided with mechanical protection to medium
severity (AG2), e.g. by use of mesh glass or by
grids which can only be removed by the use of
a tool.

An electric pump shall comply with the   WR-702.55.3
requirements of BS EN 60335-2-41.

5.9.12 Luminaires and lighting installations in agricultural and horticultural premises

Luminaires shall comply with the BS EN 60598   WR-705.559
series and be selected regarding their degree of
protection against the ingress of dust, solid objects
and moisture (e.g. IP54), suitability for mounting on
a normally flammable surface (e.g. ▽F▽) and
limited temperature of luminaire surface (e.g. ▽D▽).

5.9.13 Luminaires and lighting installations in exhibitions shows and stands

Stand installations containing a concentration of   WR-711.422.4.2
electrical equipment, luminaires or lamps liable
to generate excessive heat shall not be installed
unless adequate ventilation provisions are made,
e.g. well-ventilated ceiling constructed of
incombustible material.

ELV lighting systems for filament lamps shall comply with BS EN 60598-2-23.

WR-711.559.4.2

Insulation piercing lampholders shall **not** be used unless the cables and lampholders are compatible, and provided the lampholders are non-removable once fitted to the cable.

WR-711.559.4.3

Luminaires mounted below 2.5 m (arm's reach) from floor level or otherwise accessible to accidental contact shall be firmly and adequately fixed, and so sited or guarded as to prevent risk of injury to persons or ignition of materials.

WR-711.559.5

5.9.14 Electric discharge lamp installations

Installations of any luminous tube, sign or lamp as an illuminated unit on a stand, or as an exhibit, with nominal power supply voltage higher than 230/400 V a.c., shall comply with the following:

WR-711.559.4.4

- location – the sign or lamp shall be installed out of arm's reach or shall be adequately protected to reduce the risk of injury to persons

WR-711.559.4.5

- installation – the facia or stand fitting material behind luminous tubes, signs or lamps shall be non-ignitable.

WR-711.559.4.6

Emergency switching devices – a separate circuit shall be used to supply signs, lamps or exhibits, which shall be controlled by an emergency switch.

WR-711.559.4.7

The switch shall be easily visible, accessible and clearly marked.

5.9.15 Luminaires in caravans and motor caravans

Each luminaire in a caravan shall preferably be fixed directly to the structure or lining of the caravan.

WR-721.55.2.4

Where a pendant luminaire is installed in a caravan, provision shall be made for securing the luminaire to prevent damage when the caravan is in motion.

WR-721.55.2.4

Accessories for the suspension of pendant luminaires shall be suitable for the mass suspended and the forces associated with vehicle movement.	WR-721.55.2.4
A luminaire intended for dual-voltage operation shall comply with the appropriate standard.	WR-721.55.2.5

5.9.16 Luminaires in temporary installations

Every luminaire and decorative lighting chain shall: • have a suitable IP rating; and • be installed so as not to impair its ingress protection; and • be securely attached to the structure or support intended to carry it.	WR-740.55.1.1
Its weight shall not be carried by the supply cable, unless it has been selected and erected for this purpose.	WR-740.55.1.1
Luminaires and decorative lighting chains mounted less than 2.5 m (arm's reach) above floor level or otherwise accessible to accidental contact, shall be firmly fixed and so sited or guarded as to prevent risk of injury to persons or ignition of materials.	WR-740.55.1.1
Access to the fixed light source shall only be possible after removing a barrier or an enclosure which shall require the use of a tool.	WR-740.55.1.1
Lighting chains shall use HO5RN-F (BS 7919) cable or equivalent.	WR-740.55.1.1
Insulation-piercing lampholders shall not be used unless the cables and lampholders are compatible and the lampholders are non-removable once fitted to the cable.	WR-740.55.1.2
All lamps in shooting galleries and other sideshows where projectiles are used shall be suitably protected against accidental damage.	WR-740.55.1.3
Where transportable floodlights are used, they shall be mounted so that the luminaire is inaccessible.	WR-740.55.1.4

| Supply cables shall be flexible and have adequate protection against mechanical damage. | WR-740.55.1.4 |

 Luminaires and floodlights shall be so fixed and protected that a focusing or concentration of heat is not likely to cause ignition of any material.

5.9.16.1 Electric discharge lamp installations

| A luminous tube, sign or lamp shall be installed out of arm's reach or shall be adequately protected to reduce the risk of injury to persons. | WR-740.55.3.1 |

5.10 Mobile equipment

All final circuits for mobile equipment connected by means of a flexible cable or cord with a current-carrying capacity up to 32 A, shall be protected by RCDs.	WR-740.415.1
In a.c. systems, additional protection by means of an RCD shall be provided for mobile equipment with a current rating not exceeding 32 A for use outdoors.	WR-411.3.3
Where the use of mobile equipment is envisaged, protection (i.e. by means of a non-conducting location):	WR-418.1.6

- shall be ensured; and
- shall be permanent; and
- it shall not be possible to make the equipment ineffective.

| Where mobile equipment is likely to be used, provision shall be made so that the equipment can be fed from an adjacent and conveniently accessible socket outlet, taking account of the length of flexible cord normally fitted to portable appliances and luminaires. | WR-553.1.7 |

5.11 Plug and socket outlets

☇ A plug and socket outlet shall **not** be selected as a device for emergency switching.	WR-537.4.2.8
A plug and socket outlet may be inserted in the main supply circuit for switching off for mechanical maintenance.	WR-537.3.2.1
If the rating of the plug and socket outlet does not exceed 16 A, it shall be capable of cutting off the full load current of the relevant part of the installation.	WR-537.3.2.6
A socket outlet on a wall or similar structure shall be mounted at a height above the floor or any working surface to minimise the risk of mechanical damage to the socket outlet or to an associated plug and its flexible cord which might be caused during insertion, use or withdrawal of the plug.	WR-553.1.6
Equipment which has a protective conductor current exceeding 3.5 mA but not exceeding 10 mA, shall either be connected by a plug and socket outlet (complying with BS EN 60309-2) or be permanently connected to the fixed wiring of the installation.	WR-543.7.1.1
Equipment which has a protective conductor current exceeding 10 mA shall be connected to the supply via a flexible cable with a plug and socket outlet; or	WR-543.7.1.2

- permanently via the wiring of the installation; or
- via a protective conductor with an earth monitoring system.

☇ It shall **not** be possible for any pin of a plug to make contact with any live contact of any socket outlet within the same installation other than the type of socket outlet for which the plug is designed.	WR-553.1.1
Every plug and socket outlet (except for SELV) shall be of the non-reversible type, with provision for the connection of a protective conductor.	WR-553.1.2

Every socket outlet for household and similar use shall be of the shuttered type and, for an a.c. installation, shall preferably be of a type complying with BS 1363.

WR-553.1.4

A plug and socket outlet not complying with BS 1363, BS 546, BS 196 or BS EN 60309-2, may be used in single-line a.c. or two-wire d.c. circuits operating at a nominal voltage not exceeding 250 volts for:

WR-553.1.5

• the connection of an electric clock
• the connection of an electric shaver
• a circuit having special characteristics such that danger would otherwise arise or it is necessary to distinguish the function of the circuit.

Except for the plug of a plug and socket outlet identified in Table 53.2 on p. 117 of BS 7671:2008 as suitable for isolation, equipment of overvoltage categories I and II should **not** be used for isolation.

WR-537.2.2.1

All low-voltage plug and socket outlets shall conform with the applicable British Standard listed in Table 5.5 (see below).

WR-553.1.3

Table 5.5 Plug and socket outlets for low-voltage circuits (courtesy BSI)

Type of plug and socket outlet	Rating (amperes)	Applicable British Standard
Fused plug and shuttered socket outlets, 2-pole and earth, for a.c.	13	BS 1363 (fuses to BS 1362)
Plug, fused or non-fused, and socket outlets, 2-pole and earth	2, 5, 15, 30	BS 546 (fuses, if any, to BS 646)
Plug, fused or non-fused, and socket outlets, protected-type, 2-pole with earthing contact	5, 15, 30	BS 196
Plug and socket outlets (industrial type)	16, 32, 63, 125	BS EN 60309-2

A plug and socket outlet shall **not** be used as a device for connecting a water heater and/or boiler to the supply.

WR-554.3.3

Where mobile equipment is likely to be used, provision shall be made so that the equipment can be fed from an adjacent and conveniently accessible socket outlet, taking account of the length of flexible cord normally fitted to portable appliances and luminaires.

WR-553.1.7

Plug and socket outlets in a SELV system shall **not** have a protective conductor contact.

WR-414.4.3

For a final circuit with a number of socket outlets or connection units intended to supply two or more items of equipment, where it is known or reasonably to be expected that the total protective conductor current in normal service will exceed 10 mA, the circuit shall be provided with a high integrity protective conductor connection.

WR-543.7.2.1

The following arrangements of the final circuit are acceptable:

WR-543.7.2.1

- a ring final circuit with a ring protective conductor
- a radial final circuit with a single protective conductor.

5.11.1 Caravan and camping parks

At least one socket outlet shall be provided for each caravan pitch.

WR-708.553.1.11

Each socket outlet and its enclosure forming part of the caravan pitch electrical supply equipment shall comply with BS EN 60309-2 and meet the degree of protection of at least IP44 in accordance with BS EN 60529.

WR-708.553.1.8

Each socket outlet shall:

WR-708.553.1.12

- be provided with individual overcurrent protection.
- be protected individually by an RCD.

WR-708.553.1.13

Socket outlet protective conductors shall not be connected to any PEN conductor of the electricity supply.	WR-708.553A.14
The current rating of socket outlets shall be not less than 16 A.	WR-708.553.1.10
The socket outlets shall be placed at a height of 0.5 m to 1.5 m from the ground to the lowest part of the socket outlet. In special cases, due to environmental conditions such as risk of flooding or heavy snowfall, the maximum height is permitted to exceed 1.5 m.	WR-708.553.1.9

5.11.2 Marinas and similar locations

A maximum of four socket outlets shall be grouped together in one enclosure.	WR-709.553.1.10
Every socket outlet shall:	WR-709.553.1.8
• meet the degree of protection of IP44 or such protection shall be provided by an enclosure;	
• be located as close as practicable to the berth to be supplied. Socket outlets shall be installed in the distribution board or in separate enclosures.	WR-709.553.1.9
In general, single-line socket outlets with rated voltage 200–250 V and rated current 16 A shall be provided.	WR-709.553.1.12
One socket outlet shall supply only one pleasure craft or houseboat.	WR-709.553.1.11
Socket outlets shall:	WR-709.553.1.8
• comply with BS EN 60309-1 above 63 A and BS EN 60309-2 up to 63 A	
• be placed at a height of not less than 1 m above the highest water level. In the case of floating pontoons or walkways only, this height may be reduced to 300 mm above the highest water level provided that appropriate additional measures are taken to protect against the effects of splashing.	WR-709.553.1.13

5.11.3 Exhibitions, shows and stands

An adequate number of socket outlets shall be installed to allow user requirements to be met safely.	WR-711.55.7
Where a floor-mounted socket outlet is installed, it shall be adequately protected from accidental ingress of water and have sufficient strength to be able to withstand the expected traffic load.	WR-711.55.7

5.11.4 Mobile and transportable units

Plugs and connectors used to connect the unit to the supply shall comply with BS EN 60309-2 and shall also meet the following requirements: WR-717.55.1

- Plug shall have an enclosure of insulating material.
- Plug and socket outlets shall afford a degree of protection of not less than 1P44, if located outside.
- Appliance inlets with their enclosures shall provide a degree of protection of at least 1P44.

The plug part shall be situated on the unit.

Socket outlets located outside the unit shall be provided with an enclosure affording a degree of protection of not less than 1P44.	WR-717.55.2

5.11.5 Temporary electrical installations

An adequate number of socket outlets shall be installed to allow the user's requirements to be met safely.	WR-740.55.7

Note: In booths and stands and for fixed installations, one socket outlet for each square metre or linear metre of wall is generally considered adequate.

Socket outlets dedicated to lighting circuits placed out of arm's reach shall be encoded or marked according to their purpose.	WR-740.55.7
When used outdoors, plugs, socket outlets and couplers shall comply with BS EN 60309.	WR-740.55.7

5.12 Protection by RCDs

If an installation includes an RCD then it shall have a notice (fixed in a prominent position) that reads as follows:

> This installation, or part of it, is protected by a device that automatically switches off the supply if an earth fault develops. Test quarterly by pressing the button marked 'T' or 'Test'. The device should switch off the supply and should then be switched on to restore the supply. If the device does not switch off the supply when the button is pressed, seek expert advice.

Figure 5.2 Inspection and testing notice

Other requirements from the Regulations include:

Every installation shall be divided into circuits, as necessary, to reduce the possibility of unwanted tripping of RCDs due to excessive protective conductor currents produced by equipment.	WR-314.1
In a.c. systems, additional protection by means of an RCD shall be provided for:	WR-411.3.3

- socket outlets with a rated current not exceeding 20 A that are used by ordinary persons unless:
  - they are used under the supervision of skilled or instructed persons (e.g. in some commercial or industrial locations); or
  - a suitably labelled and/or identified socket outlet is provided for connection of a particular item of equipment
- mobile equipment with a current rating not exceeding 32 A for use outdoors.

5.12.1 Construction

The magnetic circuit of the transformer of an RCD shall enclose all the live conductors of the protected circuit. The associated protective conductor shall be outside the magnetic circuit.	WR-531.2.2
The rated residual operating current of the protective device shall comply with the requirements appropriate to the type of system earthing.	WR-531.2.3
Where an RCD is used for fault protection with, but separately from, an overcurrent protective device, it shall be verified that the residual current operated device is capable of withstanding, without damage, the thermal and mechanical stresses to which it is likely to be subjected in the case of a fault occurring on the load side of the point at which it is installed.	WR-531.2.8
Where two or more RCDs are in series and where discrimination in their operation is necessary to prevent danger, the characteristics of the devices shall be such that the intended discrimination is achieved.	WR-531.2.9

 Where the installation is not intended to be under the supervision of a skilled or instructed person (and the installation is liable to impact, consideration shall be given to providing additional protection by means of an RCD.

An RCD shall be so selected and the electrical circuits so subdivided that any protective conductor current which may be expected to occur during normal operation of the connected load(s) will be unlikely to cause unnecessary tripping of the device.	WR-531.2.4
The use of an RCD associated with a circuit normally expected to have a protective conductor shall **not** be considered sufficient for fault protection if there is no such conductor.	WR-531.2.5

5.12.2 Installation

An RCD shall be capable of disconnecting all the line conductors of the circuit at substantially the same time.	WR-531.2.1

An RCD shall be located so that its operation will not be impaired by magnetic fields caused by other equipment.	WR-531.2.7
Where an RCD may be operated by a person other than a skilled or instructed person, it shall be designed or installed so that it is not possible to modify or adjust the setting or the calibration of its rated residual operating current or time delay mechanism without a deliberate act involving the use of either a key or a tool and resulting in a visible indication of its setting or calibration.	WR-531.2.10
Where selectivity between RCDs is necessary to prevent danger and where required for proper functioning of the installation, the manufacturer's instructions shall be taken into account.	WR-536.3
Where an installation incorporates an RCD a notice shall be fixed in a prominent position at or near the origin of the installation and shall read as follows:	WR-514.12.2

This installation, or part of it, is protected by a device that automatically switches off the supply if an earth fault develops. Test quarterly by pressing the button marked 'T' or 'Test'. The device should switch off the supply and should then be switched on to restore the supply. If the device does not switch off the supply when the button is pressed, seek expert advice.

Figure 5.3 Warning notice – RCD protection

For installations and locations where there is an increased risk of shock (such as agricultural and horticultural premises, building sites, bathrooms, swimming pools) additional measures may be required, such as automatic disconnection of supply by means of an RCD with a rated residual operating current not exceeding 30 mA.

5.12.3 Locations containing a bath or shower

Additional protection shall be provided for all circuits of the location, by the use of one or more RCDs.	WR-701.411.3.3

Where the location containing a bath or shower   WR-701.415.2
is in a building with a protective equipotential
bonding system, supplementary equipotential
bonding may be omitted when all of the following
conditions are met:

- all final circuits of the location have additional
  protection by means of an RCD; and
- all final circuits of the location comply with the
  requirements for automatic disconnection; and
- all extraneous conductive parts of the location
  are effectively connected to the protective
  equipotential bonding.

5.12.4 Swimming pools and other basins

Equipment for use in the interior of basins which   WR-702.410.3.4.1
is only intended to be in operation when people
are not inside zone 0 shall be supplied by a circuit
protected by automatic disconnection of the supply
using an RCD (or SELV, or electrical separation).

 Note: The same ruling applies to fountains in zone 1.

In zone 0 or 1:   WR-702.53

- switchgear or controlgear shall **not** be installed;
  and
- a socket outlet shall **not** be installed.

In zone 2, a socket outlet or a switch is permitted   WR-702.53
only if the supply circuit is protected by an RCD
(or SELV, or electrical separation).

5.12.5 Power supply

An RCD which is powered from an independent   WR-531.2.6
auxiliary source and which does not operate
automatically in the case of failure of the
auxiliary source shall **only** be used if:

- fault protection is maintained even in the
  case of failure of the auxiliary source

- the device is incorporated in an installation intended to be supervised by an instructed person or a skilled person and inspected and tested by a competent person.

The generating set shall be connected so that any provision within the installation for protection by RCDs remains effective for every intended combination of sources of supply.

WR-551.4.2

In a TN, TT or IT system an RCD with a rated residual operating current of not more than 30 mA shall be installed to protect every circuit.

WR-551.4.4.2

Where RCDs are also used for protection against fire, the conditions for protection by automatic disconnection of the supply shall be verified.

WR-612.8

Where RCDs are required for additional protection, the effectiveness of automatic disconnection of supply by RCDs shall be verified using suitable test equipment according to BS EN 61557-6 to confirm that the relevant requirements are met.

WR-612.10

5.12.6 TN systems

In a TN system, the integrity of the earthing of the installation depends on the reliable and effective connection of the PEN or PE conductors to earth.

An RCD may be used as a protective device for fault protection.

WR-411.4.4

 Note: Where an RCD is used for earth fault protection the circuit should also incorporate an overcurrent protective device.

Where an RCD is used, the maximum values of earth fault loop impedance (see Table 41.5 on p. 50 of BS 7671:2008) may be applied for non-delayed RCDs to BS EN 61008-1 and BS EN 61009-1 for final circuits not exceeding 32 A.

WR-411.4.9

An RCD shall **not** be used in a TN-C system.

 Where an RCD is used in a TN-C-S system, a PEN conductor shall **not** be used on the load side.

Except for mineral insulated cables, busbar trunking systems or powertrack systems, a wiring system shall be protected against insulation faults in a TN system by an RCD having a rated residual operating current (I_A,) not exceeding 300 mA.

WR-422.3.9

In a TN system, where, for certain equipment in a certain part of the installation, the characteristics of a protective device do not satisfy the requirements, that part may be protected by an RCD. In these sorts of circumstance, the exposed conductive parts of that part of the installation shall be connected to the TN earthing system protective conductor or to a separate earth electrode which affords an impedance appropriate to the operating current of the RCD.

WR-531.3.1

In this latter case the circuit shall be treated as a TT system.

For a TN-S system where the neutral is not isolated, any RCD shall be positioned to avoid incorrect operation due to the existence of any parallel neutral–earth path.

WR-551.6.2

5.12.7 TT system

Where an RCD is used for earth fault protection, the circuit should also incorporate an overcurrent protective device.

WR-411.5.2

Except for mineral insulated cables, busbar trunking systems or powertrack systems, a wiring system shall be protected against insulation faults in a TT system by an RCD having a rated residual operating current (I_A,) not exceeding 300 mA.

WR-422.3.9

> If an installation which is part of a TT system is
> protected by a single RCD, this shall be placed
> at the origin of the installation unless the part of
> the installation between the origin and the device
> complies with the requirements for protection by the
> use of Class II equipment or equivalent insulation.
>
> WR-531.4.1

Note: Where there is more than one origin this requirement applies to each origin.

5.12.8 IT system

> Where fault protection is provided by an RCD, the
> product of the rated residual operating current (I_M) in
> amperes and the earth fault loop impedance in ohms
> shall not exceed 50.
>
> WR-411.8.3

> In an IT system, the neutral conductor shall not be
> distributed unless:
>
> - the circuit is protected by an RCD with a
>   rated residual operating current not exceeding
>   0.2 times the current-carrying capacity of the
>   corresponding neutral conductor; or
> - overcurrent detection is provided for the neutral
>   conductor of every circuit; or
> - the neutral conductor is effectively protected
>   against short-circuit by a protective device
>   installed on the supply side.
>
> WR-431.2.2

In an IT system without a neutral conductor it is permitted to omit the overload protective device in one of the line conductors **if** an RCD is installed in each circuit.

> Where protection is provided by an RCD and
> disconnection following a first fault is not envisaged,
> the non-operating residual current of the device shall
> be at least equal to the current which circulates on the
> first fault to earth of negligible impedance affecting a
> line conductor.
>
> WR-531.5.1

5.12.9 Devices for protection against the risk of fire

Where it is necessary to limit the consequence of   WR-532.1
fault currents in a wiring system from the point of
view of fire risk, either:

- the circuit shall be protected by an RCD for
  fault protection; and
- the RCD shall be installed at the origin of the
  circuit to be protected; and
- the RCD shall switch all live conductors; and
- the rated residual operating current of the
  RCD shall not exceed 300 mA; **or**
- the circuit will need to be continuously
  monitored by an insulation monitoring device
  which initiates an alarm on the occurrence of
  an insulation fault.

Where fault protection and/or additional protection   WR-612.13.1
is to be provided by an RCD, the effectiveness of
any test facility incorporated in the device shall be
verified.

5.12.10 Electrode water heaters and boilers

If an electrode water heater or electrode boiler is   WR-554.1.4
directly connected to a supply at a voltage exceeding
low voltage, the installation shall include an RCD.

5.12.11 Outdoor lighting installation

It is recommended that equipment such as lighting   WR-559.10.3.2
arrangements in places such as telephone kiosks, bus
shelters and town plans are provided with additional
protection by an RCD.

5.13 Residual current monitor (RCM)

An RCM is **not** intended to provide protection against electric shock.

A residual current monitor (RCM) permanently monitors any leakage current in the downstream installation or part of it.	WR-538.4
RCMs for use in a.c. systems shall comply with BS EN 62020.	WR-538.4
Where an RCM is used in an a.c. IT system, it is recommended to use a directionally discriminating RCM in order to avoid inopportune signalling of leakage current where high leakage capacitances are liable to exist downstream from the point of installation of the RCM.	WR-538.4
In supply systems, RCMs may be installed to reduce the risk of operation of the protective device in the event of excessive leakage current of the installation or the connected appliances.	WR-538.4.1
An RCM is intended to alert the user of the installation before the protective device is activated thus.	WR-538.4.1

 Note: Where an RCD is installed upstream of the RCM, it is recommended that the RCM has a rated residual operating current not exceeding a third of that of the RCD.

In all cases, the RCM shall have a rated residual operating current not higher than the first fault current level intended to be detected.	
In an IT system where interruption of the supply in the case of a first insulation fault to earth is not required or not permitted, an RCM may be installed to facilitate the location of a fault. It is recommended to install the RCM at the beginning of the outgoing circuits.	WR-538.4.2

5.14 Rotating machines and motors

All equipment, including cable, of every circuit carrying the starting, accelerating and load currents of a motor shall be suitable for a current at least equal to the full-load current rating of the motor when rated in accordance with the appropriate British Standard.	WR-552.1.1

Where the motor is intended for intermittent duty and for frequent starting and stopping, account shall be taken of any cumulative effects of the starting or braking currents upon the temperature rise of the equipment of the circuit.	WR-552.1.1
Every electric motor having a rating exceeding 0.37 kW shall be provided with control equipment incorporating means of protection against overload of the motor.	WR-552.1.2
Except where failure to start after a brief interruption would be likely to cause greater danger, every motor shall be provided with means to prevent automatic restarting after a stoppage due to a drop in voltage or failure of supply, where unexpected restarting of the motor might cause danger.	WR-552.1.3

 Note: These requirements do not preclude arrangements for starting a motor at intervals by an automatic control device, where other adequate precautions are taken against danger from unexpected restarting.

Lighting for premises where machines with moving parts are in operation should consider the stroboscopic effects which can give a misleading impression of moving parts being stationary. Such effects may be avoided by selecting luminaires with suitable lamp controlgear, such as high frequency controlgear, or by distributing lighting loads across all the phases of a three-line supply.	WR-559.9

5.15 Supplies

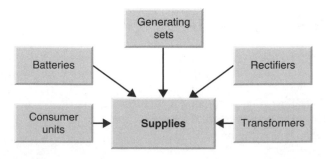

Figure 5.4 Types of supplies

5.15.1 Consumer units

A consumer unit (sometimes known as a consumer control unit or electricity control unit) is a particular type of distribution board for the control and distribution of electrical energy, primarily in domestic premises. It includes a manually operated method for isolation (both poles of the incoming circuit(s)) and assemblies of one or more fuses, circuit-breakers, RCDs, and signalling and other devices that have been manufactured for this purpose.

All circuits and final circuits **shall** be provided with a means of switching for interrupting the supply on load.   476-01-02

Note: This Regulation particularly applies to circuits and parts of an installation that (for safety reasons) need to be switched independently of other circuits and/or installations. It does not apply to short connections between the origin of the installation and the consumer's main switchgear.

A group of circuits may be switched by a common device.

5.15.2 Batteries

Stationary batteries shall be installed so that they are accessible **only** to skilled or instructed persons.   WR-551.8.1

Note: This generally requires the battery to be installed in a secure location or, for smaller batteries, a secure enclosure. The location or enclosure shall be adequately ventilated.

Battery connections shall have: WR-551.8.2

- basic protection by insulation and/or enclosures; or
- shall be arranged so that two bare conductive parts having between them a potential difference exceeding 120 volts cannot be inadvertently touched simultaneously.

Basic protection and fault protection is deemed to be provided where: the nominal voltage cannot exceed the upper limit of voltage Band I; and the supply is from a recognised source such as a battery.   WR-414.2

A battery may be used as a source for SELV and PELV systems.   WR-414.3

5.15.2.1 Protection against fault current

A device for protection against fault current need not be provided for a conductor connecting an accumulator battery to the associated control panel where the protective device is, **provided** that the wiring: • is carried out in such a way as to reduce the risk of fault to a minimum; and • is installed in such a manner as to reduce to a minimum the risk of fire or danger to persons.	WR-434.3

5.15.2.2 Central power supply sources

When operating in an emergency condition from a converter supplied by a battery, a static uninterruptible power supply source shall be able to: • operate distribution circuit protective devices; and • start the safety devices.	WR-560.6.9
There is no upper limit for the supply capacity of a central supply source such as a central battery system. However: • the batteries shall be of the vented or valve-regulated type; and • the minimum design life of the batteries shall be a minimum declared life of 10 years.	WR-560.6.9

5.15.2.3 Low-power supply sources

The power output of a low-power supply system is limited to 500 W for 3-hour duration or 1500 W for 1-hour duration. However: • the batteries may be of the gastight or valve-regulated maintenance-free type; and • the minimum design life of the batteries shall be 5 years.	WR-560.6.10

5.15.2.4 Uninterruptible power supply sources (UPS)

A static, uninterruptible power supply source shall be able to:

- operate distribution circuit protective devices; and
- start the safety devices

when operating in the emergency condition from the convertor supplied by the battery.

WR-560.6.11

5.15.2.5 Temporary electrical installations

In a temporary electrical installation, the supply to a battery-operated emergency lighting circuit shall be connected to the same RCD protecting the lighting circuit.

WR-740.415.1

5.15.3 Generating sets

5.15.3.1 As power sources

Generator sets (independent of the normal supply) are a recognised supply source for safety services such as emergency escape lighting, fire alarm systems, installations for fire pumps, fire rescue service lifts, and smoke and heat extraction equipment.

WR-351

The source of supply to a reduced low-voltage circuit shall be either:

- a motor-generator with windings that provide isolation equivalent to that provided by the windings of an isolating transformer; or
- a source independent of other supplies (such as an engine driven generator).

WR-411.8.4.1

The neutral (star) point of the secondary windings of three-line generators (or the midpoint of the secondary windings of single-line transformers and generators) shall be connected to earth.

WR-411.8.4.2

A generator supply source shall comply with BS 7698-12.

Basic and fault protection

The following sources may be used for SELV and PELV systems:

WR-414.3

- a source independent of a higher-voltage circuit (such as a diesel-driven generator)
- a motor-generator with windings providing equivalent isolation to a safety isolating transformer.

 Connection of live parts of the generator with earth may affect the protective measure.

Basic protection and fault protection is deemed to be provided where the nominal voltage cannot exceed the upper limit of voltage Band I and the supply is from a recognised source such as a diesel-driven generator or motor-generator.

WR-414.2

A device for protection against fault current need not be provided for a conductor connecting a generator to the associated control panel where the protective device is placed in the panel, **provided** that both of the following conditions are simultaneously fulfilled:

WR-434.3

- the wiring is carried out in such a way as to reduce the risk of fault to a minimum; and
- the wiring is installed in such a manner as to reduce to a minimum the risk of fire or danger to persons.

The generating set shall be connected so that any provision within the installation for protection by RCDs remains effective for every intended combination of sources of supply.

WR-551.4.2

5.15.3.2 Low-voltage generating sets

Low-voltage generating sets can be:

- combustion engines
- turbines
- electric motors
- photovoltaic cells
- electrochemical accumulators
- other suitable sources.

Where a generating set with an output not exceeding 16 A is to be connected in parallel with a system for distribution of electricity to the public, procedures for informing the electricity distributor are given in the Electricity Safety, Quality and Continuity Regulations (ESQCR) (2002).   WR-551.1

In addition to the ESQCR requirements, where a generating set with an output exceeding 16 A is to be connected in parallel with a system for distribution of electricity to the public, requirements of the electricity distributor should be ascertained before the generating set is connected.   WR-551.1

Requirements of the distributor for the connection of units rated up to 16 A are given in BS EN 50438.   WR-551.1

The proper functioning of other sources of supply shall not be impaired by the generating set.   WR-551.2.1

The prospective short-circuit current and prospective earth fault current shall be assessed for each source of supply or combination of sources which can operate independently of other sources or combinations.   WR-551.2.2

The short-circuit rating of protective devices within the installation and, where appropriate, connected to a system for distribution of electricity to the public, shall **not** be exceeded for any of the intended methods of operation of the sources.   WR-551.2.2

Where the generating set is intended to provide a supply to an installation which is not connected to a system for distribution of electricity to the public (or to provide a supply as a switched alternative to such a system) the capacity and operating characteristics of the generating set shall be such that there will be no danger or damage to equipment after the connection or disconnection of any intended load as a result of any voltage or frequency deviaton.   WR-551.2.3

Means shall be provided to automatically disconnect such parts of the installation as may be necessary if the capacity of the generating set is exceeded.   WR-551.2.3

5.15.3.3 Protection against overcurrent

Where overcurrent protection of the generating set is required, it shall be located as near as practicable to the generator terminals.	WR-551.5.1
Where a generating set is intended to operate in parallel with a system for distribution of electricity to the public, or where two or more generating sets may operate in parallel, circulating harmonic currents shall be limited so that the thermal rating of conductors is not exceeded.	WR-551.5.2

 Note: The effects of circulating harmonic currents may be limited by one or more of the following:

- the selection of a generating set with compensated windings;
- the provision of a suitable impedance in the connection to the generator star points;
- the provision of switches which interrupt the circulatory circuit but which are interlocked so that at all times fault protection is not impaired;
- the provision of filtering equipment.

5.15.3.4 Standby systems and switched alternatives to public supplies

Protection by automatic disconnection of supply shall not rely upon the connection to the earthed point of the system for distribution of electricity to the public when the generator is operating as a switched alternative to a TN system.	WR-551.4.3.2.1
Precautions for isolation shall be taken so that the generator cannot operate in parallel with the system for distribution of electricity to the public.	WR-551.6.1

5.16 Switches

Switches shall be fitted to a mounting box.	WR-530.4.2
A switch with a lock or removable handle (or other effective means of isolation) shall be provided for every electric discharge lighting installation which has an open circuit voltage exceeding low voltage.	WR-537.2.1.6

Persons and livestock shall be protected against injury, and property shall be protected against damage, as a consequence of overvoltages such as those originating from switching.

WR-131.6.2

Disconnecting devices shall be provided so as to allow electrical installations, circuits or individual items of equipment to be switched off or isolated for the purposes of operation, inspection, fault detection, testing, maintenance and repair.

WR-132.10

Note: Electrical equipment shall not have harmful effects on other equipment or impair the supply during normal service, including switching operations.

No means of isolation or switching shall be inserted in the outer conductor of a concentric cable.

WR-543.4.7

Every circuit shall be provided with a means of isolation from all live supply conductors by a linked switch or a linked circuit-breaker.

WR-422.3.13

A label or other suitable means of identification shall be provided to indicate the purpose and identification of each item of switchgear.

WR-514.1.1

WR-514.9.1

5.16.1 Protective devices

Switches shall:

WR-132.14.1

- only be inserted in the line conductor.
- be inserted in an earthed neutral conductor

WR-132.14.2

Note: Any linked switch or linked circuit-breaker inserted in an earthed neutral conductor shall be capable of breaking all of the related line conductors.

Equipment shall be selected and erected so that it will neither have harmful effects on other equipment nor impair the supply during normal service including switching operations.

WR-512.1.5

The effects of circulating harmonic currents may be limited by the provision of switches which interrupt the circulatory circuit, but which are interlocked so that at all times fault protection is not impaired.	WR-551.5.2

5.16.2 Switchgear

Switchgear:	WR-422.2.2

• shall be accessible only to authorised persons
• that is placed in an escape route, shall be enclosed in a cabinet or an enclosure constructed of non-combustible or not readily combustible material
• shall be installed outside the location unless:
  ○ it is suitable for the location; or
  ○ it is installed in an enclosure providing a degree of protection of at least IP4X or, in the presence of dust, IP5X
	WR-422.3.3

• shall not be connected to conductors intended to operate at a temperature exceeding 70°C at the equipment in normal service	WR-512.1.2
• shall be subjected to a functional test to show that it is properly mounted, adjusted and installed in accordance with the relevant requirements of these Regulations.	WR-612.13.2
At installations with a 230 V single-line supply rated up to 100 A that is under the control of ordinary persons, switchgear assemblies shall either comply with BS EN 60439-3 or be a consumer unit incorporating components and protective devices specified by the manufacturer as complying with BS EN 60439-3.	WR-530.3.4
Means of access to all live parts of switchgear where different nominal voltages exist shall be marked to indicate the voltages present.	WR-514.10.1
Where a metal enclosure or frame of a low-voltage switchgear is used as a protective conductor: • its electrical continuity shall be protected against mechanical, chemical or electrochemical deterioration	WR-543.2.4

- its cross-sectional area shall be in accordance
  with BS EN 60439-1
- it shall permit the connection of other
  protective conductors at every predetermined
  tap-off point.

5.16.3 Switching devices

A switching device:

- shall be protected against overcurrent   WR-536.5.1
- without integral overcurrent protection   WR-536.5.1
  shall be co-ordinated with an appropriate
  overcurrent protective device
- shall **not** be inserted in a protective conductor   WR-543.3.4
  unless the switch:
  - has been inserted in the connection between
    the neutral point and the means of earthing
  - is a linked switch arranged to disconnect
    and connect the earthing conductor at
    substantially the same time as the related
    live conductors
  - is a multipole linked switch or plug-in device
    in which the protective conductor circuit
    has not been interrupted before the live
    conductors and re-established not later than
    when the live conductors are reconnected.

5.16.4 Isolation and switching

Every device provided for isolation or switching   WR-537.1.1
shall comply with the relevant requirements of this
section. Table 53.2 on p. 117 of BS 7671:2008
provides information on selection.

5.16.5 Devices for isolation and switching

Isolation shall preferably be provided by a multipole   WR-537.2.2.5
switching device which disconnects all applicable
poles of the relevant supply.

Where it is intended that isolation and switching is carried out **only** by instructed persons, for TN systems, the means of switching the supply on load and the means of isolation may be provided by a suitably rated fuse carrier.	WR-559.10.6.1
Where the distributor's cut-out is used as the means of isolation of a highway power supply the approval of the distributor shall be obtained.	WR-559.10.6.2

 Note: Combined protective and neutral (PEN) conductors shall not be isolated or switched.

5.16.6 Single-pole switching devices

Single-pole switching devices: • shall not be inserted in the neutral conductor of a multiphase circuit; or • alone in the neutral conductor single-line circuits.	WR-530.3.2

5.16.7 Main switches

Where an installation is supplied from more than one source: • a main switch shall be provided for each source of supply; and • a warning notice shall be permanently fixed in such a position that any person seeking to operate any of these main switches will be warned of the need to operate all such switches to achieve isolation of the installation; or • a suitable interlock system shall be provided.	WR-537.1.6
Where an installation is supplied from more than one source of energy, one of which requires a means of earthing independent of the means of earthing of other sources and it is necessary to ensure that not more than one means of earthing is applied at any	WR-537.1.5

time, a switch may be inserted in the connection between the neutral point and the means of earthing, provided that the switch is a linked switch arranged to disconnect and connect the earthing conductor for the appropriate source, at substantially the same time as the related live conductors.

5.16.8 Main linked switches

A main linked switch shall be provided as near as practicable to the origin of every installation as a means of switching the supply on load and as a means of isolation.

WR-537.1.4

Note: In a TN-S or TN-C-S system the neutral conductor need not be isolated or switched where it can be regarded as being reliably connected to earth by a suitably low impedance.

A main switch intended for operation by ordinary persons, e.g. of a household or similar installation, shall interrupt both live conductors of a single-line supply.

WR-537.1.4

5.16.9 Emergency switching

Emergency switching may be emergency switching ON or emergency switching OFF.

Means shall be provided for emergency switching of any part of an installation where it may be necessary to control the supply to remove an unexpected danger.

WR-537.4.1.1

Other than where a risk of electric shock is involved, the emergency switching device shall be an isolating device and shall interrupt **all** live conductors.

WR-537.4.1.2

Except where the neutral conductor can be regarded as being reliably connected to earth in a TN-S or TN-C-S system, the neutral conductor need not be isolated or switched.

The execution of emergency switching shall ensure that only one single action is required to interrupt the appropriate supply conductors.   WR-537.4.1.3

The operation of an emergency switching device shall **not** introduce a further danger or interfere with the complete operation necessary to remove the danger.   WR-537.4.1.4

5.16.9.1 Devices for emergency switching

A device for emergency switching shall be capable of breaking the full load current of the relevant part(s) of the installation taking account of stalled motor currents where appropriate.

Hand-operated switching devices for direct interruption of the main circuit shall be selected where practicable and these shall be clearly identified, preferably by colour.

An emergency switching device may consist of: WR-537.4.2.2

- a switch in the main circuit (or pushbuttons in the control (auxiliary) circuit) that is capable of directly cutting off the appropriate supply; or
- a combination of equipment activated by a single action for the purpose of cutting off the appropriate supply.

The means of operating (handle, push-button, etc.) an emergency switching device:

- shall be readily accessible at places where a danger might occur and/or at any additional remote position from which that danger can be removed   WR-537.4.2.5
- shall be capable of latching or being restrained in the 'OFF' or 'STOP' position   WR-537.4.2.6
- shall **not** re-energise the relevant part of the installation   WR-537.4.2.6
- shall be so placed and marked as to be readily identifiable and convenient for the intended use.   WR-537.4.2.7

5.16.10 Functional switching devices

In general, all current-using equipment requiring   WR-537.5.1.3
control shall be controlled by an appropriate
functional switching device.

Note: A single functional switching device may control two or more items of
equipment intended to operate simultaneously.

A functional switching device that is designed to   WR-537.5.1.4
ensure the changeover of supply from alternative
sources:

- shall affect all live conductors; and
- shall not be capable of putting the sources in
  parallel, unless the installation is specifically
  designed for this condition.

Note: In these cases, no provision shall be made for isolation of the PEN or
protective conductors

- shall be suitable for the most onerous   WR-537.5.2.1
  duty they are intended to perform
- shall be provided for each part of a circuit   WR-537.5.1.1
  which may require to be controlled
  independently of other parts of the
  installation.

but it need **not** necessarily control all live conductors of a circuit

- may control the current without necessarily   WR-537.5.2.2
  opening the corresponding poles.

Semiconductor switching devices are examples of devices capable of interrupting the current in the circuit but not opening the corresponding poles.

5.16.11 Firefighter's switches

A firefighter's switch shall be provided in the low-
voltage circuit supplying:

WR-537.6.1

- exterior electrical installations operating at a
  voltage exceeding low voltage; and
- interior discharge lighting installations operating
  at a voltage exceeding low voltage.

Every firefighter's switch shall comply with the
following:

WR-537.6.3
and

- For an exterior installation, the switch shall
  be outside the building and adjacent to the
  equipment (or alternatively a notice indicating the
  position of the switch shall be placed adjacent to
  the equipment and a notice shall be fixed near the
  switch so as to render it clearly distinguishable).
- For an interior installation, the switch shall
  be independent of the switch for any exterior
  installation and in the main entrance to the
  building.
- The switch shall be placed in a conspicuous
  position, reasonably accessible to firefighters,
  and at not more than 2.75 m from the ground or
  the standing beneath the switch.
- Where more than one switch is installed on
  any one building, each switch shall be clearly
  marked to indicate the installation or part of the
  installation which it controls.

WR-537.6.2

A firefighter's switch shall:

WR-537.6.4

- be coloured red and have fixed on or near it a
  permanent nameplate marked with the words
  'FIREFIGHTER'S SWITCH'; and
- have its ON and OFF positions clearly indicated
  by lettering legible to a person standing on the
  ground at the intended site, with the OFF position
  at the top; and
- be provided with a device to prevent the switch
  being inadvertently returned to the ON position,
  and
- be arranged to facilitate operation by a
  firefighter.

Note: An IMD shall be used on a circuit comprising safety equipment which is normally de-energised by a switching means disconnecting all live poles and which is only energised in the event of an emergency (provided that the IMD is automatically deactivated whenever the safety equipment is activated).

5.16.12 Switchboards

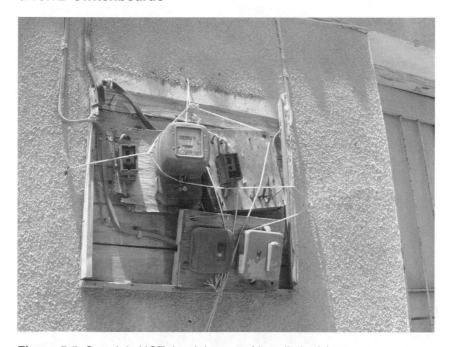

Figure 5.5 Certainly NOT the right sort of installation! (courtesy StingRay)

Passageways and working platforms which have access to an open type switchboard or an item of equipment that has dangerous exposed live parts, need to allow persons, without hazard, to:

- operate and maintain the equipment;
- pass one another as necessary with ease; and
- back away from the equipment.

Where equipment carrying current of different types (or at different voltages) is grouped in a common assembly such as a switchboard, all equipment belonging to any one type of current or any one type of voltage shall be effectively segregated so as to avoid mutual detrimental influence.   WR-515.2

Any identification of a switchboard busbar or conductor shall comply with the requirements of Table 51 on p. 92 of BS 7671:2008 so far as these are applicable.	WR-514.3.3

5.16.13 Mechanical maintenance

The capability of switching off for mechanical maintenance shall be provided where mechanical maintenance could involve a risk of physical injury.	WR-537.3.1.1 and WR-537.3.1.2

 Suitable means shall be provided to prevent electrically powered equipment from becoming unintentionally reactivated during mechanical maintenance

5.16.13.1 Devices for switching off for mechanical maintenance

A device such as a: • multipole switch • circuit-breaker • control and protective switching device (CPS) • control switch operating a contactor • plug and socket outlet may be inserted in the main supply circuit for switching off for mechanical maintenance.	WR-537.3.2.1

 The open position of the contacts of the device shall be visible or be clearly and reliably indicated by the use of the symbols '**0**' and '**I**' to indicate the open and closed positions respectively.

A device for switching off for mechanical maintenance: • shall require manual operation • shall be designed and/or installed so as to prevent inadvertent or unintentional switching on • shall be so placed, readily identifiable and convenient for the intended use • shall be capable of cutting off the full load current of the relevant part of the installation when required.	WR-537.3.2.2 WR-537.3.2.3 WR-537.3.2.4 WR-537.3.2.5

Note: A plug and socket outlet or similar device of rating not exceeding 16 A may be used as a device for switching off for mechanical maintenance.

Heaters and/or boilers

A heater or boiler shall be permanently connected to the electricity supply through a double-pole linked switch which is either separate from and within easy reach of the heater or boiler or is incorporated therein. The wiring from the heater or boiler shall be connected directly to that switch without the use of a plug and socket outlet.	WR-554.3.3

If the heater or boiler is installed in a room containing a fixed bath, the switch shall comply with Section 701 of BS 7671:2008.

Where a step-up transformer is used, a linked switch shall be provided for disconnecting the transformer from all live conductors of the supply.	WR-555.1.3

5.16.14 Lighting installations

A lighting installation shall be appropriately controlled by a switch, combination of switches or by an automatic control system, which is suitable for discharge lighting circuits.	WR-559.6.1.9

5.16.15 Special installations and locations

5.16.15.1 Swimming pools and other basins

In zone 0 or 1, switchgear or controlgear shall not be installed.	WR-702.53
In zone 2, a switch is permitted **only** if the supply circuit is protected by one of the following protective measures: • SELV • automatic disconnection of supply using an RCD • electrical separation.	WR-702.53

5.16.15.2 Rooms and cabins containing sauna heaters

Switchgear which forms part of the sauna heater equipment or of other fixed equipment installed in zone 2, may be installed within the sauna room or cabin.	WR-703.537.5
Other switchgear (e.g. for lighting) shall be placed outside the sauna room or cabin.	WR-703.537.5

5.16.15.3 Construction and demolition site installations

Each Assembly for Construction Sites (ACS) shall incorporate suitable devices for the switching and isolation of the incoming supply.	WR-704.537.2.2
Theses devices shall be suitable for securing in the off position by padlock (or by location) inside a lockable enclosure.	WR-704.537.2.2

5.16.15.4 Agricultural and horticultural premises

The electrical installation of each building or part of a building shall be isolated by a single isolation device.	WR-705.537.2
• These isolation devices shall be clearly marked according to the part of the installation to which they belong.	WR-705.537.2
• Emergency stopping devices shall **not** be erected where they are accessible to livestock or in any position where access may be impeded by livestock.	WR-705.537.2

5.16.15.5 Marinas and similar locations

At least one means of isolation: • shall be installed in each distribution cabinet • shall disconnect all live conductors including the neutral conductor.	WR-709.537.2.1.1

Note: There shall be one isolating switching device for a maximum of four socket outlets.

5.16.15.6 Exhibitions, shows and stands

Switchgear shall be placed in closed cabinets which can only be opened by the use of a key or a tool, except for those parts designed and intended to be operated by ordinary persons.	WR-711.51
A separate circuit shall be used to supply signs, lamps or exhibits, which shall be controlled by an emergency switch.	WR-711.559.4.7
The switch shall be easily visible, accessible and clearly marked	

5.16.15.7 Solar photovoltaic (PV) power supply systems

Switchgear assemblies shall be in compliance with BS EN 60439-1.	WR-712.511.1
To allow maintenance of the PV converter, means of isolating the PV converter from the d.c. side and the a.c. side shall be provided.	WR-712.537.2.1.1

In the selection and erection of devices for isolation and switching to be installed between the PV installation and the public supply, the public supply shall be considered the source and the PV installation shall be considered the load.

A switch-disconnector shall be provided on the d.c. side of the PV converter.	WR-712.537.2.2.5

Note: All junction boxes (PV generator and PV array boxes) shall carry a warning label indicating that parts inside the boxes may still be live after isolation from the PV converter.

5.16.15.8 Electrical installations in caravans and motor caravans

In an installation consisting of only one final circuit, the isolating switch may be the overcurrent protective device fulfilling the requirements for isolation.	WR-721.537.2.1.1

A notice shall be permanently fixed near the main isolating switch inside the caravan.	WR-721.537.2.1.1.1

5.16.15.9 Temporary electrical installations

Switchgear shall be placed in cabinets which can be opened only by the use of a key or a tool, except for those parts designed and intended to be operated by ordinary persons.	WR-740.51
Every electrical installation of a booth, stand or amusement device shall have its own means of isolation and switching, which shall be readily accessible.	WR-740.537.1
Every separate temporary electrical installation for amusement devices and each distribution circuit supplying outdoor installations shall be provided with its own readily accessible and properly identified means of isolation.	WR-740.537.2.1.1
A device for isolation shall disconnect all live conductors (line and neutral conductors).	WR-740.537.2.2
A separate circuit shall be used to supply luminous tubes, signs or lamps, which shall be controlled by an emergency switch.	WR-740.55.3.2
The switch shall be easily visible, accessible and marked in accordance with the requirements of the local authority.	WR-740.55.3.2

5.17 Rectifiers

A device for protection against fault current need not be provided for a conductor connecting a rectifier where the protective device is placed in the panel **provided** that both of the following conditions are simultaneously fulfilled: • The wiring is carried out in such a way as to reduce the risk of fault to a minimum. • The wiring is installed in such a manner as to reduce to a minimum the risk of fire or danger to persons.	WR-434.3

5.17.1 Temporary installations

Enclosures containing rectifiers shall be adequately ventilated and the vents shall not be obstructed when in use.	WR-740.55.5

5.18 Transformers

Enclosures containing transformers shall be adequately ventilated and the vents shall not be obstructed when in use.	WR-740.55.5
The magnetic circuit of the transformer of an RCD shall enclose all the live conductors of the protected circuit. The associated protective conductor shall be outside the magnetic circuit.	WR-531.2.2
The source of supply to a reduced low-voltage circuit may be a double-wound isolating transformer (complying with BS EN 61558-1 and BS EN 61558-2-23).	WR-411.8.4.1

5.18.1 Safety isolating transformers

A safety isolating transformer for an extra-low-voltage lighting installation shall comply with BS EN 61558-2-6 and either   WR-559.11.3.1

- the transformer shall be protected on the primary side by a protective device complying with the requirements of Regulation 559.11.4.2; **or**
- the transformer shall be short-circuit proof (both inherently and non-inherently) and shall be marked with the symbol shown here.
- Safety isolating transformers shall comply with BS EN 61558-2-6 or provide an equivalent degree of safety.   WR-740.55.5
- A manually reset protective device shall protect the secondary circuit of each transformer or electronic convertor.   WR-740.55.5

- Safety isolating transformers shall be mounted out of arm's reach or be mounted in a location that provides equal protection, and shall have adequate ventilation.   WR-740.55.5

5.18.2 Autotransformers and step-up transformers

 A step-up autotransformer shall **not** be connected to an IT system.

Where an autotransformer is connected to a circuit having a neutral conductor, the common terminal of the winding shall be connected to the neutral conductor.   WR-555.1.1

Where a step-up transformer is used, a linked switch shall be provided for disconnecting the transformer from all live conductors of the supply.   WR-555.1.3

5.18.3 ELV transformers and electronic converters

ELV (extra-low-voltage) transformers shall be mounted out of arm's reach of the public and shall have adequate ventilation.   WR-711.55.6

A manual reset protective device shall protect the secondary circuit of each transformer or electronic converter.   WR-711.55.6

Electronic converters shall conform with BS EN 61347-1.   WR-711.55.6

5.18.4 IT systems

An IT system can be provided by: WR-717.411.6.2

- an isolating transformer or a low-voltage generating set, with an insulation monitoring device installed; or

- a transformer providing simple separation
  and providing:
  - automatic disconnection of the supply in
    the case of a first fault between live parts
    and the frame of the unit, or
  - an RCD.

5.18.5 Electric dodgems

Electric dodgems shall only be operated at voltages not exceeding 50 V a.c. or 120 V d.c.	WR-740.55.9
The circuit shall be electrically separated from the supply mains by means of a transformer or a motor-generator set.	WR-740.55.9

6

Cables and conductors

Within the Wiring Regulations there is frequent reference to different types of cable (e.g. single-core, multicore, fixed, flexible), conductors (e.g. live supply, protective, bonding) and conduits, cable ducting, cable trunking and so on. Unfortunately, similarly to equipment and components, the requirements for these items are liberally sprinkled throughout the Standard. The aim of this chapter, therefore, is to provide a catalogue of all the different types identified and referred to in the Wiring Regulations in three main headings (e.g. cables, conductors and conduits) and then make a list of their essential requirements.

 Similarly to other chapters, please remember that these lists of requirements are **only** the author's impression of the most important aspects of the Wiring Regulations and electricians should **always** consult BS 7671 to satisfy compliance!

6.1 Cables

An electric power cable is defined as:

> *an assembly of two or more electrical conductors consisting of a core protected by twisted wire strands held together with (and typically covered by) an overall sheath. The conductors may be of the same or different sizes, each with their own insulation. A bare conductor is normally used for the equipment safety earth.*

There are five main types of cable found in electrical installations. These are:

1. single-core cables
2. multicore cables
3. flexible cables
4. heating cables
5. fixed wiring.

6.1.1 General

Cables and cords shall comply with the requirements of BS EN 50265-2-1 or 2-2.	482-03-03
Cables with a non-metallic sheath or non-metallic enclosure are not considered to be a Class II construction.	471-09-04

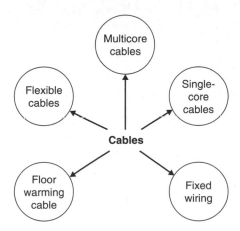

Figure 6.1 Cables

6.1.2 Single-core cables

The general requirements for single-core cables include the following:

Metallic sheaths and/or the non-magnetic armour of single-core cables that are in the same circuit must either be bonded together:

- at both ends of their run (solid bonding); or
- at one point in their run (single point bonding);

provided that, at full load, voltages from sheaths and/or armour to earth:

- do not exceed 25 volts; and
- do not cause corrosion; and
- do not cause danger or damage to property.

Single-core cables may be used as protective conductors but shall be coloured green-and-yellow throughout their length.

The conductance of the outer conductor of a concentric single-core cable shall not be less than that of the internal conductor.

When non-twisted single-core cables with a cross-sectional area greater than 50 mm² in copper (or 70 mm² in aluminium) are connected in parallel, the load current shall be shared equally between them. (This ruling does not apply to final ring circuits.)

Owing to possible electromagnetic effects, single-core cables that are armoured with steel wire or tape shall **not** be used for a.c. circuits.

In ferromagnetic enclosures:

> Single-core cables armoured with steel wire or steel   WR-521.5.2
> tape shall **not** be used for an a.c. circuit.

6.1.3 Multicore cables

The general requirements for single-core cables include the requirements that:

for telecommunication circuits, data transfer circuits and similar, consideration shall also be given to electrical interference, both electromagnetic and electrostatic. (See BS EN 50081 and BS EN 50082.)

A Band I circuit shall **not** be contained in the same wiring system as a Band II voltage circuit unless it is in a multicore cable (or cord) and the cores of the Band I circuit are:

- insulated for the highest voltage present in the Band II circuit;
- separated from the cores of the Band II circuit by an earthed metal screen;

and the cables:

- are insulated for their system voltage;
- are installed in a separate compartment of a cable ducting or trunking system;
- are installed on a tray or ladder separated by a partition;
- use a separate conduit, trunking or ducting system.

SELV circuit conductors that are contained in a multicore cable with other circuits should be insulated for the highest voltage present in that cable.

Separated circuits shall, preferably, use a separate wiring system. If this is not feasible, multicore cables (without a metallic sheath or insulated conductors) may be used.

The conductance of the outer conductor of a concentric cable for a multicore cable:

- serving a number of points contained within one final circuit (or where the internal conductors are connected in parallel) should not be less than that of the internal conductors;

- in a multiphase or multipole circuit should not be less than that of one internal conductor.

Two or more circuits are allowed in the same cable (but see Section 528 for specific requirements).	WR-521.7
Each part of a circuit shall be arranged such that the conductors are **not** distributed over different multicore cables, conduits, ducting systems, franking systems or tray or ladder systems.	WR-521.8.1
Where multicore cables are installed in parallel each cable shall contain one conductor of each line.	WR-521.8.1
A voltage Band I circuit shall not be contained in the same wiring system as a Band II circuit unless:	WR-528.1

- each conductor of a multicore cable is insulated for the highest voltage present in the cable
- for a multicore cable or cord, the cores of the Band I circuit are separated from the cores of the Band II circuit by an earthed metal screen of equivalent current-carrying capacity to that of the largest core of a Band II circuit.

All protective conductors shall be incorporated in a multicore cable or in a conduit together with the live conductors.	WR-721.543.2.1

6.1.4 Flexible cables

The general requirements for flexible cables include the following:
Flexible cables and flexible cords:

- shall be of a heavy-duty type with a voltage rating of not less than 45 or 50 V, or
- shall be suitably protected against mechanical damage;
- that are liable to mechanical damage shall be visible throughout its length;
- that are used as an overhead, low-voltage, line shall comply with the relevant British or Harmonised Standard;
- shall only be used for fixed wiring where the relevant Regulations permit.

Insulated flexible cables and cords may include a flexible metallic armour, braid or screen.

Non-flexible cables (and flexible cords not forming part of a portable appliance or Iuminaire) that are sheathed with lead, PVC or an elastomeric material may include a catenary wire (or hard-drawn copper conductor) for aerial use or when suspended.

Provided that all flexible equipment cables (other than Class II equipment) have a protective conductor for use as an equipotential bonding conductor, then source supplies may supply more than one item of equipment.

Flexible cables and cords shall be visible throughout any part of their length liable to mechanical damage.	WR-413.3.4

For separated circuits the use of separate wiring systems is recommended.

All flexible cables (unless they supply equipment with double or reinforced insulation) shall include a protective bonding conductor.	WR-418.3.6
Flexible cords shall **not** be laid in areas accessible to the public unless they are protected against mechanical damage.	WR-711.52
A flexible cable or flexible cord shall be used for fixed wiring only where the relevant provisions of the Regulations are met.	WR-521.9.3
Equipment that is intended to be moved in use shall be connected by flexible cables or cords, except equipment supplied by contact rails.	WR-521.9.1
Stationary equipment which is moved temporarily for the purposes of connecting, cleaning etc. (e.g. a cooker or a flush-mounting unit for installations in a false floor) may be connected with a non-flexible cable; however, if it is subject to vibration whilst in use it shall be connected by a flexible cable or cord.	WR-521.9.2
Flexible cables and flexible cords shall be either: • a heavy-duty type (with a voltage rating of not less than 450/750 V); or • suitably protected against mechanical damage.	WR-422.3.14

In locations subject to an external heat source (e.g. from solar gain of the wiring system or its surrounding medium):

parts of a cable or flexible cord within an accessory, WR-522.2.2
appliance or luminaire shall be suitable for the
temperatures likely to be encountered, or shall be
provided with additional insulation suitable for those
temperatures.

6.1.5 Heating cables

The general requirements for heating cables include the following:

Heating cables:

- passing through (or in close proximity to) a fire hazard:
  - shall be enclosed in material with an ignitability characteristic 'P' as specified in BS 476 Part 5;
  - shall be protected from any mechanical damage;
- that are going to be laid (directly) in soil, concrete, cement screed, or other material used for road and building construction shall be:
  - capable of withstanding mechanical damage;
  - constructed of material that will be resistant to damp and/or corrosion;
- that are going to be laid (directly) in soil, a road, or the structure of a building shall be installed so that they:
  - are completely embedded in the substance intended to be heated;
  - are not damaged by movement (by the embedding substance)
  - comply with the maker's instructions and recommendations.

The maximum loading of floor-warming cable under operating conditions is shown in Table 6.1.

Table 6.1 Maximum conductor operating temperatures for a floor-warming cable

Type of cable	Maximum conductor operating temperature (°C)
General-purpose PVC over conductor	70
Enamelled conductor, polychlorophene over enamel, PVC overall	70
Enamelled conductor, PVC overall	70
Enamelled conductor, PVC over enamel, lead-alloy 'E' sheath overall	70
Heat-resisting PVC over conductor	85
Nylon over conductor, heat-resisting PVC overall	85
Synthetic rubber or equivalent elastomeric insulation over conductor	85
Mineral insulation over conductor, copper sheath overall	Temperature dependent on type of seal employed, outer covering etc.
Silicone-treated woven-glass sleeve over conductor	180

6.1.6 Electric floor heating systems

For electric floor heating systems, only heating cables or thin sheet flexible heating elements shall be erected **provided** that they have either a metal sheath or a metal enclosure or a fine mesh metallic grid.	WR-701.753

6.1.7 Armoured single-core cables

The metallic sheaths and/or non-magnetic armour of single-core cables in the same circuit shall normally be bonded together at both ends of their run (solid bonding).	WR-523.10

6.1.8 Building design

Where risks due to structural movement exist (CB3), the cable support and protection system employed shall be capable of permitting relative movement so that conductors and cables are not subjected to excessive mechanical stress.	WR-522.15.1

6.1.9 Cable conduit, ducting, trunking, tray and ladder systems

Cable conduits shall comply with the BS EN 61386 series.	WR-521-6
Cable trunking or ducting shall comply with the appropriate part of the BS EN 50085 series.	WR-521-6
Cable tray and ladder systems shall comply with BS EN 61537.	WR-521-6
Non-sheathed cables for fixed wiring shall be enclosed in conduit, ducting or trunking.	WR-521.10.1
Non-sheathed cables are permitted in a cable trunking system which provides a minimum of IP4X or IPXXD protection, **or** if the cover can only be removed by means of a tool or a deliberate action.	WR-521.10.2

6.1.10 Cable couplers

A cable coupler shall be arranged so that the connector   WR-553.2.2
of the coupler is fitted at the end of the cable remote
from the supply.

Every cable coupler in a FELV system: WR-411.7.5

- shall have a protective conductor contact; and
- shall **not** be dimensionally compatible with those
  used for any other system in use in the same
  premises.

Except for a SELV or a Class II circuit, a cable coupler   WR-553.2.1
shall be non-reversible and shall have provision for the
connection of a protective conductor.

6.1.11 Cables in thermal insulation

A cable should preferably not be installed in a location   WR-523.7
where it is liable to be covered by thermal insulation.

Where a cable is to be run in a space to which thermal   WR-523.7
insulation is likely to be applied it shall, wherever
practicable, be fixed in a position such that it will **not**
be covered by the thermal insulation.

For a single cable likely to be totally surrounded by   WR-523.7
thermally insulating material over a length of 0.5 m or
more, the current-carrying capacity shall be taken, in the
absence of more precise information, as 0.5 times the
current-carrying capacity for that cable clipped direct to
a surface and open.

Where a cable is to be totally surrounded by thermal   WR-523.7
insulation for less than 0.5 m the current-carrying
capacity of the cable shall be reduced appropriately
depending on the size of cable, length in insulation
and thermal properties of the insulation.

6.1.12 Connection of multiwire, fine wire and very fine wire conductors

Cores of sheathed cables from which the sheath has been   WR-526.9
removed and non-sheathed cables at the termination of
conduit, ducting or trunking shall be enclosed.

6.1.13 Cross-sectional areas of conductors of cables

> The cross-sectional area of each conductor in an a.c.   WR-524.1
> circuit or of a conductor in a d.c. circuit shall be not
> less than the values given in Table 52.3 on p. 105 of BS
> 7671:2008.

6.1.14 Current-carrying capacities of cables

> The current, including any harmonic current, to be carried   WR-523.1
> by any conductor for sustained periods during normal
> operation, shall be such that the appropriate temperature
> limit specified in Table 6.2 (see below) is not exceeded.

Table 6.2 Maximum operating temperatures for types of cable insulation (reproduced with permission of IET)

Type of insulation	Temperature limit
Thermoplastic	70°C at the conductor
Thermosetting	90°C at the conductor
Mineral (thermoplastic covered or bare exposed to touch)	70°C at the sheath
Mineral (bare not exposed to touch and not in contact with combustible)	105°C at the sheath

6.1.15 Earth electrodes

As previously covered in chapter 3 (i.e. earthing), the following types of earth electrode are recognised as being suitable for the purposes of these Regulations:

- earth rods or pipes;
- earth tapes or wires;
- earth plates;
- underground structural metalwork embedded in foundations;
- welded metal reinforcement of concrete (except pre-stressed concrete) embedded in the earth;
- lead sheaths and other metal coverings of cables;
- other suitable underground metalwork.

 Note: Further information on earth electrodes can be found in BS 7430.

The use, as an earth electrode, of the lead sheath or other   WR-542.2.5
metal covering of a cable shall be subject to all of the
following conditions:

- Adequate precautions shall be taken to prevent
  excessive deterioration by corrosion.
- The sheath or covering shall be in effective contact
  with earth.
- The consent of the owner of the cable shall be
  obtained.
- Arrangements shall exist for the owner of the
  electrical installation to be warned of any proposed
  change to the cable which might affect its suitability
  as an earth electrode.

6.1.16 Electrical connections to bare connectors and/or busbars

Where a cable is to be connected to a bare conductor   WR-526.4
or busbar its type of insulation and/or sheath shall be
suitable for the maximum operating temperature of the
bare conductor or busbar.

6.1.17 Electrode water heaters and boilers

The shell of the electrode water heater or electrode boiler   WR-554.1.3
shall be bonded to the metallic sheath and armour, if any,
of the incoming supply cable.

6.1.18 Electromechanical stresses

Every cable shall have adequate strength and be so   WR-521.5.1
installed as to withstand the electromechanical forces that
may be caused by any current, including fault current, it
may have to carry in service.

6.1.19 Fire propagating structures

In the selection and erection of installations in locations of national, commercial,
industrial or public significance, the following measures may be considered:

- installation of mineral insulated cables according to BS EN 60702;

- installation of cables with improved fire-resisting characteristics in case of a fire hazard;
- installation of cables in non-combustible solid walls, ceilings and floors;
- installation of cables in areas with constructional partitions having a fire-resisting capability for a time of 30 minutes or 90 minutes.

 Note: Where these measures are not practicable improved fire protection may be possible by the use of reactive fire protection systems.

6.1.20 Fault current protective devices

Every fault current protective device shall ensure that:

A fault occurring at any point in a circuit is interrupted within a time such that the fault current does not cause the permitted limiting temperature of any conductor or cable to be exceeded.	WR-434.5.2

6.1.21 Groups containing more than one circuit

The group rating factors (see Tables 4C1 to 4C5 of Appendix 4 to BS 7671:2008) are applicable to groups of non-sheathed or sheathed cables having the same maximum operating temperature.	WR-523.5
For groups containing non-sheathed or sheathed cables having different maximum operating temperatures, the current-carrying capacity of all the non-sheathed or sheathed cables in the group shall be based on the lowest maximum operating temperature of any cable in the group together with the appropriate group rating factor.	WR-523.5

6.1.22 Heating conductors and cables

Where a heating cable is required to pass through, or be in close proximity to, material which presents a fire hazard, the cable: - shall be enclosed in material having the ignitability characteristic '13' as specified in BS 476-12; and - shall be adequately protected from any mechanical damage reasonably foreseeable during installation and use.	WR-554.4.1

A heating cable intended for laying directly in soil, concrete, cement screed or other material used for road and building construction shall be:

- capable of withstanding mechanical damage under the conditions that can reasonably be expected to prevail during its installation; and
- constructed of material that will be resistant to damage from dampness and/or corrosion under normal conditions of service.

WR-554.4.2

A heating cable laid directly in soil, a road or the structure of a building, shall be installed so that it:

- is completely embedded in the substance it is intended to heat; and
- does not suffer damage in the event of movement normally to be expected in it or the substance in which it is embedded; and
- complies in all respects with the manufacturer's instructions and recommendations.

WR-554.4.3

The load of every floor-warming cable (under operation) shall be limited to a value such that the manufacturer's stated conductor temperature is not exceeded.

WR-554.4.4

6.1.23 Identification of conductors

Cores of cables shall be identified at their terminations (and preferably throughout the cable length) by:

- colour; (see BS 7671:2008 Regulation 514.4); and/or
- lettering and/or numbering; (see BS 7671:2008 Regulation 514.5).

WR-514.3.1

WR-514.3.2

6.1.24 Impact

A cable installed under a floor or above a ceiling shall be run in such a position that it is not liable to be damaged by contact with the floor or the ceiling or their fixings.

WR-522.6.5

A cable passing through a joist within a floor or ceiling   WR-522.6.5
construction or through a ceiling support (e.g. under
floorboards), shall:

- be at least 50 mm measured vertically from the
  top, or bottom as appropriate, of the joist or
  batten; or
- incorporate an earthed metallic covering; or
- be enclosed in earthed conduit; or
- be enclosed in earthed trunking or ducting; or
- be mechanically protected against damage
  sufficient to prevent penetration of the cable by
  nails, screws etc.

A cable concealed in a wall or partition at a depth of   WR-522.6.6
less than 50 mm from a surface of the wall or partition
shall:

- incorporate an earthed metallic covering; or
- be enclosed in earthed conduit; or
- be enclosed in earthed trunking or ducting
  complying; or
- be mechanically protected against damage
  sufficient to prevent penetration of the cable by
  nails, screws etc.; or
- be installed in a zone within 150 mm from the top
  of the wall or partition or within 150 mm of an
  angle formed by two adjoining walls or partitions.

The cables of an installation not intended to be under   WR-522.6.8
the supervision of a skilled or instructed person and
which are concealed in a wall or partition, the internal
construction of which includes metallic parts, other
than metallic fixings such as nails, screws and the like,
shall:

- incorporate an earthed metallic covering; or
- be enclosed in earthed conduit; or
- be enclosed in earthed trunking or ducting; or
- be mechanically protected sufficiently to avoid
  damage to the cable during construction of the
  wall or partition and during installation of the
  cable; or
- be provided with additional protection by means
  of an RCD.

Where the installation is not intended to be under the supervision of a skilled or instructed person, consideration shall be given to providing additional protection by means of an RCD.

6.1.25 Inspection

 Note: Inspection shall always precede testing and shall normally be done with that part of the installation under inspection disconnected from the supply.

The inspection shall include at least the checking of the routing of cables in safe zones (or protection against mechanical damage), where relevant to the installation and, where necessary, during erection.	WR-611.3

6.1.26 Lifts and/or the proximity to non-electrical services

No cable shall be run in a lift or hoist shaft unless it forms part of the lift installation.	WR-528.3.5

6.1.27 Locations with risks of fire due to the nature of processed or stored materials

Where BE2 conditions exist and where there is a risk of fire due to the manufacture, processing or storage of flammable materials such as:

- barns (due to the accumulation of dust and fibres)
- woodworking facilities
- paper mills and textile factories (due to the storage and processing of combustible materials)

a fire risk will be present and in such locations, and in these circumstances:

• cables shall, as a minimum, satisfy the test under fire conditions specified in BS EN 60332-1-2 • cables not completely embedded in non-combustible material such as plaster or concrete or otherwise protected from fire shall meet the flame propagation characteristics as specified in BS EN 60332-1-2	WR-422.3.4

- a cable trunking system or cable ducting system shall satisfy the test under fire conditions specified in BS EN 50085
- precautions shall be taken such that a cable or wiring system cannot propagate flame
- where the risk of flame propagation is high the cable shall meet the flame propagation characteristics specified in the appropriate part of the BS EN 50266 series.

Except for mineral insulated cables, a wiring system shall be protected against insulation faults: WR-422.3.9

- in a TN or TT system, by an RCD having a rated residual operating current (I_a) not exceeding 300 mA
- in an IT system, by an insulation monitoring device with audible and visual signals.

Flexible cables and flexible cords shall be either: WR-422.3.14

- a heavy-duty type (with a voltage rating of not less than 450/750 V); or
- suitably protected against mechanical damage.

6.1.28 Luminaires

Any cable or cord between the fixing means and the luminaire shall be installed so that any expected stresses in the conductors, terminals and terminations will not impair the safety of the installation. WR-559.6.1.5

6.1.29 Mechanical stresses

A wiring system shall be selected and erected to avoid during installation, use, or maintenance, damage to the sheath or insulation of cables and their terminations. WR-522.8.1

A conduit system or cable ducting system (other than a pre-wired conduit assembly that as been specifically designed for the installation) that is going to be buried in the structure, shall be completely erected between access points before any cable is drawn in. WR-522.8.2

The radius of every bend in a wiring system shall be such that the cables do not suffer damage and terminals are not stressed.

WR-522.8.3

Where cables are not supported continuously they shall be supported by suitable means at appropriate intervals in such a manner that the conductors or cables do not suffer damage by their own weight.

WR-522.8.4

Every cable shall be supported in such a way that it is not exposed to undue mechanical strain and so that there is no appreciable mechanical strain on the terminations of the conductors.

WR-522.8.5

A wiring system intended for the drawing in or out of conductors or cables shall have adequate means of access to allow this operation.

WR-522.8.6

A cable buried in the ground (that is not installed in a conduit or duct) shall incorporate an earthed armour or metal sheath or both, suitable for use as a protective conductor.

WR-522.8.10

The location of buried cables shall be marked by cable covers or a suitable marking tape.

Buried cables shall be at a sufficient depth to avoid being damaged by any reasonably foreseeable disturbance of the ground.

 Note: See IEC 61386-24 for further details concerning underground conduits.

Cable supports and enclosures shall not have sharp edges liable to damage the wiring system.

WR-522.8.11

A cable shall not be damaged by the means of fixing.

WR-522.8.12

Cables which pass across expansion joints shall be so selected and/or erected that anticipated movement does not cause damage to the electrical equipment.

WR-522.8.13

6.1.30 Omission of identification by colour or marking

Identification by colour or marking is not required for:

WR-514.6.1

- concentric conductors of cables
- the metal sheath or armour of cables when used as a protective conductor.

6.1.31 Precautions within a fire-segregated compartment

The risk of spread of fire shall be minimised by the selection of appropriate materials and erection.

 Note: In installations where particular risk is identified, cables shall meet the flame propagation requirements given in the relevant part of the BS EN 50266 series.

Cables complying with the requirements of BS EN 60332-1-2 may be installed without special precautions.	WR-527.1.3
Cables **not** complying with the flame propagation requirements of BS EN 60332-1-2 shall be limited to short lengths for connection of appliances to the permanent wiring system and shall not pass from one fire-segregated compartment to another.	WR-527.1.4

6.1.32 Protective conductor

Single-core cables that are coloured green-and-yellow throughout their length shall **only** be used as a protective conductor and shall **not** be over-marked at their terminations.	WR-514.4.2

6.1.33 Reduced low-voltage system

Every cable coupler of a reduced low-voltage system: WR-411.8.5

- shall have a protective conductor contact; and
- shall **not** be dimensionally compatible with those used for any other system in use in the same premises.

6.1.34 Requirements for SELV and PELV circuits

Protective separation of wiring systems of SELV or PELV circuits from the live parts of other circuits (which have at least basic insulation) shall be achieved by one of the following arrangements: WR-414.4.2

- SELV and PELV circuit conductors

- circuit conductors contained in a multi-conductor
  cable or other grouping of conductor
- the rated voltage of the cable(s) not being less than
  the nominal voltage of the system, mechanical
  protection, and basic insulation etc.

6.1.35 Rotating machines

All equipment, including cable, of every circuit   WR-552.1.1
carrying the starting, accelerating and load currents of
a motor shall be suitable for a current at least equal to
the full-load current rating of the motor when rated in
accordance with the appropriate British Standard.

6.1.36 Telecommunication cables

Protective equipotential bonding shall be applied to   WR-411.3.1.2
any metallic sheath of a telecommunication cable.

6.1.37 Temperature

Cables and wiring accessories shall only be installed   WR-522.1.2
or handled at temperatures within the limits stated in
the relevant product specification or as given by the
manufacturer.

Parts of a cable or flexible cord within an accessory,   WR-522.2.2
appliance or luminaire shall be suitable for the
temperatures likely to be encountered, or shall be
provided with additional insulation suitable for those
temperatures.

6.1.38 Types of wiring system

The installation method of a wiring system in relation   WR-521.1
to the type of conductor or cable used shall be in
accordance with Table 4A1 of Appendix 4 of BS
7671:2008.

6.1.39 Underground cables

In the event of crossing (or being in the proximity of) underground telecommunication cables and underground power cables, a minimum clearance of 100 mm shall be maintained.	WR-528.2

- A fire-retardant partition shall be provided between the cables.
- For crossings, mechanical protection between the cables shall be provided.

6.1.40 Vibration

The cables and cable connections of a wiring system supported by (or fixed to) a structure or equipment that is subject to vibration of medium severity (AII2) or high severity (AH3) shall be suitable for such conditions.	WR-522.7.1
The connection of suspended current-using equipment (such as a luminaire) shall be made by cable with flexible cores.	WR-522.7.2

 Where no vibration or movement can be expected, cable with non-flexible cores may be used.

6.1.41 Wiring systems

The installation of wiring systems will meet the requirements if:	WR-412.2.4.1

- the rated voltage of the cable(s) is not less than the nominal voltage of the system and at least 300/500 V; and
- adequate mechanical protection of the basic insulation is provided by one or more of the following:
  - the non-metallic sheath of the cable
  - non-metallic trunking or ducting (complying with the BS EN 50085)

A voltage Band I circuit shall not be contained in the same wiring system as a Band II circuit unless:

WR-528.1

- every cable is insulated for the highest voltage present
- each conductor of a multicore cable is insulated for the highest voltage present in the cable
- the cables are insulated for their system voltage and installed in a separate compartment of a cable ducting or cable trunking system
- the cables are installed on a cable tray system where physical separation is provided by a partition
- for a multicore cable or cord, the cores of the Band I circuit are separated from the cores of the Band II circuit by an earthed metal screen of equivalent current-carrying capacity to that of the largest core of a Band II circuit.

The minimum cross-sectional area of the extra-low-voltage conductors shall normally be $1.5\,mm^2$ copper, but:

WR-559.11.5.2

- for flexible cables with a maximum length of $3\,m$, a cross-sectional area of $1\,mm^2$ copper may be used
- for suspended flexible cables, for mechanical reasons, $4\,mm^2$ copper should be used
- for composite cables consisting of braided tinned copper outer sheath, having a material of high tensile strength inner core, $4\,mm^2$ copper should be used.

A cable for through wiring:

shall be selected in accordance with the temperature information on the luminaire or on the manufacturer's instruction sheet.

WR-559.6.2.2

6.1.42 Special installations and locations

6.1.42.1 Agricultural and horticultural premises

The particular requirements of this section apply to fixed electrical installations indoors and outdoors in agricultural and horticultural premises. Some of the requirements are also applicable to other locations that are in common buildings belonging to the agricultural and horticultural premises.

 Note: Section 705 does not cover electric fence installations. Refer to BS EN 60335-2-76 and BS EN 6100-1.

> Where vehicles and mobile agricultural machines are   WR-705.522
> operated, the following methods of installation shall
> be applied:
>
> • Cables shall be buried in the ground at a depth of
>   at least 0.6 m with added mechanical protection.
> • Cables in arable or cultivated ground shall be
>   buried at a depth of at least 1 m.
> • Self-supporting suspension cables shall be
>   installed at a height of at least 6 m.

 Special attention shall be given to the presence of different kinds of fauna (e.g. rodents).

> The following documentation shall be provided to   WR-705.514.9.3
> the user of the installation:
>
> • the routing of all concealed cables
> • a single-line distribution diagram.

6.1.42.2 Caravan and camping parks

> Underground cables shall be buried at a depth of at   WR-708.521.1.1
> least 0.6 m and, unless having additional mechanical
> protection, be placed outside any caravan pitch or
> away from any surface where tent pegs or ground
> anchors are expected to be present.

 Not more than four socket outlets should be grouped in one location, in order to avoid the supply cable crossing a pitch other than the one intended to be supplied.

6.1.42.3 Caravans and motor caravans

 Note: In order not to mix regulations on different subjects, such as those for electrical installation of caravan parks with those for electrical installation inside caravans, two sections have been created:

- Section 708, which concerns electrical installations in caravan parks, camping parks and similar locations; and
- Section 721, which concerns electrical installations in caravans and motor caravans.

Types of wiring system

The wiring systems shall be installed using one or more of the following:	WR-721.521.2
• insulated single-core cables, with flexible Class 5 conductors, in non-metallic conduit • insulated single-core cables, with stranded Class 2 conductors (minimum of 7 strands), in non-metallic conduit • sheathed flexible cables.	
All cables shall, as a minimum, meet the requirements of BS EN 60332-1-2.	WR-721.521.2
Cable management systems shall comply with BS EN 61386.	WR-721.521.2

All cables, unless enclosed in rigid conduit and all flexible conduit shall be supported at intervals not exceeding 0.4 m for vertical runs and 0.25 m for horizontal runs.	WR-721.522.8.1.3
Cables of low-voltage systems shall be run separately from the cables of extra-low-voltage systems, so that there is no risk of physical contact between the two wiring systems.	WR-721.528.1
Where cables have to run through such a compartment, they shall be run at a height of less than 500 mm above the base of the cylinder(s), and such cables shall be protected against mechanical damage by installation within a continuous gas-tight conduit or duct passing through the compartment.	WR-721.528.3.5

All cables shall meet the requirements of BS EN 60332-1-2.	WR-740.521.1
Cables shall have a minimum rated voltage of 450/750 V.	WR-740.521.1
The routes of cables buried in the ground shall be marked at suitable intervals.	WR-740.521.1
Buried cables shall be protected against mechanical damage.	WR-740.521.1
Armoured cables or cables protected against mechanical damage shall be used wherever there is a risk of mechanical damage due to external influence, e.g. >AG2.	WR-740.521.1
Joints shall not be made in cables except where necessary as a connection into a circuit. Where joints are made, these shall either use connectors in accordance with the relevant British Standard or the connection shall be made in an enclosure with a degree of protection of at least IP4X or IPXXD.	WR-740.526
Where strain can be transmitted to terminals the connection shall incorporate cable anchorage(s).	WR-740.526
Insulation-piercing lampholders shall not be used unless the cables and lampholders are compatible and the lampholders are non-removable once fitted to the cable.	WR-740.55.1.2
Supply cables shall be flexible and have adequate protection against mechanical damage.	WR-740.55.1.4

 Luminaires and floodlights shall be so fixed and protected that a focusing or concentration of heat is not likely to cause ignition of any material.

6.1.42.4 Construction and demolition site installations

Cables shall **not** be installed across a site road or a walkway unless adequate protection of the cable against mechanical damage is provided.	WR-704.522.8.10
For reduced low-voltage systems, low temperature 300/500 V thermoplastic (BS 7919) or equivalent flexible cables shall be used.	WR-704.522.8.11

For applications exceeding reduced low voltage, flexible cable shall be HO7RN-F (BS 7919) type or equivalent having 450/750 V rating and resistant to abrasion and water.	WR-704.522.8.11

6.1.42.5 Exhibitions, shows and stands

The particular requirements of this section apply to the temporary electrical installations in exhibitions, shows and stands (including mobile and portable displays and equipment) to protect users.

A cable intended to supply temporary structures shall be protected at its origin by an RCD whose rated residual operating current does not exceed 300 mA.	WR-711.410.3.4
Armoured cables or cables protected against mechanical damage shall be used wherever there is a risk of mechanical damage.	WR-711.52
Wiring cables shall be copper, have a minimum cross-sectional area of 1.5 mm², and shall comply with an appropriate British Standard for either thermoplastic or thermosetting insulated electric cables.	WR-711.52
Flexible cords shall **not** be laid in areas accessible to the public unless they are protected against mechanical damage.	WR-711.52
Where no fire alarm system is installed in a building used for exhibitions etc., cable systems shall be either: • flame retardant to BS EN 60332-1-2 and low smoke to BS EN 61034-2; or • single-core or multicore unarmoured cables enclosed in metallic or non-metallic conduit or trunking, providing a degree of fire protection of at least IP4X.	WR-711.521
Joints shall not be made in cables except where necessary as a connection into a circuit. Where joints are made, these shall either use connectors in accordance with relevant standards or be in enclosures with a degree of protection of at least IP4X or IPXXD.	WR-711.526.1

Where strain can be transmitted to terminals the connection shall incorporate suitable cable anchorage(s).	WR-711.526.1
Insulation piercing lampholders shall **not** be used unless the cables and lampholders are compatible, and provided the lampholders are non-removable once fitted to the cable.	WR-711.559.4.3
On the a.c. side, a photovoltaic power supply system (PV) supply cable shall be connected to the supply side of the protective device for automatic disconnection of circuits supplying current-using equipment.	WR-2.411.3.2.1.1
Overload protection may be omitted to PV string and PV array cables when the continuous current-carrying capacity of the cable is equal to or greater than 1.25 times Isc STC at any location.	WR-712.433.1
Overload protection may be omitted to the PV main cable if the continuous current-carrying capacity is equal to or greater than 1.25 times Isc STC of the PV generator.	WR-712.433.2
The PV supply cable on the a.c. side shall be protected against fault current by an overcurrent protective device installed at the connection to the a.c. mains.	WR-712.434.1
PV string cables, PV array cables and PV d.c. main cables shall be selected and erected so as to minimise the risk of earth faults and short-circuits.	WR-712.522.8.1
Wiring systems shall withstand the expected external influences such as wind, ice formation, temperature and solar radiation.	WR-712.522.8.3
Where protective bonding conductors are installed, they shall be parallel to and in as close contact as possible with d.c. cables and a.c. cables and accessories.	WR-712.54

6.1.42.6 Marinas and similar locations

The particular requirements of this section are applicable only to circuits intended to supply pleasure craft or houseboats in marinas and similar locations. They do not apply to the supply to houseboats that are directly supplied from the public network or to the internal electrical installations of pleasure craft or houseboats.

The following wiring systems are suitable for distribution circuits of marinas:

WR-709.521.1.4

- underground cables
- overhead cables or overhead insulated conductors
- cables with copper conductors and thermoplastic or elastomeric insulation and sheath installed within an appropriate cable management system taking into account external influences such as movement, impact, corrosion and ambient temperature
- mineral-insulated cables with a PVC protective covering
- cables with armouring and serving of thermoplastic or elastomeric material
- other cables and materials that are no less suitable than those listed above.

The following wiring systems shall **not** be used on or above a jetty, wharf, pier or pontoon:

WR-709.521.1.5

- cables in free air suspended from or incorporating a support wire
- non-sheathed cables in conduit, trunking etc.
- cables with aluminium conductors
- mineral insulated cables.

Underground distribution cables shall, unless provided with additional mechanical protection, be buried at a sufficient depth to avoid being damaged, e.g. by heavy vehicle movement.

WR-709.521.1.7

 Note: A depth of 0.5 m is generally considered as a minimum depth to fulfil this requirement.

Cables shall be selected and installed so that mechanical damage due to tidal and other movement of floating structures is prevented.

WR-709.521.1.6

Cable management systems shall be installed to allow the drainage of water by drainage holes and/or installation of the equipment on an incline.

WR-709.521.1.6

6.1.42.7 Mobile or transportable units

For the purposes of this section, the term 'unit' is intended to mean a vehicle and/or mobile (self-propelled or towed) or transportable structure (such as a container or cabin) in which all or part of an electrical installation is contained and which is provided with a temporary supply by means of, for example, a plug and socket outlet.

Flexible cables (for connecting the unit to the supply), or cables of equivalent design, having a minimum cross-sectional area of 2.5 mm^2 copper shall be used.	WR-717.52.1
The flexible cable shall enter the unit by an insulating inlet in such a way as to minimise the possibility of any insulation damage or fault which might energise the exposed conductive parts of the unit.	WR-717.52.1
The following or other equivalent cable types are permitted for the internal wiring of the unit: <ul><li>thermoplastic- or thermosetting-insulated-only cable (BS 6004, BS 7211, BS 7919) installed in conduits in accordance with BS EN 61386-1</li><li>thermoplastic or thermosetting insulated and sheathed cable (BS 6004, BS 7211, BS 7919), if precautionary measures are taken to prevent mechanical damage due to any sharp-edged parts or abrasion.</li></ul>	WR-717.52.2
Where cables have to run through such a compartment, they shall be run at a height of less than 500 mm above the base of the cylinder(s), and such cables shall be protected against mechanical damage by installation within a continuous gas-tight conduit or duct passing through the compartment.	WR-717.528.3.5
Where installed, this conduit or duct shall be able to withstand an impact equivalent to AG3 without visible physical damage.	WR-717.528.3.5

6.1.42.8 Swimming pools and other basins

In zones 0, 1 and 2, any metallic sheath or metallic covering of a wiring system shall be connected to the supplementary equipotential bonding.	WR-702.522.21

Note: Cables should preferably be installed in conduits made of insulating material.

Additional requirements for the wiring of fountains

For a fountain, the following additional requirements shall be met: • A cable for electrical equipment in zone 0 shall be installed as far outside the basin rim as is reasonably practicable and run to the electrical equipment inside zone 0 by the shortest practicable route. • In zone 1, a cable shall be selected, installed and provided with mechanical protection to medium severity (AG2) and the relevant submersion in water depth (AD8).	WR-702.522.23

6.1.42.9 Rooms and cabins containing sauna heaters

The particular requirements of this section apply to:

• sauna cabins erected on site (e.g. in a location or in a room);
• the room where the sauna heater is, or where the sauna heating appliances are installed.

Zone 3

In zone 3: the insulation and sheaths of cables shall withstand a minimum temperature of 170°C.

6.2 Conductors

 Conductors intended to operate at temperatures above 70°C shall **not** be connected to switchgear, protective devices, accessories or other types of equipment.

Conductors shall **not** be subjected to excessive mechanical stress.

Conductors **shall** be capable of withstanding all foreseen electromechanical forces (including fault current) during service.

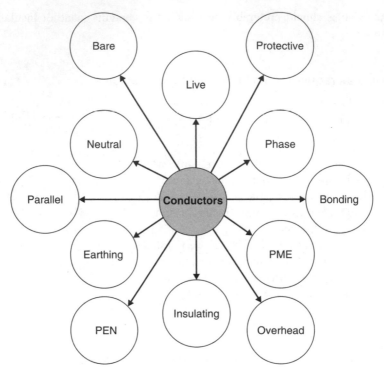

Figure 6.2 Conductors

6.2.1 General

The number of conductors to be considered in a circuit are those carrying load current less those conductors which only serve the purpose of protective conductors.	WR-523.6.1 WR-523.6.4
PEN conductors need to be taken into consideration in the same way as neutral conductors.	WR-523.6.4

6.2.2 Types of protective conductor

A protective conductor may consist of one or more of the following:

- a single-core cable;
- a conductor in a cable;
- an insulated or bare conductor in a common enclosure with insulated live conductors;
- a fixed bare or insulated conductor;

- a metal covering (for example, the sheath, screen or armouring of a cable);
- a metal conduit or other enclosure or electrically continuous support system for conductors.

Note: If a protective conductor is formed by conduit, trunking, ducting or the metal sheath and/or armour of a cable, the earthing terminal of each accessory shall be connected by a separate protective conductor to an earthing terminal incorporated in the associated box or enclosure.

6.2.3 Automatic disconnection in the case of a fault

A protective device shall automatically interrupt the supply to the line conductor of a circuit or equipment in the event of a fault of negligible impedance between the line conductor and an exposed conductive part or a protective conductor in the circuit or equipment within the disconnection time required in Table 41.1 on p. 46 of BS 7671:2008.	WR-411.3.2.1

Disconnection is not required for protection against electric shock but may be required for other reasons, such as protection against thermal effects.

In a TN system, a disconnection time not exceeding 5 s is permitted for a distribution circuit and for a circuit not covered by Table 41.1 on p. 46 of BS 7671:2008.	WR-411.3.2.3
In a TT system, a disconnection time not exceeding 1 s is permitted for a distribution circuit and for a circuit not covered by Table 41.1 on p. 46 of BS 7671:2008.	WR-411.3.2.4

Where automatic disconnection cannot be achieved in the time required, supplementary equipotential bonding shall be provided.

6.2.4 Bare conductors

The minimum cross-sectional area of line conductors in a.c. circuits and of live conductors in d.c. circuits shall be as shown in Table 6.3.

Table 6.3 Minimum cross-sectional area of bare conductors

Type of wiring system	Use of circuit	Conductor	
		Material	Minimum cross-sectional area (mm)
Bare conductors	Power circuits	Copper	10
		Aluminium	16
	Signalling and control circuits	Copper	4

Bare conductors shall be painted or identified by a coloured tape, sleeve or disc as per Table 6.4.

Table 6.4 Identification of conductors

Function	Alphanumeric	Colour
Protective conductors		Green-and-yellow
Functional earthing conductor		Cream
a.c. power circuit (Note 1)		
Phase of single-line circuit	L	Brown
Neutral of single- or three-line circuit	N	Blue
Phase 1 of three-line a.c. circuit	L1	Brown
Phase 2 of three-line a.c. circuit	L2	Black
Phase 3 of three-line a.c. circuit	L3	Grey
Two-wire unearthed d.c. power circuit		
Positive of two-wire circuit	L+	Brown
Negative of two-wire circuit	L−	Grey
Two-wire earthed d.c. power circuit		
Positive (of negative earthed) circuit	L+	Brown
Negative (of negative earthed) circuit (Note 2)	M	Blue
Positive (of positive earthed) circuit (Note 2)	M	Blue
Negative (of positive earthed) circuit	L−	Grey
Three-wire d.c. power circuit		
Outer positive of two-wire circuit derived from three-wire system	L+	Brown
Outer negative of two-wire circuit derived from three-wire system	L−	Grey
Positive of three-wire circuit	L+	Brown
Mid-wire of three-wire circuit (Notes 2 & 3)	M	Blue
Negative of three-wire circuit	L−	Grey
Control circuits, ELV and other applications		
Phase conductor	L	Brown, Black, Red, Orange, Yellow, Violet, Grey, White, Pink or Turquoise
Neutral or mid-wire (Note 4)	N or M	Blue

Note:
(1)   Power circuits include lighting circuits.
(2)   M identifies either the mid-wire of a three-wire d.c. circuit, or the earthed conductor of a two-wire earthed d.c. circuit.
(3)   Only the middle wire of three-wire circuits may be earthed.
(4)   An earthed PELV conductor is blue.

Colour or marking is not required for bare conductors (where permanent identification is not practicable).

If the nominal voltage does not exceed 25 V a.c. or 60 V d.c., bare conductors may be used for extra-low-voltage lighting installations, provided that:	WR-559.11.5.3

- the lighting installation has been designed, installed or enclosed in such a way that the risk of a short-circuit is reduced to a minimum
- the conductors used have a cross-sectional area of at least 4 mm^2
- the conductors are not placed directly on combustible material.

For suspended bare conductors, at least one conductor and its terminals shall be insulated for that part of the circuit between the transformer and the short-circuit protective device to prevent a short-circuit.	WR-559.11.5.3

A permanent label/warning notice – with the words shown in Figure 6.3 – shall be permanently fixed at or near the bonding conductor's connection point to an extraneous part.

Safety Electrical Connection – Do Not Remove

Figure 6.3 Earthing and bonding notice

In agricultural and horticultural premises, protective bonding conductors shall be protected against mechanical damage and corrosion, and shall be selected to avoid electrolytic effects.	WR-705.544.2

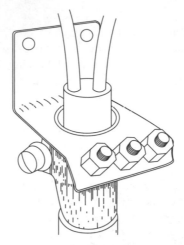

Figure 6.4 Typical earthing conductor (courtesy Stingray)

In solar photovoltaic power supply systems, where protective bonding conductors are installed, the conductors shall be parallel to and in as close contact as possible with d.c. cables and a.c. cables and accessories.	WR-712.54

6.2.5 Equipotential bonding conductor

• all socket outlets shall be provided with a protective conductor contact (that is connected to the equipotential bonding conductor); • all flexible equipment cables (other than Class II equipment) shall have a protective conductor for use as an equipotential bonding conductor.	413-06-05

6.2.6 Non-earthed equipotential bonding conductor

Source supplies may supply more than one item of equipment provided that: • all exposed conductive parts of the separated circuit are connected together by an insulated and non-earthed equipotential bonding conductor	413-06-05

- the non-earthed equipotential bonding conductor
  is not connected to a protective conductor (or to an
  exposed conductive part of any other circuit or to any
  extraneous conductive part).

6.2.7 Main protective bonding conductors

Except where protective multiple earthing (PME)
conditions apply, a main protective bonding conductor
shall have a cross-sectional area not less than half the
cross-sectional area required for the earthing conductor
of the installation and not less than $6\,mm^2$.   WR-544.1.1

The cross-sectional area need not exceed $25\,mm^2$ WR-544.1.1
if the bonding conductor is of copper or a cross-
sectional area affording equivalent conductance in
other metals.

Except for highway power supplies and street   WR-544.1.1
furniture, where PME conditions apply, the main
protective bonding conductor shall be selected in
accordance with the neutral conductor of the supply
and Table 54.8 on p. 134 of BS 7671:200.

Where an installation has more than one source   WR-544.1.1
of supply to which PME conditions apply, a main
protective bonding conductor shall be selected
according to the largest neutral conductor of
the supply.

6.2.8 Supplementary equipotential bonding

The use of supplementary bonding does not exclude the need to disconnect the
supply for other reasons, for example protection against fire, thermal stresses
in equipment.

Supplementary equipotential bonding system shall   WR-415.2.1
be connected to the protective conductors of all
equipment including those of socket outlets.

In locations containing a bath and/or a shower,   WR-701.415.2
local supplementary equipotential bonding shall be
established connecting together the terminals of the
protective conductor of each circuit supplying Class
I and Class II equipment to the accessible extraneous
conductive parts, within a room containing a bath or
shower, including the following:

- metallic pipes supplying services and metallic
  waste pipes (e.g. water, gas)
- metallic central heating pipes and air conditioning
  systems
- accessible metallic structural parts of the building
  (metallic door architraves, window frames and
  similar parts are not considered to be extraneous
  conductive parts unless they are connected to
  metallic structural parts of the building).

Supplementary equipotential bonding may be installed outside or inside rooms
containing a bath or shower, preferably close to the point of entry of extraneous conductive parts into such rooms.

All extraneous conductive parts in zones 0, 1 and   WR-702.411.3.3
2 of a swimming pool (and/or other basins) shall
be connected by supplementary protective bonding
conductors to the protective conductors of exposed
conductive parts of equipment situated in these zones.

A supplementary bonding conductor connecting two   WR-544.2.1
exposed conductive parts shall have a conductance
(if sheathed or otherwise provided with mechanical
protection) not less than that of the smaller protective
conductor connected to the exposed conductive parts.

Note: If mechanical protection is not provided, its cross-sectional area shall
be not less than 4 mm².

A supplementary bonding conductor connecting an   WR-544.2.2
exposed conductive part to an extraneous conductive
part shall have a conductance (if sheathed or
otherwise provided with mechanical protection)
not less than half that of the protective conductor
connected to the exposed conductive part.

A supplementary bonding conductor connecting two   WR-544.2.3
extraneous conductive parts shall have a cross-sectional
area not less than 2.5 mm^2 if sheathed or otherwise
provided with mechanical protection or 4 mm^2 if
mechanical protection is not provided.

Supplementary bonding shall be provided by a   WR-544.2.4
supplementary conductor, a conductive part of a
permanent and reliable nature, or by a combination
of these.

Where supplementary bonding is to be applied to a fixed   WR-544.2.5
appliance (which is supplied via a short length of flexible
cord from an adjacent connection unit or other accessory,
incorporating a flex outlet) the circuit protective
conductor within the flexible cord shall be deemed to
provide the supplementary bonding connection to the
exposed conductive parts of the appliance.

6.2.9 Building design

Where risks due to structural movement exist (CB3),   WR-522.15.1
the cable support and protection system employed
shall be capable of permitting relative movement
so that conductors are not subjected to excessive
mechanical stress.

6.2.9.1 Suspended conductors

Suspension devices for extra-low-voltage luminaires,   WR-559.11.6
including supporting conductors, shall be capable
of carrying five times the mass of the luminaires
(including their lamps) intended to be supported, but
not less than 5 kg.

Terminations and connections of conductors shall   WR-559.11.6
be made by screw terminals or screwless clamping
devices complying with BS EN 60998-2-1 or BS EN
60998-2-2.

Insulation piercing connectors and termination wires   WR-559.11.6
which rely on counterweights hung over suspended
conductors to maintain the electrical connection shall
not be used.

The suspended system shall be fixed to walls or
ceilings by insulated distance cleats and shall be
continuously accessible throughout the route.

WR-559.11.6

6.2.10 Cable couplers

A cable coupler shall be arranged so that the connector of the coupler is fitted
at the end of the cable remote from the supply.

Except for a SELV or a Class II circuit, a cable coupler
shall be non-reversible and shall be capable of being
connected to a protective conductor.

WR-553.2.1

Every cable coupler in a SELV system:

WR-411.7.5

* shall have a protective conductor contact; and
* shall **not** be dimensionally compatible with those
  used for any other system in use in the same
  premises.

6.2.11 Cross-sectional area of conductors

The cross-sectional area of conductors shall be
determined for both normal operating conditions
and, where appropriate, for fault conditions
according to:

WR-132.6

* the admissible maximum temperature
* the voltage drop limit
* the electromechanical stresses likely to occur due
  to short-circuit and earth fault currents
* other mechanical stresses to which the conductors
  are likely to be exposed
* the maximum impedance for operation of short-
  circuit and earth fault protection
* the method of installation
* harmonics
* thermal insulation.

The cross-sectional area of a line conductor in an a.c.
circuit or of a live conductor in a d.c. circuit shall be
as shown in Table 6.5.

WR-524.1

Table 6.5 Minimum nominal cross-sectional area of conductor

Type of wiring system	Use of circuit	Conductor	
		Material	Minimum cross-sectional area (mm)
Cables and insulated conductors	Power and lighting circuits	Copper Aluminium	1.0 16.0
	Signalling and control circuits	Copper	0.5
Bare conductors	Power circuits	Copper Aluminium	10 16
	Signalling and control circuits	Copper	4
Flexible connections with insulated conductors and cables	For a specific appliance	Copper	See relevant British Standard
	For any other application	Copper	0.5
	Extra-low-voltage circuits for special applications	Copper	0.5

The neutral conductor (if any) shall have a cross-sectional area not less than that of the line conductor. **WR-524.2**

For a polyphase circuit where each line conductor has a cross-sectional area greater than $16\,mm^2$ for copper or $25\,mm^2$ for aluminium, the neutral conductor is permitted to have a smaller cross-sectional area than that of the line conductors provided that: **WR-524.3**

- the expected maximum current including harmonics (if any) in the neutral conductor during normal service is not greater than the current-carrying capacity of the reduced cross-sectional area of the neutral conductor; and
- the neutral conductor is protected against overcurrents; and
- the size of the neutral conductor is at least equal to $16\,mm^2$ for copper or $25\,mm^2$ for aluminium.

In caravans and motor caravans, the cross-sectional area of every conductor shall be not less than $1.5\,mm^2$. **WR-721.524.1**

6.2.12 Conductors in parallel

Where two or more live conductors or PEN conductors are connected in parallel in a system: **WR-523.8**

- measures shall be taken to achieve equal load current sharing between them and either:

- the conductors in parallel are multicore cables or twisted single-core cables or non-sheathed cables; or
- the conductors in parallel are non-twisted single-core cables or non-sheathed cables in trefoil or flat formation and where the cross-sectional area is greater than 50 mm^2 in copper or 70 mm^2 in aluminium.

Note: This regulation does not preclude the use of ring final circuits with or without spur connections. Where adequate current sharing is not possible (or where four or more conductors have to be connected in parallel) consideration shall be given to the use of busbar trunking.

6.2.13 Current-carrying capacities of conductors

The current (including any harmonic current) to be carried   WR-523.1
by any conductor for sustained periods during normal
operation shall be such that the appropriate temperature
limit specified in Table 6.6 is not exceeded.

Table 6.6 Maximum operating temperatures for types of cable insulation

Type of insulation	Temperature limit
Thermoplastic	70°C at the conductor
Thermosetting	90°C at the conductor
Mineral (thermoplastic covered or bare exposed to touch)	70°C at the sheath
Mineral (bare not exposed to touch and not in contact with combustible material)	105°C at the sheath

6.2.14 Earthing conductors

 Earthing conductors **shall** be capable of being disconnected to enable the resistance of the earthing arrangements to be measured.

In every installation a main earthing terminal shall   WR-542.4.1
be provided to connect the following to the earthing
conductor:

- the circuit protective conductors
- the protective bonding conductors

- functional earthing conductors (if required)
- lightning protection system bonding conductor, if any.

Buried earthing conductors shall have a cross-sectional area not less than that stated in Table 6.7.

Table 6.7 Minimum cross-sectional areas of a buried earthing conductor (reproduced with permission of IET)

	Protected against mechanical damage	Not protected against mechanical damage
Protected against corrosion by a sheath	See Note A below	16 mm² copper 16 mm² coated steel
Not protected against corrosion	25 mm² copper 50 mm² steel	25 mm² copper 50 mm² steel

Note A: The cross-sectional area of protective conductors shall not be less than:

$$S = \sqrt{I^2 t / k}$$

where:

S = the nominal cross-sectional area of the conductor in mm²

I = the value in amperes (rms for a.c.) of fault current for a fault of negligible impedance, which can flow through the associated protective device

t = the operating time of the disconnecting device in seconds

k = a factor taking account of the resistivity, temperature coefficient and heat capacity of the conductor material.

6.2.15 Earthing requirements

 If a protective conductor forms part of a cable, then this **shall only** be earthed in the installation containing the associated protective device.

Protective conductors buried in the ground shall have a cross-sectional area not less than that stated in Table 6.8.

Table 6.8 Minimum cross-sectional areas of a buried earthing conductor (reproduced with permission of IET)

	Protected against mechanical damage	Not protected against mechanical damage
Protected against corrosion by a sheath	See Note A below	16 mm² copper 16 mm² coated steel
Not protected against corrosion	25 mm² copper 50 mm² steel	25 mm² copper 50 mm² steel

Note A: The cross-sectional area of protective conductors shall not be less than:

$$S = \sqrt{I^2 t}/k$$

where:

S = the nominal cross-sectional area of the conductor in mm^2

I = the value in amperes (rms for a.c.) of fault current for a fault of negligible impedance, which can flow through the associated protective device

t = the operating time of the disconnecting device in seconds

k = is a factor taking account of the resistivity, temperature coefficient and heat capacity of the conductor material.

Where overcurrent protective devices are used for fault protection, the protective conductor shall be incorporated in the same wiring system as the live conductors or in their immediate proximity.	WR-543.6.1

6.2.16 Main earthing terminals or bars

To enable the resistance of the earthing arrangements to be measured, the earthing conductor needs to be capable of being easily disconnected.	WR-542.4.2
Any joint shall be capable of disconnection only by means of a tool.	WR-542.4.2

6.2.16.1 Earthing requirements for the installation of equipment having high protective conductor currents

Equipment having a protective conductor current exceeding 3.5 mA but not exceeding 10 mA, shall be either permanently connected to the fixed wiring of the installation without the use of a plug and socket outlet or connected by means of a plug and socket outlet complying with BS EN 60309-2.	WR-543.7.1.1
Equipment having a protective conductor current exceeding 10 mA shall be connected to the supply:	WR-543.7.1.2

- permanently via the wiring of the installation; or
- via a flexible cable with a plug and socket outlet; or
- via an protective conductor with an earth monitoring system.

The wiring of every final circuit and distribution circuit where the total protective conductor current is likely to exceed 10 mA, shall have a high integrity protective connection complying with one or more of the following:

WR-543.7.1.3

- a single protective conductor with a cross-sectional area greater than 10 mm^2
- a single copper protective conductor having a cross-sectional area of not less than 4 mm^2
- two individual protective conductors
- an earth monitoring system which, in the event of a continuity fault occurring in the protective conductor, automatically disconnects the supply to the equipment
- connection of the equipment to the supply by means of a double-wound transformer or equivalent unit, such as a motor–alternator set.

Where two protective conductors are used, the ends of the protective conductors shall be terminated independently of each other, at all connection points (such as at the distribution board, junction boxes and socket outlets) throughout the circuit.

WR-543.7.1.4

At the distribution board, information shall be provided indicating those circuits having a high protective conductor current.

WR-543.7.1.5

 Note: This information shall be positioned so as to be visible to a person who is modifying or extending the circuit.

6.2.1.6.2 Earthing arrangements and protective conductors

For a TN-S system, means shall be provided for the main earthing terminal of the installation to be connected to the earthed point of the source of energy. Part of the connection may be formed by the distributor's lines and equipment.

WR-542.1.2

For a TN-C-S system, where protective multiple earthing is provided, means shall be provided for the main earthing terminal of the installation to be connected by the distributor to the neutral of the source of energy.

WR-542.1.3

For a TT or IT system, the main earthing terminal shall be connected via an earthing conductor to an earth electrode.	WR-542.1.4

The earthing arrangements may be used jointly or separately for protective and functional purposes, according to the requirements of the installation.	WR-542.1.5

The earthing arrangements shall be such that:

- the value of impedance from the consumer's main earthing terminal to the earthed point of the supply for TN systems, or to earth for TT and IT systems, is in accordance with the protective and functional requirements of the installation, and considered to be continuously effective; and
- earth fault currents and protective conductor currents which may occur are carried without danger, particularly from thermal, thermomechanical and electromechanical stresses; and
- they are adequately robust or have additional mechanical protection appropriate to the assessed conditions of external influence.

Precautions shall be taken against the risk of damage to other metallic parts through electrolysis.	WR-542.1.7

Where a number of installations have separate earthing arrangements, any protective conductor common to any of these installations shall either:	WR-542.1.8

- be capable of carrying the maximum fault current likely to flow through them; or
- be earthed within one installation only and insulated from the earthing arrangements of any other installation.

6.2.16.3 Earthing conductors

Where buried in the ground, the earthing conductor shall have a cross-sectional area not less than that stated in Table 6.9	WR-542.3.1

Table 6.9 Minimum cross-sectional area of a buried earthing conductor

	Protected against mechanical damage	Not protected against mechanical damage
Protected against corrosion by a sheath	2.5 mm² copper 10 mm² steel	16 mm² copper
Not protected against corrosion	25 mm² copper 50 mm² steel	16 mm² coated steel

For a tape or strip conductor, the thickness shall be such as to withstand mechanical damage and corrosion.

WR-542.3.2

The connection of an earthing conductor to an earth electrode or other means of earthing shall be:

WR-542.3.2

- soundly made
- electrically and mechanically satisfactory
- labelled; and
- suitably protected against corrosion.

The earthing conductor of a street electrical fixture shall have a minimum copper equivalent cross-sectional area not less than that of the supply neutral conductor at that point or not less than 6 mm², whichever is the smaller.

WR-559.10.3.4

6.2.17 Electric floor heating systems

The load of every floor-warming cable under operation shall be limited to a value such that the manufacturer's stated conductor temperature is not exceeded.

WR-554.4.4

In locations containing a bath and/or shower:

WR-701.753

- only heating cables or thin sheet flexible heating elements shall be erected (**provided** that they have, a metal sheath or a metal enclosure or a fine mesh metallic grid)
- the fine mesh metallic grid, metal sheath or metal enclosure shall be connected to the protective conductor of the supply circuit.

Compliance with the latter requirement is not required if the protective measure SELV is provided for the floor heating system.

For electric floor heating systems the protective measure 'protection by electrical separation' is **not** permitted.	WR-701.753

6.2.18 Electrical connections

Connections between conductors or between a conductor and other equipment shall provide durable electrical continuity and adequate mechanical strength and protection.	WR-526.1
The type of connector chosen shall take account of:	WR-526.2

- the material of the conductor and its insulation
- the number and shape of the wires forming the conductor
- the cross-sectional area of the conductor
- the number of conductors to be connected together
- the temperature attained at the terminals in normal service
- the provision of adequate locking arrangements in situations subject to vibration or thermal cycling.

Where a soldered connection is used, the design shall take account of creep, mechanical stress and temperature rise under fault conditions.	WR-526.2
Every connection shall be accessible for inspection, testing and maintenance, except for the following:	WR-526.3

- a joint designed to be buried in the ground
- a compound filled or encapsulated joint
- a connection between a cold tail and the heating element as in ceiling heating, floor heating or a trace heating system
- a joint made by welding, soldering, brazing or appropriate compression tool
- a joint forming part of the equipment complying with the appropriate product standard.

Where necessary, precautions shall be taken so that the temperature attained by a connection in normal service shall not impair the effectiveness of the insulation of the conductors connected to it or any insulating material used to support the connection.   WR-526.4

Where a cable is to be connected to a bare conductor or busbar its type of insulation and/or sheath shall be suitable for the maximum operating temperature of the bare conductor or busbar.   WR-526.4

Every termination and joint in a live conductor or a PEN conductor shall be made within one of the following or a combination thereof: WR-526.5

- a suitable accessory
- an equipment enclosure
- an enclosure partially formed or completed with non-combustible building material.

There shall be no appreciable mechanical strain on the connections of conductors.   WR-526.6

Where a connection is made in an enclosure the enclosure shall provide adequate mechanical protection and protection against relevant external influences.   WR-526.7

6.2.19 Electrical installations

 The characteristics of the electrical equipment shall **not** be impaired by the process of erection.

Conductors and terminals shall be identified in accordance with Section 514.   WR-134.1.3

Every installation shall be divided into circuits, as necessary, to reduce the possibility of unwanted tripping of RCDs due to excessive protective conductor currents produced by equipment.   WR-314.1

The number of final circuits required, and the number of points supplied by any final circuit, shall be such as to enable compliance with the requirements for overcurrent protection, isolation and switching and with due regard to the current-carrying capacities of conductors.   WR-314.3

6.2.20 Electrode water heaters and boilers

If an electrode water heater or electrode boiler is connected to a three-line low-voltage supply, the shell of the electrode water heater or electrode boiler shall be connected to the neutral of the supply as well as to the earthing conductor.	WR-554.1.5
If the supply to an electrode water heater or electrode boiler is single-line and one electrode is connected to a neutral conductor earthed by the distributor, the shell of the electrode water heater or electrode boiler shall be connected to the neutral of the supply as well as to the earthing conductor.	WR-554.1.6
If the electrode water heater or electrode boiler is not piped to a water supply or in physical contact with any earthed metal (and where the electrodes and the water in contact with the electrodes are so shielded in insulating material that they cannot be touched while the electrodes are live) a fuse in the line conductor may be substituted for the circuit-breaker and the shell of the electrode water heater or electrode boiler need not be connected to the neutral of the supply.	WR-554.1.7

6.2.20.1 Water heaters having immersed and uninsulated heating elements

All metal parts of the heater or boiler which are in contact with the water (other than current-carrying parts) shall be solidly and metallically connected to a metal water pipe through which the water supply to the heater or boiler is provided and that water pipe shall be connected to the main earthing terminal by means independent of the circuit protective conductor.	WR-554.3.2
Single-line water heaters and boilers with an uninsulated heating element immersed in the water shall not have a single-pole switch, non-linked circuit-breaker or fuse fitted in the neutral conductor, in any part of the circuit between the heater or boiler, or in the origin of the installation.	WR-554.3.1 and 4

6.2.21 Electromechanical stresses

Every conductor or cable shall have adequate strength   WR-521.5.1
and be installed so as to withstand the electromechanical
forces that may be caused by any current, including
fault current, it may have to carry whilst in service.

6.2.22 Emergency switching

Emergency switching may be emergency switching ON or emergency switching OFF.

Emergency switching for any part of an installation   WR-537.4.1.1
where it may be necessary to control the supply to
remove an unexpected danger.

Other than where a risk of electric shock is involved,   WR-537.4.1.2
the emergency switching device shall be an isolating
device and shall interrupt all live conductors.

Except where the neutral conductor can be regarded   WR-537.4.1.2
as being reliably connected to earth in a TN-S or
TN-C-S system the neutral conductor need not be
isolated or switched.

Means for emergency switching shall act as directly   WR-537.4.1.3
as possible on the appropriate supply conductors.

6.2.23 Enclosures

No conductive part that is enclosed in an insulating   WR-412.2.2.4
enclosure shall be connected to a protective conductor

No exposed conductive part or intermediate part shall   WR-412.2.2.4
be connected to a protective conductor unless the
specification for the equipment concerned, allows for
this to happen.

6.2.24 Fault protection

The exposed conductive parts of the equipment of   WR-411.7.3
the FELV circuit shall be connected to the protective
conductor of the primary circuit of the source:

provided that the primary circuit is subject to
protection by automatic disconnection of supply.

6.2.25 Fault protection

Fault protection by automatic disconnection of supply   WR-411.8.3
shall be provided by means of an overcurrent protective
device in each line conductor or by an RCD.

Live parts of the separated circuit shall **not** be   WR-413.3.3
connected at any point to another circuit or to earth or
to a protective conductor.

For separated circuits the use of separate wiring systems is recommended.

No exposed conductive part of the separated circuit   WR-413.3.6
shall be connected either to the protective conductor or
exposed conductive parts of other circuits, or to earth.

6.2.26 Ferromagnetic enclosures: electromagnetic effects

The conductors of an a.c. circuit installed in a   WR-413
ferromagnetic enclosure shall be arranged so that the
line conductors, the neutral conductor (if any) and the
appropriate protective conductor are all contained in the
same enclosure.

Where such conductors enter a ferrous enclosure, they   WR-413
shall be arranged such that the conductors are only
collectively surrounded by ferrous material.

6.2.27 Final circuits

A final circuit with a number of socket outlets or   WR-543.7.2.1
connection units that is intended to supply two or
more items of equipment (and where it is known that
the total protective conductor current in normal service
will exceed 10 mA) shall be provided with a high
integrity protective conductor connection.

The following arrangements of the final circuit are acceptable:

WR-543.7.2.1

- a ring final circuit with a ring protective conductor
- a radial final circuit with a single protective conductor.

Each final circuit shall be protected by an overcurrent protective device which disconnects all live conductors of that circuit.

WR-721.43.1

6.2.28 Fire risk

In locations where a fire risk exists, conductors of circuits supplied at extra-low voltage shall be protected either:

WR-705.422.8

- by barriers or enclosures affording a degree of protection of IPXXD or IP4X; or
- in addition to their basic insulation, by an enclosure of insulating material.

6.2.29 Fuses

A fuse base shall be arranged so as to exclude the possibility of the fuse carrier making contact between conductive parts belonging to two adjacent fuse bases.

WR-533.1.1.1

A fuse base using screw-in fuses shall be connected so that the centre contact is connected to the conductor from the supply and the shell contact is connected to the conductor to the load.

WR-533.1.1.1

6.2.30 Functional switching

A functional switching device shall be provided for each part of a circuit which may require to be controlled independently of other parts of the installation.

WR-537.5.1.1

Functional switching devices need not necessarily control all live conductors of a circuit.	WR-537.5.1.2
Functional switching devices ensuring the changeover of supply from alternative sources shall affect all live conductors and shall **not** be capable of putting the sources in parallel, unless the installation is specifically designed for this condition.	WR-537.5.1.4

Note: In these cases, no provision shall be made for isolation of the PEN or protective conductors.

6.2.31 Harmonic currents

Overcurrent detection shall be provided for the neutral conductor in a multiphase circuit where the harmonic content of the line currents is such that the current in the neutral conductor may exceed the current-carrying capacity of that conductor.	WR-431.2.3

Note: Overcurrent detection shall cause disconnection of the line conductors but not necessarily the neutral conductor.

6.2.32 Identification of conductors

The identification of any conductor shall comply with the requirements of Table 51 on p. 92 of BS 7671:2008 so far as these are applicable.	WR-514.3.3
Unambiguous marking shall be provided at the interface between conductors.	WR-514.1.3
The single colour green shall **not** be used.	WR-514.4.5

Note:
(1) Power circuits include lighting circuits.
(2) M identifies either the mid-wire of a three-wire d.c. circuit, or the earthed conductor of a two-wire earthed d.c. circuit.
(3) Only the middle wire of three-wire circuits may be earthed.
(4) An earthed PELV conductor is blue.

6.2.32.1 Neutral or midpoint conductor

Where a circuit includes a neutral or midpoint conductor, WR-514.4.1
the colour used shall be blue.

6.2.32.2 Protective conductor

The bi-colour combination green-and-yellow shall be used WR-514.4.2
exclusively for identification of a protective conductor and
this combination shall **not** be used for any other purpose.

Single-core cables that are coloured green-and-yellow WR-514.4.2
throughout their length shall **only** be used as a protective
conductor and shall **not** be over-marked at their
terminations.

One of the colours shall cover at least 30% and at most WR-514.4.2
70% of the surface being coloured, while the other
colour shall cover the remainder of the surface.

A bare conductor or busbar used as a protective WR-514.4.2
conductor shall be identified, where necessary, by equal
green and yellow stripes, each not less than 15 mm
and not more than 100 mm wide, close together, either
throughout the length of the conductor or in each
compartment and unit and at each accessible position. If
adhesive tape is used, it shall be bi-coloured.

6.2.32.3 PEN conductor

A PEN conductor shall be marked by one of the WR-514.4.3
following methods:

- green-and-yellow throughout its length with, in
  addition, blue markings at the terminations
- blue throughout its length with, in addition, green-
  and-yellow markings at the terminations.

6.2.32.4 Other conductors

All other conductors shall be identified by colour in WR-514.4.4
accordance with Table 51 on p. 92 of BS 7671:2008.

6.2.32.5 Bare conductors

A bare conductor shall be identified by the application   WR-514.4.6
of tape, sleeve or disc of the appropriate colour
prescribed in Table 51 on p. 92 of BS 7671:2008 or
by painting it with such a colour.

6.2.33 Identification of conductors by letters and/or numbers

The lettering or numbering system applies to identification of individual conductors and of conductors in a group. All numerals:

- shall be clearly legible and durable;
- shall be in strong contrast to the colour of the insulation;
- shall be given in letters or Arabic numerals (in order to avoid confusion, unattached numerals 6 and 9 shall be underlined).

6.2.33.1 Numeric

Conductors may be identified by numbers, the number 0   WR-514.5.4
being reserved for the neutral or midpoint conductor.

6.2.33.2 Protective conductor

Conductors with green-and-yellow colour identification   WR-514.5.2
shall **not** be numbered other than for the purpose of
circuit identification.

6.2.33.3 Omission of identification by colour or marking

Identification by colour or marking is not required for: WR-514.6.1

- concentric conductors of cables
- metal sheath or armour of cables when used as a
  protective conductor
- bare conductors where permanent identification is
  not practicable
- extraneous conductive parts used as a protective
  conductor
- exposed conductive parts used as a protective
  conductor.

6.2.34 Notices

6.2.34.1 Safety earth

A warning notice for the safety earth shall be   WR-514.13.1
permanently fixed at:

- the point of connection of every earthing conductor
  to an earth electrode; and
- the point of connection of every bonding conductor
  to an extraneous conductive part; and
- the main earth terminal, where separate from main
  switchgear.

The warning notice shall be worded as shown in Figure 6.5.

Safety Electrical Connection – Do Not Remove

Figure 6.5 Warning notice – earthing and bonding

6.2.34.2 Electrical separation

Where electrical separation to the supply to more than   WR-514.13.2
one item of current-using equipment is used (Regulation
418.2.5 or 418.3), the warning notice shall read as
follows:

The protective bonding conductors associated with the electrical
installation in this location MUST NOT BE CONNECTED TO

EARTH

Equipment having exposed-conductive-parts connected to earth must
not be brought into this location

Figure 6.6 Warning notice – protective bonding conductors (reproduced
with permission of IET)

6.2.34.3 Non-standard colours

If wiring additions or alterations are made to an installation such that some of the wiring complies with the current Regulations but there is also wiring to previous versions of these Regulations, a warning notice shall be affixed at or near the appropriate distribution board with the following wording:	WR-514.14.1

CAUTION

This installation has wiring colours to two versions of BS 7671. Great care should be taken before undertaking extension, alteration or repair that all conductors are correctly identified.

Figure 6.7 Warning notice – non-standard colours (reproduced with permission of IET)

6.2.35 Installation

A bare live conductor shall be installed on insulators.	WR-521.10.2
A circuit supplying one or more items of Class II equipment shall have a circuit protective conductor run to and terminated at each point in wiring and at each accessory.	WR-412.2.3.2
An IMD shall be connected between earth and a live conductor of the monitored equipment.	WR-538.3
The degree of protection of electrical equipment shall be maintained after installation of conductors.	WR-522.6.3
The connection of conductors shall not affect the protection being supplied by an enclosure.	WR-412.2.3.1
The 'line' terminal(s) of an IMD shall be connected as close as practicable to the origin of the system to:	WR-538.1.2

- the neutral point of the power supply; or
- an artificial neutral point with impedances connected to the line conductors; or
- a line conductor or two or more line conductors.

For d.c. installations, the 'line' terminal(s) of the IMD   WR-538.1.2
shall be connected either directly to the midpoint, if
any, or to one or all of the supply conductors.

In some particular d.c. IT two-conductor installations,   WR-538.1.4
a passive IMD that does not inject current into the
system may be used, provided that:

- the insulation of all live distributed conductors is
  monitored; and
- all exposed conductive parts of the installation are
  interconnected; and
- circuit conductors are selected and installed so as
  to reduce the risk of an earth fault to a minimum.

6.2.36 Inspection

Note: Inspection shall precede testing and shall normally be done with that
part of the installation under inspection disconnected from the supply.

The inspection shall include at least the checking of   WR-611.3
the following items, where relevant to the installation
and, where necessary, during erection:

- connection of conductors
- identification of conductors
- selection of conductors for current-carrying capacity
  and voltage drop, in accordance with the design
- connection of single-pole devices for protection
  or switching in line conductors only.

6.2.37 Isolation

Isolation is intended, for reasons of safety, to make dead a circuit by separating an installation or section from every source of electric energy.

Every circuit shall be capable of being isolated from   WR-537.2.1.1
each of the live supply conductors.

In a TN-S or TN-C-S system, it is not necessary   WR-537.2.1.1
to isolate or switch the neutral conductor where it
is regarded as being reliably connected to earth by
suitably low impedance.

Provision shall be made for disconnecting the neutral conductor. Where this is a joint it shall be such that it is in an accessible position which can only be disconnected by means of a tool, is mechanically strong and will reliably maintain electrical continuity.	WR-537.2.1.7
In marinas and similar locations, this switching device shall disconnect all live conductors including the neutral conductor.	WR-709.537.2.1.1
In temporary electrical installations, devices for isolation shall disconnect all live conductors (line and neutral conductors).	WR-740.537.2.2

6.2.37.1 Devices for isolation

An isolation device shall isolate all live supply conductors from the circuit concerned.	WR-537.2.2.1
Where a link is inserted in the neutral conductor, the link shall:	WR-537.2.2.4

- not be capable of being removed without the use of a tool
- only be accessible to skilled persons.

Semiconductor devices shall **not** be used as isolating devices.

6.2.38 Line conductors

Detection of overcurrent shall be provided for all line conductors and shall cause the disconnection of the conductor in which the overcurrent is detected.	WR-431.1.1
In a TN or TT system, for a circuit supplied between line conductors and in which the neutral conductor is not distributed, overcurrent detection need not be provided for one of the line conductors, provided that:	WR-431.1.2

- there exists, in the same circuit or on the supply side, differential protection intended to detect unbalanced loads and cause disconnection of all the line conductors; **and**

- the neutral conductor is not distributed from an artificial neutral point of the circuits situated on the load side of this differential protective device.

6.2.39 Live conductors

Bare live conductors **shall** be installed on insulators.

The supply to all live conductors **shall** be automatically interrupted in the event of overload current and fault current.

Conductors **shall** be able to carry fault current without overheating.

Persons and livestock **shall** be protected against injury, and property shall be protected against damage, due to excessive temperatures or electromechanical stresses caused by any overcurrents likely to arise in live conductors.

Live supply conductors **shall** be capable of being isolated from circuits.

A main switch intended for operation by ordinary persons (such as a household or similar installation) shall interrupt **both** live conductors of a single-line supply.   WR-537.1.4

Where an installation is supplied from more than one source of energy (one of which requires a means of earthing independent of the means of earthing of other sources and it is necessary to ensure that not more than one means of earthing is applied at any time) a switch may be inserted in the connection between the neutral point and the means of earthing, provided that the switch is a linked switch arranged to disconnect and connect the earthing conductor for the appropriate source, at substantially the same time as the related live conductors.   WR-537.1.5

Where no neutral point or midpoint exists, a line conductor may be connected to earth.

After the occurrence of a first fault – and in the event of a second fault occurring on a different live conductor – automatic disconnection of supply shall occur.   WR-411.6.4

Note: Where the exposed conductive parts are interconnected by a protective conductor and collectively earthed to the same earthing system, the conditions similar to a TN system shall apply.

The number and type of live conductors (e.g. single-line two-wire a.c., three-line four-wire a.c.) shall be determined, both for the source of energy and for each circuit to be used within the installation.	WR-312.2.1
In agricultural and horticultural premises:	WR-705.537.2
• the electrical installation of each building or part of a building shall be isolated by a single isolation device	
• means of isolation of all live conductors, including the neutral conductor, shall be provided for circuits used occasionally (such as during harvest time).	WR-705.537.2
Where both the live circuit conductors are uninsulated, either:	WR-559.11.4.1
• they shall be provided with a protective device complying with the requirements of Regulation 559.11.4.2; or	
• the system shall comply with BS EN 60598-2-23.	
In caravans and motor caravans, each installation shall be provided with a main disconnector which shall disconnect all live conductors and which shall be suitably placed for ready operation within the caravan.	WR-721.537.2.1.1
In locations containing a bath and/or shower, the equipment shall only be accessible via a hatch (or a door) by means of a key or a tool which shall disconnect all live conductors and the supply cable. In addition, the main disconnecting means shall be installed in a way which provides protection of Class II or equivalent insulation.	WR-702.55.4
The supply circuit of the equipment shall be protected by:	
• SELV; or	
• an RCD; or	
• electrical separation.	

In agricultural and horticultural premises, RCDs shall disconnect all live conductors.   WR-705.422.7

In solar photovoltaic power supply systems, earthing of one of the live conductors of the d.c. side is permitted, if there is at least simple separation between the a.c. side and the d.c. side.   WR-712.312.2

 Note: Any connections with earth on the d.c. side should be electrically connected so as to avoid corrosion (see BS 7361-1:1991).

6.2.40 Low-voltage generating sets

Where a generating set is intended to operate in parallel with a system for distribution of electricity to the public (or where two or more generating sets may operate in parallel) circulating harmonic currents shall be limited so that the thermal rating of conductors is not exceeded.   WR-551.5.2

6.2.41 Luminaires

Any cable or cord between the fixing means and the luminaire shall be installed so that any expected stresses in the conductors, terminals and terminations will not impair the safety of the installation.   WR-559.6.1.5

In circuits of a TN or TT system (except for E14 and E27 lampholders complying with BS EN 60238) the outer contact of every Edison screw or single centre bayonet cap type lampholder shall be connected to the neutral conductor.   WR-559.6.1.8

 Note: This Regulation also applies to track-mounted systems.

Groups of luminaires divided between the three line conductors of a three-line system with only one common neutral conductor shall be provided with at least one device that simultaneously disconnects all line conductors.   WR-559.6.2.3

6.2.42 Mechanical stresses

The radius of every bend in a wiring system shall be such that conductors do not suffer damage and terminals are not stressed.	WR-522.8.3
Every conductor shall be supported in such a way that it is not exposed to undue mechanical strain and so that there is no appreciable mechanical strain on the terminations of the conductors.	WR-522.8.5
Where the conductors are not supported continuously, they shall be supported by suitable means at appropriate intervals in such a manner that the conductors do not suffer damage by their own weight.	WR-522.8.4
Cables, busbars and other electrical conductors which pass across expansion joints shall be so selected and/ or erected that anticipated movement does not cause damage to the electrical equipment.	WR-522.8.13
A wiring system intended for the drawing in or out of conductors shall have adequate means of access to allow this operation.	WR-522.8.6
A cable buried in the ground (that is not installed in a conduit or duct) shall incorporate an earthed armour or metal sheath or both, suitable for use as a protective conductor.	WR-522.8.10
A conductor shall not be damaged by the means of fixing.	WR-522.8.12

6.2.43 Multiwire, fine wire and very fine wire conductors

In order to avoid inappropriate separation or spreading of individual wires of multiwire, fine wire or very fine wire conductors, suitable terminals shall be used or the conductor ends shall be suitably treated.	WR-526.8.1
Soldering (tinning) of the whole conductor end of multiwire, fine wire and very fine wire conductors is **not** allowed if screw terminals are used.	WR-526.8.2
Soldered (tinned) conductor ends on fine wire and very fine wire conductors are **not allowed** at connection and junction points which are subject in service to a relative movement between the soldered and the non-soldered part of the conductor.	WR-526.8.3

6.2.44 Neutral conductor

Consideration shall be given to the fact that: WR-442.3

- if the neutral conductor in a three-line TN or   WR-442.4
TT system is interrupted, basic, double and   WR-442.5
reinforced insulation (as well as components
rated for the voltage between line and neutral
conductors) can be temporarily stressed with the
line-to-line voltage
- if a line conductor of an IT system is earthed
accidentally, insulation or components rated for
the voltage between line and neutral conductors
can be temporarily stressed with the line-to-line
voltage
- if a short-circuit occurs in the low-voltage
installation between a line conductor and the
neutral conductor, the voltage between the other
line conductors and the neutral conductor can
reach the value of $1.45 \times U_o$ for a time up to 5 s.

If the total harmonic distortion due to third harmonic   WR-523.6.3
current (or multiples of the third harmonic) is greater
than 15% of the fundamental line current the neutral
conductor shall not be smaller than the line conductors.

In a TN or TT system, the neutral conductor shall be   WR-431.2.1
protected against short-circuit current.

In an IT system, the neutral conductor shall not be   WR-431.2.2
distributed unless:

- overcurrent detection is provided for the neutral
conductor of every circuit; or
- the neutral conductor is effectively protected
against short-circuit by a protective device
installed on the supply side; or
- the circuit is protected by an RCD with a
rated residual operating current not exceeding
0.2 times the current-carrying capacity of the
corresponding neutral conductor.

In a TN-S or TN-C-S system the neutral conductor   WR-537.1.2
need not be isolated or switched where it can be
regarded as being reliably connected to earth by a
suitably low impedance.

In temporary electrical installations, the neutral   WR-740.551.8
conductor of the star-point of the generator shall,
except for an IT system, be connected to the exposed
conductive-parts of the generator.

Where conductors in polyphase circuits carry balanced currents, the associated neutral conductor need not be taken into consideration.	WR-523.6.1
Where the neutral conductor carries current without a corresponding reduction in load of the line conductors, the neutral conductor shall be considered when ascertaining the current-carrying capacity of the circuit.	WR-523.6.3
The neutral conductor shall not be disconnected before the line conductors and shall be reconnected at the same time as (or before) the line conductors.	WR-431.3

6.2.45 Non-conducting location

This protective measure is intended to prevent simultaneous contact with parts which may be at different potentials through failure of the basic insulation of live parts. It is **not** recognised for general application.

In a non-conducting location there is no need for a protective conductor, provided that the location has an insulating floor and walls and: • all exposed conductive parts and extraneous conductive parts are well spaced; or • effective obstacles exist between all exposed conductive parts and extraneous conductive parts; or • all extraneous conductive parts are insulated.	WR-418.1.3

6.2.46 Multicore cables, conduits, ducting systems, franking systems or tray or ladder systems

Each part of a circuit shall be arranged such that the conductors are **not** distributed over different multicore cables, conduits, ducting systems, franking systems or tray or ladder systems.	WR-521.8.1
Where multicore cables are installed in parallel each cable shall contain one conductor of each line.	WR-521.8.1

6.2.47 Multiphase circuits

In multiphase circuits an independently operated   WR-530.3.2
single-pole switching device or protective device
shall not be inserted in the neutral conductor.

In single-line circuits an independently operated
single-pole switching or protective device shall not
be inserted in the neutral conductor alone.

6.2.48 Operational conditions

Switchgear, protective devices, accessories and   WR-512.1.2
other types of equipment shall not be connected
to conductors intended to operate at a temperature
exceeding 70°C at the equipment in normal service.

6.2.49 Overhead conductors

In caravan and camping parks, all overhead   WR-708.521.1.2
conductors:

- shall be insulated
- shall be at a height above ground of not less than   WR-708.521.1.2
  6 m in all areas subject to vehicle movement and
  3.5 m in all other areas.

In marinas and similar locations all overhead   WR-709.521.1.8
conductors shall be:

- insulated
- at a height above ground of not less than 6 m   WR-709.521.1.8
  in all areas subjected to vehicle movement and
  3.5 m in all other areas.

6.2.50 Parallel conductors

Except for a ring final circuit (where spurs are   WR-433.4.1
permitted) where a single device protects conductors
in parallel and the conductors are sharing currents
equally, the value of the current-carrying capacity
of the conductor is the sum of the current-carrying
capacities of the parallel conductors.

Where the use of a single conductor is impractical   WR-433.4.2
and the currents in the parallel conductors are unequal,
the design current and requirements for overload
protection for each conductor shall be considered
individually.

Note: Currents in parallel conductors are considered to be unequal if the difference between the currents is more than 10% of the design current for each conductor.

6.2.51 Plug and socket outlets

Except for SELV, every plug and socket outlet shall   WR-553.1.2
be of the non-reversible type, with provision for the
connection of a protective conductor.

6.2.52 Protective conductors

 Gas pipes, oil pipes and flexible (or pliable) conduit may **not** be used as a protective conductor.

Exposed-conductive parts of equipment shall **not** be used as a protective conductor for other equipment.

In installations and locations where the risk of an electric shock is increased by a reduction in body resistance and/or by contact with earth potential, all plugs, socket outlets and cable couplers of a reduced low-voltage system **shall** have a protective conductor contact.

If the protective conductor: WR-543.1

- is not an integral part of a cable; or
- is not formed by conduit, ducting or trunking; or
- is not contained in an enclosure formed by a
  wiring system

then the cross-sectional area shall be not less than:

- 2.5 mm^2 copper equivalent if protection against
  mechanical damage is provided; and
- 4 mm^2 copper equivalent if mechanical protection
  is not provided.

Where a protective conductor is common to two or more   WR-543.1.2
circuits, its cross-sectional area shall be:

- calculated for the most onerous of the values of
  fault current and operating time encountered in
  each of the various circuits; or
- selected so as to correspond to the cross-sectional
  area of the largest line conductor of the circuits.

6.2.52.1 Types of protective conductor

 A gas pipe, an oil pipe, flexible or pliable conduit,   WR-543.2.1
support wires or other flexible metallic parts, or
constructional parts subject to mechanical stress in
normal service, shall **not** be selected as a protective
conductor.

A protective conductor may consist of one or more of the   WR-543.2.2
following:

- a single-core cable
- a conductor in a cable
- an insulated or bare conductor in a common
  enclosure with insulated live conductors
- a fixed bare or insulated conductor
- a metal covering, for example, the sheath, screen or
  armouring of a cable
- a metal conduit, metallic cable management system
  or other enclosure or electrically continuous support
  system for conductors
- an extraneous conductive part.

The metal covering (including the sheath – bare or   WR-543.2.5
insulated) of a cable, trunking, ducting and metal
conduit, may be used as a protective conductor for the
associated circuit.

An extraneous conductive part may be used as a   WR-543.2.6
protective conductor if:

- electrical continuity can be assured with either
  construction or connection so that it is protected
  against mechanical, chemical or electrochemical
  deterioration

- precautions have been taken against its removal
- it has been considered for such a use and, if necessary, suitably adapted.

Where a metal enclosure or frame of a low-voltage switchgear or controlgear assembly or busbar trunking system is used as a protective conductor: WR-543.2.4

- its electrical continuity shall be assured, either by construction or by suitable connection, in such a way as to be protected against mechanical, chemical or electrochemical deterioration
- its cross-sectional area shall be in accordance with BS EN 60439-1
- it shall permit the connection of other protective conductors at every predetermined tap-off point.

Where the protective conductor is formed by conduit, trunking, ducting or the metal sheath and/or armour of a cable, the earthing terminal of each accessory shall be connected by a separate protective conductor to an earthing terminal incorporated in the associated box or other enclosure.   WR-543.2.7

An exposed conductive part of equipment shall not be used to form a protective conductor for other equipment.   WR-543.2.8

Except where the circuit protective conductor is formed by a metal covering or enclosure containing all of the conductors of the ring, the circuit protective conductor of every ring final circuit shall also be run in the form of a ring having both ends connected to the earthing terminal at the origin of the circuit.   WR-543.2.9

A separate metal enclosure for cable shall not be used as a PEN conductor.   WR-543.2.10

6.2.52.2 Preservation of electrical continuity of protective conductors

A protective conductor shall be suitably protected against mechanical and chemical deterioration and electrodynamic effects.   WR-543.3.1

A protective conductor with a cross-sectional area up to and including $6\,mm^2$ shall be protected throughout by a covering at least equivalent to that provided by   WR-543.3.2

the insulation of a single-core non-sheathed cable of appropriate size having a voltage rating of at least 450/750 V unless it is:

- a protective conductor forming part of a multicore cable
- cable trunking or conduit used as a protective conductor.

Where the sheath of a cable incorporating an uninsulated protective conductor of cross-sectional area up to and including 6 mm^2 is removed adjacent to joints and terminations, the protective conductor shall be protected by insulating sleeving complying with BS EN 60684 series.

WR-543.3.2

Every connection and joint shall be accessible for inspection, testing and maintenance.

A switching device shall **not** be inserted in a protective conductor unless:

WR-543.3.4

- the switch has been inserted in the connection between the neutral point and the means of earthing, and the switch is a linked switch arranged to disconnect and connect the earthing conductor for the appropriate source, at substantially the same time as the related live conductors
- it is a multipole linked switch or plug-in device in which the protective conductor circuit has not been interrupted before the live conductors, and re-established not later than when the live conductors are reconnected.

Joints intended to be disconnected for test purposes are permitted in a protective conductor circuit.

WR-543.3.4

Where electrical monitoring of earthing is used, no dedicated devices (e.g. operating sensors, coils) shall be connected in series with the protective conductor.

WR-543.3.5

Every joint in metallic conduit shall be mechanically and electrically continuous.

WR-543.3.6

6.2.53 PEN conductors

 A separate metal cable enclosure shall **not** be used as a PEN conductor.

PEN conductors shall **not** be isolated or switched.

Automatic disconnection using an RCD shall **not** be applied to a circuit incorporating a PEN conductor.

Combined protective and neutral (PEN) conductors shall **not** be isolated or switched.

PEN conductors may only be used within an installation:	WR-543.4.1 and 543.4.2

- where any necessary authorisation for use of a PEN conductor has been obtained and where the installation complies with the conditions for that authorisation; or
- where the installation is supplied by a privately owned transformer or converter in such a way that there is no metallic connection (except for the earthing connection) with the distributor's network; or
- where the supply is obtained from a private generating plant.

For a fixed installation, a conductor of a cable not subject to flexing and having a cross-sectional area not less than $10\,\text{mm}^2$ for copper or $16\,\text{mm}^2$ for aluminium may serve as a PEN conductor provided that the part of the installation concerned is not supplied through an RCD.	WR-543.4.3
The outer conductor of a concentric cable shall not be common to more than one circuit.	WR-543.4.4
The conductance of the outer conductor of a concentric cable (measured at a temperature of 20°C) shall:	WR-543.4.5

- for a single-core cable, be not less than that of the internal conductor
- for a multicore cable serving a number of points contained within one final circuit or having the internal conductors connected in parallel, be not less than that of the internal conductors connected in parallel.

At every joint in the outer conductor of a concentric cable and at a termination, the continuity of that joint shall be supplemented by a conductor additional to any means used for sealing and clamping the outer conductor.

WR-543.4.6

No means of isolation or switching shall be inserted in the outer conductor of a concentric cable.

Other than a cable conforming to BS EN 60702-1, all PEN conductors of every cable shall be insulated or have an insulating covering suitable for the highest voltage to which it may be subjected.

WR-543.4.8

If, from any point of the installation, the neutral and protective functions are provided by separate conductors, those conductors shall not then be reconnected together beyond that point.

WR-543.4.9

At the point of separation, separate terminals or bars shall be provided for the protective and neutral conductors.

WR-543.4.9

The PEN conductor shall be connected to the terminals or bar intended for the protective earthing conductor and the neutral conductor.

WR-543.4.9

In marinas and similar locations, the final circuits for the supply to pleasure craft or houseboats shall **not** include a PEN conductor.

WR-709.411.4

In temporary electrical installations (where the type of system earthing is TN) a PEN conductor shall not be used downstream of the origin of the temporary electrical installation.

WR-740.411.4

In the UK for a TN system, the final circuits for the supply to caravans or similar shall **not** include a PEN conductor.

WR-740.411.4

In caravan and camping parks, socket outlet protective conductors shall **not** be connected to any PEN conductor of the electricity supply.

WR-708.553A.14

6.2.54 Protection against fault current

A device providing protection against fault current shall be installed at the point where a reduction in the cross-sectional area or other change causes a reduction in the current-carrying capacity of the conductors of the installation (except installations situated in locations presenting a fire risk or risk of explosion and where the requirements for special installations and locations specify different conditions).

WR-434.2

The device protecting a conductor may be installed on the supply side of the point where a change occurs provided that it possesses an operating characteristic such that it protects the wiring situated on the load side against fault current.

WR-434.2.2

A device for protection against fault current need not be provided for:

WR-434.3

- a conductor connecting a generator, transformer, rectifier or an accumulator battery to the associated control panel where the protective device is placed in the panel
- a circuit where disconnection could cause danger for the operation of the installation concerned
- certain measuring circuits
- the origin of an installation where the distributor installs one or more devices that provide protection against fault current

provided that both of the following conditions are simultaneously fulfilled:

- the wiring is carried out in such a way as to reduce the risk of fault to a minimum; and
- the wiring is installed in such a manner as to reduce to a minimum the risk of fire or danger to persons.

A single protective device may protect conductors in parallel against the effects of fault currents provided that the operating characteristic of the

WR-434.4

device results in its effective operation should a
fault occur at the most onerous position in one of
the parallel conductors.

Conductors other than live conductors, and any other parts intended to carry a fault current, shall be capable of carrying that current without attaining an excessive temperature.	WR-131.5
Conductors, shall be provided with mechanical protection against electromechanical stresses of fault currents as necessary to prevent injury or damage to persons, livestock or property.	WR-131.5
A fault occurring at any point in a circuit shall be interrupted within a time such that the fault current does not cause the permitted limiting temperature of any conductor or cable to be exceeded.	WR-434.5.2
Conductors are considered to be protected against overload current and fault current where they are supplied from a source incapable of supplying a current exceeding the current-carrying capacity of the conductors (e.g. certain bell transformers, certain welding transformers and certain types of thermoelectric generating set).	WR-436

6.2.55 Protection against fire

Every termination of a live conductor or connection or joint between live conductors shall be contained within an enclosure.	WR-421.7
A wiring system which passes through the location but is not intended to supply electrical equipment in the location shall:	WR-422.3.5

- have no connection or joint within the location, unless the connection or joint is installed in an enclosure; and
- is protected against overcurrent; and
- does **not** use bare live conductors.

A PEN conductor shall **not** be used unless it is a circuit traversing the location.	WR-422.3.12
Every circuit shall be provided with a means of isolation from all live supply conductors by a linked switch or a linked circuit-breaker.	WR-422.3.13

6.2.56 Protection against overcurrent

A protective device shall be provided to break any overcurrent in the circuit conductors before such a current could cause a danger due to thermal or mechanical effects detrimental to insulation, connections, joints, terminations or the surroundings of the conductors.

 Note: Protection of conductors according to these Regulations does not necessarily protect the equipment connected to the conductors.

In an IT system, in the event of a second fault, an overcurrent protective device shall disconnect all corresponding live conductors, including the neutral conductor, if any.	WR-531.1.3

6.2.57 Protection against overload current

The rated current or current setting of a device protecting a conductor against overload shall not be less than the design current of the circuit; and	WR-433.1.1
• shall not exceed the lowest of the current-carrying capacities of any of the conductors of the circuit; and • the current causing effective operation of the protective device shall not exceed 1.45 times the lowest of the current-carrying capacities of any of the conductors of the circuit.	
A device for protection against overload shall be installed at the point where a reduction occurs in the value of the current-carrying capacity of the conductors of the installation.	WR-433.2.1

 Note: A reduction in current-carrying capacity may be due to a change in cross-sectional area, method of installation, or type of cable or conductor, or in environmental conditions.

The device protecting a conductor against overload may be installed along the run of that conductor provided that part of the run (i.e. between the point where a change occurs and the position of the protective device), has neither	WR-433.2.2

branch circuits nor outlets for connection of current-using equipment and is protected against fault current; or

- its length does not exceed 3 m
- it is installed so as to reduce the risk of fault to a minimum; and
- it is installed in so as to reduce to a minimum the risk of fire or danger to persons.

A device for protection against overload need not be provided: WR-433.3.1

- for a conductor situated on the load side of the point where a reduction occurs in the value of current-carrying capacity, where the conductor is effectively protected against overload by a protective device installed on the supply side of that point
- for a conductor which, because of the characteristics of the load or the supply, is not likely to carry overload current
- at the origin of an installation where the distributor provides an overload device.

In an IT system without a neutral conductor it is permitted to omit the overload protective device in one of the line conductors if an RCD is installed in each circuit.   WR-433.3.2.2

Where a single protective device protects two or more conductors in parallel there shall be no branch circuits or devices for isolation or switching in the parallel conductors.   WR-443.1

6.2.58 Protection by earth-free local equipotential bonding

Earth-free local equipotential bonding is intended to prevent the appearance of a dangerous touch voltage and shall **only** be used in special circumstances.

Protective bonding conductors shall interconnect every simultaneously accessible exposed conductive part and extraneous conductive part.   WR-418.2.2

Unless protection by automatic disconnection of supply   WR-418.2.3
can be applied, local protective bonding conductors
shall **not:**

- be in electrical contact with earth directly; or
- through exposed conductive parts; or
- through extraneous conductive parts.

The exposed conductive parts of the separated circuit   WR-418.3.4
shall be connected together by insulated, non-earthed
protective bonding conductors.

These conductors shall **not** be connected to the protective conductor or exposed
conductive parts of any other circuit or to any extraneous conductive parts.

Every socket outlet shall be provided with a protective   WR-418.3.5
conductor contact which shall be connected to the
equipotential bonding system.

All flexible cables (unless they supply equipment   WR-418.3.6
with double or reinforced insulation) shall include a
protective bonding conductor.

If two faults affect two exposed conductive parts at   WR-418.3.7
the same time and these are fed by conductors with a
different polarity, a protective device shall disconnect
the supply in a disconnection time conforming to
Table 41.1 on p. 46 of BS 7671:2008.

6.2.59 Protective conductors

Every plug, socket outlet, luminaire supporting coupler   WR-411.8.5
(LSC), device for connecting a luminaire (DCL) and
cable coupler of a reduced low-voltage system:

- shall have a protective conductor contact; and
- shall **not** be dimensionally compatible with those
  used for any other system in use in the same
  premises.

In a TN system, the integrity of the earthing of the installation depends on the
reliable and effective connection of the PEN or PE conductors to earth.

Each exposed conductive part of the installation shall be connected by a protective conductor to the main earthing terminal of the installation, which shall be connected to the earthed point of the power supply system.

WR-411.4.2

In a fixed installation, a single conductor may serve both as a protective conductor and as a neutral conductor (PEN conductor).

WR-411.4.3

Where an RCD is used in a TN-C-S system, a PEN conductor shall not be used on the load side.

WR-411.4.4

For an outdoor lighting installation (where the protective measure for the whole installation is by double or reinforced insulation) no protective conductor shall be provided and the conductive parts of the lighting column shall not be intentionally connected to the earthing system.

WR-559.10.4

An extra-low-voltage luminaire without provision for the connection of a protective conductor shall be installed **only as part of** a SELV system.

WR-559.11.2

In caravans and motor caravans, all protective conductors shall be incorporated in a multicore cable or in a conduit together with the live conductors.

WR-721.543.2.1

6.2.60 Protective devices and switches

Single-pole fuses, switches or circuit-breakers shall only be inserted in the line conductor.

WR-132.14.1

Switches, circuit-breaker (except where linked) or fuses shall be inserted in an earthed neutral conductor.

WR-132.14.2

Any linked switch or linked circuit-breaker inserted in an earthed neutral conductor shall be capable of breaking all of the related line conductors.

WR-132.14.2

6.2.61 Protective earthing

Exposed conductive parts shall be connected to a protective conductor.

WR-411.3.1.1

Simultaneously accessible exposed conductive parts shall be connected to the same earthing system individually, in groups or collectively.

Conductors for protective earthing shall comply with Chapter 54 of BS 7671:2008.	WR-411.3.1.1
A circuit protective conductor shall be run to and terminated at each point in wiring and at each accessory (except a lampholder having no exposed conductive parts and suspended from such a point).	WR-411.3.1.1

6.2.62 Protective equipotential bonding

In each installation, main protective bonding conductors shall connect extraneous conductive parts to the main earthing terminal including the following: WR-411.3.1.2

- central heating and air conditioning systems
- exposed metallic structural parts of the building
- gas installation pipes
- water installation pipes
- other installation pipework and ducting.

Where an installation serves more than one building the above requirement shall be applied to each building.

In exhibitions and showgrounds, structural metallic parts which are accessible from within the stand, vehicle, wagon, caravan or container shall be connected through the main protective bonding conductors to the main earthing terminal within the unit.   WR-711.411.3.1.2

In mobile or transportable units:

- accessible conductive parts of the unit, such as the chassis, shall be connected through the main protective bonding conductors to the main earthing terminal within the unit;   WR-717.411.3.1.2   WR-717.411.3.1.2
- the main protective bonding conductors shall be finely stranded.

In caravans and motor caravans, where protection by   WR-721.411.1
automatic disconnection of supply is used, an RCD
shall be used and the wiring system shall include a
circuit protective conductor which shall be connected to:

- the protective contact of the inlet; and
- the exposed conductive parts of the electrical
  equipment; and
- the protective contacts of the socket outlets.

 Note: Structural metallic parts which are accessible from within the caravan shall be connected through main protective bonding conductors to the main earthing terminal within the caravan.

6.2.63 Protective multiple earthing

Where protective multiple earthing (PME) exists, the cross-sectional area of the main equipotential bonding conductor) shall be in accordance with Table 6.10.

Table 6.10 Minimum cross-sectional area of the main equipotential bonding conductor in relation to the neutral

Copper equivalent cross-sectional area of the supply neutral conductor	Minimum copper equivalent cross-sectional area of the main equipotential bonding conductor
$35\,mm^2$ or less	$10\,mm^2$
over $35\,mm^2$ up to $50\,mm^2$	$16\,mm^2$
over $50\,mm^2$ up to $95\,mm^2$	$25\,mm^2$
over $95\,mm^2$ up to $150\,mm^2$	$35\,mm^2$
over $150\,mm^2$	$50\,mm^2$

 Note: Local distributors' network conditions may require a larger conductor.

6.2.64 RCDs

An RCD shall be capable of disconnecting all the line   WR-531.2.1
conductors of the circuit at substantially the same time.

An RCD shall be so selected and the electrical circuits   WR-531.2.4
so subdivided that any protective conductor current
which may be expected to occur during normal
operation of the connected load(s) will be unlikely to
cause unnecessary tripping of the device.

In an IT system, where protection is provided by an RCD and disconnection following a first fault is not envisaged, the non-operating residual current of the device shall be at least equal to the current which circulates on the first fault to earth of negligible impedance affecting a line conductor.	WR-531.5.1
The use of an RCD associated with a circuit normally expected to have a protective conductor shall **not** be considered sufficient for fault protection if there is no such conductor.	WR-531.2.5
The magnetic circuit of the transformer of an RCD shall enclose all the live conductors of the protected circuit. The associated protective conductor shall be outside the magnetic circuit.	WR-531.2.2
Where it is necessary to limit the consequence of fault currents in a wiring system from the point of view of fire risk, the circuit shall:	WR-532.1

- be protected by an RCD for fault protection; and
- the RCD shall be installed at the origin of the circuit to be protected; and
- the RCD shall switch all live conductors.

6.2.65 SELV and PELV circuits

The earthing of PELV circuits may be achieved by a connection to earth or to an earthed protective conductor within the source itself.	WR-414.4.1
Protective separation of wiring systems of SELV or PELV circuits from the live parts of other circuits (which have at least basic insulation) shall be achieved by one of the following arrangements:	WR-414.4.2

- SELV and PELV circuit conductors:
  - enclosed in a non-metallic sheath or insulating enclosure in addition to basic insulation
  - separated from conductors of circuits at voltages higher than Band I by an earthed metallic sheath or earthed metallic screen
- circuit conductors (at voltages higher than Band I) contained in a multi-conductor cable or other grouping of conductor (but only if the SELV and

PELV conductors are insulated for the highest
voltage present).

Plug and socket outlets in a SELV system shall **not**   WR-414.4.3
have a protective conductor contact.

Exposed conductive parts of a SELV circuit shall not   WR-414.4.4
be connected to protective conductors.

6.2.66 Supply source

In caravans and motor caravans, the means of   WR-721.55.2.6
connection to the caravan pitch socket outlet shall
be supplied with the caravan and shall comprise the
following:

- a plug complying with BS EN 60309-2; and
- a flexible cord or cable of 25 m (±2 m) length
  incorporating a protective conductor.

Annex A to BS 7671:2008 provides guidance for extra-low-voltage d.c.
installations:

In temporary electrical installations (irrespective   WR-740.313.3
of the number of sources of supply) the line and
neutral conductors from different sources shall not
be interconnected downstream of the origin of the
temporary electrical installation.

6.2.67 Transformers

Where an autotransformer is connected to a circuit   WR-555.1.1
having a neutral conductor, the common terminal
of the winding shall be connected to the neutral
conductor.

A step-up autotransformer shall **not** be connected to an IT system.

Where a step-up transformer is used, a linked switch   WR-555.1.3
shall be provided for disconnecting the transformer
from all live conductors of the supply.

6.2.68 Testing

A test shall be made to verify the continuity of each conductor, including the protective conductor, of every ring final circuit.	WR-612.2.2
The insulation resistance shall be measured between live conductors and between live conductors and the protective conductor connected to the earthing arrangement.	WR-612.3.1
The insulation resistance measured with the test voltages indicated in Table 61 on p. 158 of BS 7671:2008 shall be considered satisfactory if the main switchboard and each distribution circuit tested separately, with all its final circuits connected but with current-using equipment disconnected, has an insulation resistance not less than the appropriate value given in Table 6.11.	WR-612.3.2

More stringent requirements are applicable for the wiring of fire alarm systems in buildings, see BS 5839-1.

Table 6.11 Minimum values of insulation resistance

Circuit nominal voltage (V)	Test voltage d.c. (V)	Minimum insulation resistance
SELV and PELV	250	>0.5
Up to and including 500 V with the exception of the above systems	500	>1.0
Above 500 V	1000	>1.0

In locations exposed to fire hazard, a measurement of the insulation resistance between the live conductors should be applied.	WR-612.3.2

Insulation resistance values are usually much higher than those of Table 6.11.

Where the circuit includes electronic devices which are likely to influence the results or be damaged, only a measurement between the live conductors connected together and the earthing arrangement shall be made.	WR-612.3.3

6.2.68.1 Polarity

A test of polarity shall be made and it shall be verified that: WR-612.6

- every fuse and single-pole control and protective device is connected in the line conductor only; and
- except for E14 and E27 lampholders to BS EN 60238, circuits which have an earthed neutral conductor, centre contact bayonet and Edison screw lampholders have the outer or screwed contacts connected to the neutral conductor; and
- wiring has been correctly connected to socket outlets and similar accessories.

6.2.68.2 Verification of voltage drop

When required, compliance to the Regulations may be confirmed by using the following options: WR-612.14

- The voltage drop may be evaluated by measuring the circuit impedance.
- The voltage drop may be evaluated by using calculations, for example, by diagrams or graphs showing maximum cable length vs load current for different conductor cross-sectional areas with different percentage voltage drops for specific nominal voltages, conductor temperatures and wiring systems.

 Note: Verification of voltage drop is not normally required during initial verification.

6.2.69 Wiring systems

The installation method of a wiring system in relation to the type of conductor or cable used shall be in accordance with Table 4A1 of Appendix 4 of BS 7671:2008.   WR-521.1

Examples of wiring systems are shown in Table 4A2 of BS 7671:2008.   WR-521.3

A voltage Band I circuit shall not be contained in the same wiring system as a Band II circuit unless: • every cable or conductor is insulated for the highest voltage present • each conductor of a multicore cable is insulated for the highest voltage present in the cable.	WR-528.1
Metallic structural parts of buildings (such as pipe systems or parts of furniture) shall **not** be used as live conductors.	WR-559.11.5.1
The minimum cross-sectional area of the extra-low-voltage conductors shall normally be $1.5\,mm^2$ copper, but: • in the case of flexible cables with a maximum length of 3 m a cross-sectional area of $1\,mm^2$ copper may be used • in the case of suspended flexible cables or insulated conductors, for mechanical reasons, $4\,mm^2$ copper should be used • in the case of composite cables consisting of braided tinned copper outer sheath, having a material of high tensile strength inner core, $4\,mm^2$ copper should be used.	WR-559.11.5.2
The following wiring systems are suitable for distribution circuits of marinas: • overhead cables or overhead insulated conductors • cables with copper conductors and thermoplastic or elastomeric insulation and sheath installed within an appropriate cable management system taking into account external influences such as movement, impact, corrosion and ambient temperature.	WR-709.521.1.4
In caravans and motor caravans, the wiring systems shall be installed using one or more of the following: • insulated single-core cables, with flexible Class 5 conductors, in non-metallic conduit	WR-721.521.2

- insulated single-core cables, with stranded
  Class 2 conductors (minimum of 7 strands), in
  non-metallic conduit
- sheathed flexible cables.

16.3 Conduits, cable ducting, cable trunking, busbar or busbar trunking

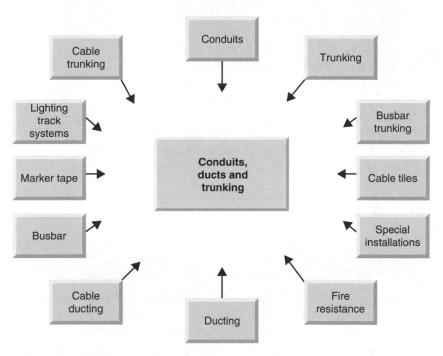

Figure 6.8 Conduits and conduit fittings

 Conduit and trunking **shall** comply with the resistance to flame propagation requirements of BS EN 50085 or BS EN 50086.

 Flexible or pliable conduit shall **not** be selected as a protective conductor.

 Note: Conduits and conduit fittings shall comply with the appropriate British Standard shown in Table 6.12.

Table 6.12 Conduits and conduit fittings (courtesy BSI)

Steel conduit and fittings	BS 31, BS EN 60423, BS EN 50086-1
Flexible steel conduit	BS 731-1, BS EN 60423, BS EN 50086-1
Steel conduit fittings with metric threads	BS 4568, BS EN 60423, BS EN 50086-1
Non-metallic conduits and fittings	BS 4607, BS EN 60423, BS EN 50086-2-1

16.3.1 Ducting and trunking

A bare conductor or busbar used as a protective   WR-514.4.2
conductor shall be identified, where necessary, by equal
green and yellow stripes, each not less than 15 mm
and not more than 100 mm wide, close together, either
throughout the length of the conductor or in each
compartment and unit and at each accessible position. If
adhesive tape is used, it shall be bi-coloured.

A busbar trunking system or a powertrack system shall   WR-521.4
be installed in accordance with the manufacturer's
instructions taking account of external influences.

Note: A busbar trunking system shall comply with BS EN 60439-2 and a powertrack system shall comply with BS EN 61534 series.

A cable concealed in a wall or partition at a depth of less   WR-522.6.6
than 50 mm from a surface of the wall or partition shall:

- incorporate an earthed metallic covering; or
- be enclosed in earthed conduit; or
- be enclosed in earthed trunking or ducting
  complying; or
- be mechanically protected against damage
  sufficient to prevent penetration of the cable by
  nails, screws etc.

A cable passing through a joist within a floor or ceiling
construction or through a ceiling support (e.g. under
floorboards), shall:

- incorporate an earthed metallic covering; or
- be enclosed in earthed conduit; or
- be enclosed in earthed trunking or ducting; or
- be mechanically protected against damage
  sufficient to prevent penetration of the cable by
  nails, screws etc.

A cable trunking system or cable ducting system shall satisfy the test under fire conditions specified in BS EN 50085.

WR-422.3.4

A conduit system or cable ducting system (other than a pre-wired conduit assembly that has been specifically designed for the installation) that is going to be buried in the structure, shall be completely erected between access points before any cable is drawn in.

WR-522.8.2

A conduit system, cable trunking system or cable ducting system classified as non-flame propagating according to the relevant product standard and having a maximum internal cross-sectional area of $710\,mm^2$ need not be internally sealed provided that:

WR-527.2.6

- the system satisfies the test of BS EN 60529 for IP33; and
- any termination of the system in one of the compartments, separated by the building construction being penetrated, satisfies the test of BS EN 60529 for IP33.

A protective conductor with a cross-sectional area up to and including $6\,mm^2$ shall be protected throughout by a covering at least equivalent to that provided by the insulation of a single-core non-sheathed cable of appropriate size having a voltage rating of at least 450/750 V unless it is:

WR-543.3.2

- a protective conductor forming part of a multicore cable
- cable trunking or conduit used as a protective conductor.

A voltage Band I circuit shall not be contained in the same wiring system as a Band II circuit unless:

WR-528.1

- every cable or conductor is insulated for the highest voltage present
- each conductor of a multicore cable is insulated for the highest voltage present in the cable
- the cables are insulated for their system voltage and installed in a separate compartment of a cable ducting or cable trunking system
- the cables are installed on a cable tray system where physical separation is provided by a partition

- a separate conduit, trunking or ducting system is employed
- for a multicore cable or cord, the cores of the Band I circuit are separated from the cores of the Band II circuit by an earthed metal screen of equivalent current-carrying capacity to that of the largest core of a Band II circuit.

A wiring system (such as a conduit system, cable ducting system, cable trunking system, busbar or busbar trunking system) which penetrates elements of building construction having specified fire-resistance shall be internally sealed to the degree of fire-resistance of the respective element before penetration as well as being externally sealed.   WR-527.2.4

Cable trunking systems and cable ducting systems shall comply with the relevant Part 2 of BS EN 50085.   WR-740.521.1

Cables, busbars and other electrical conductors which pass across expansion joints shall be so selected and/or erected that anticipated movement does not cause damage to the electrical equipment.   WR-522.8.13

Conduit and trunking systems shall be in accordance with BS EN 61386-1 and BS EN 50085-1 respectively and shall meet the fire-resistance tests within these Standards.   WR-422.4.6

Conduit systems shall comply with BS EN 61386 series.   WR-740.521.1

Except for mineral insulated cables, busbar trunking systems or powertrack systems, a wiring system shall be protected against insulation faults: WR-422.3.9

- in a TN or TT system, by an RCD having a rated residual operating current (I_a) not exceeding 300 mA
- in an IT system, by an insulation monitoring device with audible and visual signals.

For locations where the wiring system may be exposed to impact and mechanical shock due to vehicles and mobile agricultural machines, etc., the external influences shall be classified as AG3 and: WR-705.522.16

- conduits shall provide a degree of protection against impact of 5 J according to BS EN 61386-2
- cable trunking and ducting systems shall provide a degree of protection against impact of 5 J according to BS EN 50085-2-1.

The cables of an installation not intended to be under the supervision of a skilled or instructed person and are concealed in a wall or partition the internal construction of which includes metallic parts, other than metallic fixings such as nails, screws and the like, shall:

WR-522.6.8

- incorporate an earthed metallic covering; or
- be enclosed in earthed conduit; or
- be enclosed in earthed trunking or ducting; or
- be mechanically protected sufficiently to avoid damage to the cable during construction of the wall or partition and during installation of the cable.

In each installation main protective bonding conductors shall connect extraneous conductive parts to the main earthing terminal including other installation pipework and ducting.

WR-411.3.1.2

Cores of sheathed cables from which the sheath has been removed and non-sheathed cables at the termination of conduit, ducting or trunking, shall be enclosed.

WR-526.9

If the protective conductor:

WR-543.1

- is not an integral part of a cable; or
- is not formed by conduit, ducting or trunking; or
- is not contained in an enclosure formed by a wiring system

the cross-sectional area shall be not less than $2.5\,\text{mm}^2$ copper equivalent if protection against mechanical damage is provided, and $4\,\text{mm}^2$ copper equivalent if mechanical protection is not provided.

In multicore cables, each part of a circuit shall be arranged such that the conductors are **not** distributed over different multicore cables, conduits, ducting systems, franking systems or tray or ladder systems.

WR-521.8.1

Non-sheathed cables are permitted in a cable trunking system which provides a minimum of IP4X or IPXXD protection, **or** if the cover can only be removed by means of a tool or a deliberate action.

WR-521.10.2

Non-sheathed cables for fixed wiring shall be enclosed in conduit, ducting or trunking.

WR-521.10.1

The following wiring systems shall **not** be used on or above a jetty, wharf, pier or pontoon: WR-709.521.1.5

- cables in free air suspended from or incorporating a support wire
- non-sheathed cables in conduit, trunking etc.
- cables with aluminium conductors
- mineral insulated cables.

The installation of wiring systems will meet the requirements if non-metallic trunking or ducting (complying with the BS EN 50085) is used.   WR-412.2.4.1

The metal covering (including the sheath – bare or insulated) of a cable, trunking, ducting and metal conduit, may be used as a protective conductor for the associated circuit.   WR-543.2.5

Two or more circuits are allowed in the same conduit, ducting or trunking system (but see Section 528 for specific requirements).   WR-521.6

Wherever equipment is fixed on or in cable trunking or skirting trunking or in mouldings it shall not be fixed on covers which can be removed inadvertently.   WR-530.4.3

Where a metal enclosure or frame of a low-voltage switchgear or controlgear assembly or busbar trunking system is used as a protective conductor: WR-543.2.4

- its electrical continuity shall be assured, either by construction or by suitable connection, in such a way as to be protected against mechanical, chemical or electrochemical deterioration
- its cross-sectional area shall be in accordance with BS EN 60439-1
- it shall permit the connection of other protective conductors at every predetermined tap-off point.

Where a protective conductor is formed by conduit, trunking, ducting or the metal sheath and/or armour of a cable, the earthing terminal of each accessory shall be connected by a separate protective conductor to an earthing terminal incorporated in the associated box or other enclosure.   WR-543.2.7

Where no fire alarm system is installed in a building   WR-711.521
used for exhibitions etc., cable systems shall be either:

- flame retardant to BS EN 60332-1-2 and low
  smoke to BS EN 61034-2; or
- single-core or multicore unarmoured cables enclosed
  in metallic or non-metallic conduit or trunking, providing a degree of fire protection of at least IP4X.

7

Special installations and locations

 These particular Regulations are **additional** to all of the other requirements and **not** alternatives to them.

Whilst the Regulations apply to all electrical installations in buildings, there are also some indoor and out-of-doors special installations and locations that are subject to special requirements due to the extra dangers they pose. This chapter considers the requirements for these special locations and installations.

In addition to the normal safety protection methods against direct and indirect contact listed in other parts of the Regulations, special installations and locations such as:

- agricultural and horticultural premises;
- conducting locations with restricted movement;
- construction and demolition sites;
- electrical installations in caravan/camping parks and similar locations;
- electrical installations in caravans and motor caravans;
- exhibitions, shows and stands;
- floor and ceiling heating systems;
- locations containing a bath or shower;
- marinas and similar locations;
- mobile and transportable units;
- rooms and cabins containing saunas;
- solar, photovoltaic (PV) power supply systems;
- swimming pools and other basins;
- temporary electrical installations for structures, amusement devices and booths at fairgrounds, amusement parks and circuses;

must also comply with the requirements for safety protection in respect of:

- electric shock
- thermal effects
- overcurrent
- undervoltage
- isolation and switching.

These requirements are described in the following sections.

7.1 General requirements

The following are intended to act as a reminder of the general requirements that are applicable to special installations and locations conductors.

7.1.1 Accessibility

Connections and joints of heating appliances must be accessible for inspection, testing and maintenance, unless:

- they are in a compound-filled or encapsulated joint;
- the connection is between a cold tail and a heating element;
- the joint is made by welding, soldering, brazing or a compression tool.

7.1.2 Electrical heating units

 The equipment, system design, installation and testing of an electric surface heating system **must** meet the requirements of BS 6351.

 Heating appliances **must** be fixed so as to minimise the risks of bums to livestock and of fire from combustible material.

Electric heating units that are embedded in the floor (and intended for heating the location) may be installed below any zone in a bathroom, provided that they are covered by an earthed metallic grid or sheath that is connected to local supplementary equipotential bonding.

If an electric heating unit is embedded in the floor in zone B or C of a swimming pool it must either:

- be connected to the local supplementary equipotential bonding by a metallic sheath; or
- be covered by an earthed metallic grid connected to the equipotential bonding.

Radiant heaters must be fixed not less than 0.5 m from livestock and from combustible material.

7.1.3 Equipotential bonding conductors

Main equipotential bonding conductors (for each installation) need to be connected to the main earthing terminal of that installation, which can include:

- water service pipes;
- gas installation pipes;
- other service pipes and ducting;
- central heating and air conditioning systems;
- exposed metallic structural parts of the building;
- the lightning protective system.

 Note: where an installation serves more than one building the above requirement shall be applied to each building.

7.1.4 Forced air heating systems

- Forced air heating systems **must** have two, independent, temperature limiting devices.
- Electric heating elements of forced air heating systems (other than those of central-storage heaters) should:
  - not be capable of being activated until the prescribed air flow has been established;
  - deactivate when the air flow is reduced or stopped.
- Frames and enclosures of electric heating elements must be of non-ignitable material.

7.1.5 Heating appliances

In the Regulations, the general requirements for heating appliances (e.g. water heaters, boilers, heating units, heating conductors and cables, surface and underfloor heating systems) are very important to special installations and locations, and the following, whilst perhaps not being a compete list, represents the most important requirements.

7.1.6 Heating conductors and cables

- Heating cables passing through (or in close proximity to) a fire hazard:
  - must be enclosed in material with an ignitability characteristic 'P' as specified in BS 476 Part 5;
  - must be protected from any mechanical damage.
- Heating cables that are going to be laid (directly) in soil, concrete, cement screed, or other material used for road and building construction must be:
  - capable of withstanding mechanical damage;
  - resistant to damage from dampness or corrosion.
- Heating cables that are going to be laid (directly) in soil, a roadway, or the structure of a building must be installed so that they are:
  - completely embedded in the substance it is intended to heat;
  - not damaged by movement (either by it, or the substance in which it is embedded;
  - compliant (in all respects) with the maker's instructions and recommendations.

7.1.7 Hot water and/or steam appliances

Electric appliances producing hot water or steam must be protected against overheating.

7.1.8 Locations with risks of fire due to the nature of processed or stored materials

 Heating appliances **must** always be fixed.

- Equipment enclosures (such as heaters and resistors) must not attain surface temperatures higher than:
  - 90°C under normal conditions; and
  - 115°C under fault conditions.
- Heat storage appliances must not ignite combustible dust and/or fibres.
- Heating appliances mounted close to combustible materials must be protected by barriers.
- Where heating and ventilation systems containing heating elements are installed:
  - the dust or fibre content and the temperature of the air must not present a fire hazard;
  - temperature limiting devices must have manual reset.

7.1.9 Overhead wiring systems

Whilst the only economic method of transmitting power from a grid station is by means of lines suspended from pylons, at lower voltages there is a choice between running them overhead or underground. The supply to most domestic buildings (particularly in towns) is predominantly underground but for electrical installations such as in agricultural buildings, the most cost-effective way is via overhead cables. The downside of this, of course, is the potential for the overhead cable to become a safety hazard and protective methods must be used to guard against this possibility.

7.1.9.1 Protection against indirect contact

Protective measures against indirect contact may only be dispensed with if:

- overhead line insulator brackets (and metal parts connected to them) are not within arm's reach;
- the steel reinforcement of concrete poles in not accessible;
- exposed conductive parts (including small isolated metal parts such as bolts, rivets, nameplates and cable clips) cannot be gripped or cannot be contacted by a major surface of the human body;

Figure 7.1 Overhead wiring systems (courtesy Stingray)

Figure 7.1 (Continued)

- there is no risk of fixing screws used for non-metallic accessories coming into contact with live parts;
- inaccessible lengths of metal conduit do not exceed 150 mm;
- metal enclosures mechanically protecting equipment comply with requirements for Class II protection;
- unearthed street furniture that is supplied from an overhead line is inaccessible whilst in normal use.

7.1.9.2 Protection against direct contact

 Bare live parts (other than overhead lines) must **not** be within arm's reach.

- Bare and/or insulated overhead lines being used for distribution between buildings and structures must be installed in accordance with the Electricity Safety, Quality and Continuity Regulations 2002.
- Conductors used as an overhead line, operating at low voltage, must comply with the relevant British and/or Harmonised Standard.

Note: If access to live equipment (from a normally occupied position) is restricted by an obstacle (such as a handrail, mesh or screen) with a degree of protection less than 1P2X or IPXXB, the extent of arm's reach must be measured **from** that obstacle.

When protection against direct contact is required for highway power supplies:

- protection by obstacles must not be used;
- protection by placing out of reach shall only apply to low-voltage overhead lines constructed in accordance with the Electricity Safety, Quality and Continuity Regulations 2002;
- except when the maintenance of equipment is to be restricted to skilled persons specially trained, items of street furniture (or street located equipment) that are within 1.5 m of a low-voltage overhead line, must have protection against direct contact provided by some means other than placing out of reach.

7.1.9.3 Protection against overvoltages

No additional protection against overvoltages of atmospheric origin is necessary for:

- installations that are supplied by low-voltage systems which do not contain overhead lines;
- installations that are supplied by low-voltage networks which contain overhead lines and their location is subject to less than 25 thunderstorm days per year;
- installations that contain overhead lines and their location is subject to less than 25 thunderstorm days per year;

provided that they meet the required minimum equipment impulse withstand voltages shown in Table 44.3 on p. 85 of BS 7671:2008.

Note: Suspended cables having insulated conductors with earthed metallic coverings are considered to be 'underground cables'.

Installations that are supplied by (or include) low-voltage overhead lines must incorporate protection against overvoltages of atmospheric origin if the location is subject to more than 25 thunderstorm days per year.

This protection must be provided either by:

- a surge protective device with a protection level not exceeding Category II; or
- other means providing at least an equivalent attenuation of overvoltages.

7.1.9.4 Overhead power lines and cables

Mains operated electric fence controllers must:

- take account of the effects of induction when in the vicinity of overhead power lines;
- not be fixed to any supporting pole of an overhead power or telecommunication line.

All overhead conductors in caravan sites must be:

- protected by insulation of live parts;
- at least 2 m away from the boundary of any caravan pitch;
- not less than 6 m in vehicle movement areas and 3.5 m in all other areas.

Note: Poles and other overhead wiring supports shall be protected against any reasonably foreseeable vehicle movement.

7.1.10 Type of demand

The number and type of circuits required for lighting, heating, power, control, signalling, communication and information technology, etc. will generally depend on:

- the location and points of power demand;
- the loads to be expected on the various circuits;
- the daily and yearly variation of demand;
- any special conditions;
- requirements for control, signalling, communication and information technology, etc.

7.1.11 Water heaters and boilers

Heaters that are intended for liquid or other substances must incorporate (or be provided with) an automatic device to prevent a dangerous rise in temperature.

Metal parts (other than the current-carrying parts of single-line water heaters and boilers) that are in contact with the water shall be solidly and metallically connected to the metal water pipe supplying that heater/boiler.

This metal water pipe should be connected to the main earthing terminal by a circuit protective conductor that is independent of the heater/boiler.

The heater/boiler must be permanently connected to the electricity supply via a double-pole linked switch that is either:

- separate from and within easy reach of the heater/ boiler; or
- part of the boiler/heater (provided that the wiring from the heater or boiler is directly connected to the switch without use of a plug and socket outlet).

7.2 Special installations and locations

7.2.1 Agricultural and horticultural premises

It is a mandatory requirement of the Regulations that all fixed agricultural and horticultural installations (outdoors and indoors) and locations where livestock is kept (such as stables, chicken houses, piggeries, feed-processing locations, lofts and storage areas for hay, straw and fertilisers) shall be inspected to confirm that they comply with Part 705 of the Regulations (see Section 7.3.1 for a list of inspections and tests that need to be completed).

Note: If these premises include dwellings that are intended solely for human habitation, then the dwellings are excluded from the scope of these particular Regulations.

7.2.2 Conducting locations with restricted movement

Fixed equipment in conducting locations (particularly where the movement of persons is restricted by the location) and supplies to mobile equipment for use in such locations shall be inspected to confirm that they comply with Part 706 of the Regulations (see Section 7.3.2 for a list of inspections and tests that need to be completed).

7.2.3 Construction and demolition sites

Installations providing an electricity supply for:

- new building construction;
- repairs, alterations, extensions or demolition of existing buildings;
- engineering construction;
- earthworks;

shall be inspected to confirm that they comply with Part 704 of the Regulations (see Section 7.3.3 for a list of requirements, inspections and tests that need to be completed).

These requirements do **not** apply to:

- construction site offices, cloakrooms, meeting rooms, canteens, restaurants, dormitories and toilets;
- installations covered by BS 6907.

7.2.4 Electrical installations in caravan/camping parks and similar locations

Electrical installations in caravan/camping parks and similar locations providing facilities for supplying leisure accommodation vehicles (including caravans) or tents, shall be inspected to confirm that they comply with Part 708 of the Regulations (see Section 7.3.4 for a list of requirements, inspections and tests that need to be completed).

7.2.5 Electrical installations in caravans and motor caravans

All electrical installations in caravans and motor caravans shall be inspected to confirm that they comply with Part 721 of the Regulations (see Section 7.3.5 for a list of requirements, inspections and tests that need to be completed).

It should be noted that the requirements of this section do **not** apply to:

- electrical circuits and equipment covered by the Road Vehicles Lighting Regulations 1989;
- installations covered by BS EN 1648-1 and BS EN 1648-2.

7.2.6 Exhibitions, shows and stands

Temporary electrical installations in exhibitions, shows and stands (including mobile and portable displays and equipment, shall be inspected to confirm that they comply with Part 711 of the Regulations (see Section 7.3.6 for a list of inspections and tests that need to be completed).

7.2.7 Floor and ceiling heating systems

Electric floor and ceiling heating systems which are erected as either thermal storage heating systems or direct heating systems, shall be inspected to confirm that they comply with Part 753 of the Regulations (see Section 7.3.7 for a list of requirements, inspections and tests that need to be completed).

7.2.8 Locations containing a bath or shower

All locations containing a bath or shower shall be inspected to confirm that they comply with Part 701 of the Regulations (see Section 7.3.8 for a list of requirements, inspections and tests that need to be completed).

 Locations containing baths or showers for medical treatment, or for disabled persons, may have special requirements.

7.2.9 Marinas and similar locations

Circuits intended to supply pleasure craft or houseboats in marinas and similar locations, shall be inspected to confirm that they comply with Part 709 of the Regulations (see Section 73.9 for a list of requirements, inspections and tests that need to be completed).

7.2.10 Mobile and transportable units

A vehicle and/or mobile (self-propelled or towed) or transportable structure (such as a container or cabin) in which all or part of an electrical installation is contained and which is provided with a temporary supply by means of, for example, a plug and socket outlet, shall be inspected to confirm compliance with Part 717 of the Regulations (see Section 7.3.10 for a list of requirements, inspections and tests that need to be completed).

7.2.11 Rooms and cabins containing saunas

Installations supplying electricity for locations within which are hot air sauna heating equipment (in accordance with BS EN 60335-2-53), shall be inspected to confirm that they comply with Part 703 of the Regulations (see Section 7.3.11 for a list of requirements, inspections and tests that need to be completed).

7.2.12 Solar, photovoltaic (PV) power supply systems

Electrical installations of PV power supply systems (including subsystems with a.c. modules) shall be inspected to confirm that they comply with Part 712 of the Regulations (see Section 7.3.12 for a list of requirements, inspections and tests that need to be completed).

7.2.13 Swimming pools and other basins

Requirements applicable to basins of swimming pools and paddling pools and other basins plus their surrounding zones shall be inspected to confirm that they comply with Part 702 of the Regulations (see Section 7.3.13 for a list of requirements, inspections and tests that need to be completed).

7.2.14 Temporary electrical installations for structures, amusement devices and booths at fairgrounds, amusement parks and circuses

Electrical installation required for the safe design, installation and operation of temporarily erected mobile or transportable electrical machines and structures which incorporate electrical equipment, shall be inspected to confirm that they comply with Part 740 of the Regulations (see Section 7.3.14 for a list of requirements, inspections and tests that need to be completed).

7.3 Requirements from the Regulations

7.3.1 Agricultural and horticultural premises

In contrast to normal domestic installations, an agricultural installation is usually prone to damp conditions and so contact with earth will be better and people and animals are more liable to electric shock. For animals (whose body resistance is much lower than in humans), this situation is worsened as their contact with earth will be greater and so even a small voltage could prove lethal to them. Animals can also cause a lot of physical damage to electrical installations and animal effluents present a greater risk of corrosion. Horticultural installations are also subject to the same wet/high earth contact conditions and for these reasons, special requirements have been introduced for all agricultural or horticultural installation.

The Regulations list a number of requirements for fixed agricultural and horticultural installations (outdoors and indoors) and for locations where livestock is kept (e.g. cow sheds, stables, chicken houses, piggeries) plus food processing stations and storage areas for hay, straw and fertilisers.

The particular requirements of this section apply to fixed electrical installations indoors and outdoors in agricultural and horticultural premises. Some of the requirements are also applicable to other locations that are in common buildings belonging to the agricultural and horticultural premises.

If these premises include dwellings that are intended solely for human habitation, then the dwellings are excluded from the scope of these particular Regulations.

7.3.1.1 Accessibility by livestock

As the possibility of animals unintentionally coming in direct contact with a live installation is greater than for humans (e.g. they cannot read the warning notices!) and as livestock cannot be protected by earthed equipotential bonding and automatic disconnection (EEBAD) (because the voltages to which they would be subjected in the event of a fault would be unsafe for them) the following protective methods have to be used.

Electrical equipment generally shall be inaccessible to livestock.	WR-705.513.2
Equipment that is unavoidably accessible to livestock (such as equipment for feeding and basins for watering) shall be adequately constructed and installed to avoid damage by, and to minimise the risk of injury to, livestock.	WR-705.513.2

7.3.1.2 Automatic disconnection of supply

In circuits, whatever the type of earthing system, the following disconnection device shall be provided: • in final circuits supplying socket outlets with rated current not exceeding 32 A, an RCD with a rated residual operating current not exceeding 30 mA	WR-705.411.1

- in final circuits supplying socket outlets with rated current more than 32 A, an RCD with a rated residual operating current not exceeding 100 mA
- in all other circuits, RCDs with a rated residual operating current not exceeding 300 mA.

7.3.1.3 Conduit systems, cable trunking systems and cable ducting systems

For locations where livestock is kept, external influences shall be classified AF4, and conduits shall have protection against corrosion of at least Class 2 (medium) for indoor use and Class 4 (high protection) outdoors according to BS EN 61386-21.	WR-705.522.16

For locations where the wiring system may be exposed to impact and mechanical shock due to vehicles and mobile agricultural machines, etc, the external influences shall be classified AG3 and: WR-705.522.16

- conduits shall provide a degree of protection against impact of 5 J according to BS EN 61386-2;
- cable trunking and ducting systems shall provide a degree of protection against impact of 5 J according to BS EN 50085-2-1.

7.3.1.4 Electric fence controllers

Electric fencing systems have been developed to stop the free movement of animals across pasture. They are semipermanent solutions that can be extended, altered, or removed to allow grazing to be divided up and/or protected from livestock access.

The system consists of plastic or wooden posts, insulators, conductive wire/rope/tape and an electrical energiser unit. This unit can be mains, battery or solar powered and it sends short electrical impulses along a conductive wire (tape, rope, etc.) so that when the conductive wire or fence is touched by an animal the current passes through it to the ground and causes the animal to feel a shock. The shock is sufficient to alarm an animal but not to harm it and in time the animal will learn to stay away from it.

Specific requirements for electric fence installations: see BS EN 60335-2-BS 7671:2008 and BS EN 6100-1.

7.3.1.5 External influences

In agricultural or horticultural premises, electrical equipment shall have a minimum degree of protection of IP44, when used under normal conditions.	WR-705.512.2
Where equipment of IP44 rating is not available, it shall be placed in an enclosure complying with IP44.	WR-705.512.2
Socket outlets shall be installed in a position where they are unlikely to come into contact with combustible material.	WR-705.512.2
Where there are conditions of external influences >AD4, >AE3 and/or >AG1, socket outlets shall be provided with the appropriate protection.	WR-705.512.2
Protection may also be provided by the use of additional enclosures or by installation in building recesses.	WR-705.512.2

Note: These requirements do not apply to residential locations, offices, shops and locations with similar external influences belonging to agricultural and horticultural premises where (i.e. for socket outlets) BS 1363-2 or BS 546 applies.

Where corrosive substances are present (e.g. in dairies or cattle sheds) the electrical equipment shall be adequately protected.	WR-705.512.2

7.3.1.6 Extra-low voltage provided by SELV or PELV

Where SELV or PELV is used, whatever the nominal voltage, basic protection shall be provided by: • basic insulation; or • barriers or enclosures affording a degree of protection of at least IPXXB or IP2X.	WR-705.414.4.5

7.3.1.7 Identification

The following documentation shall be provided to the user of the installation:	WR-705.514.9.3

- a plan indicating the location of all electrical equipment
- the routing of all concealed cables
- a single-line distribution diagram
- an equipotential bonding diagram indicating locations of bonding connections.

7.3.1.8 Isolation and switching

The electrical installation of each building or part of a building shall be isolated by a single isolation device.	WR-705.537.2
Means of isolation of all live conductors, including the neutral conductor, shall be provided for circuits used occasionally (e.g. during harvest time).	WR-705.537.2
The isolation devices shall be clearly marked according to the part of the installation to which they belong.	WR-705.537.2
Devices for isolation and switching and devices for emergency stopping or emergency switching shall not be erected where they are accessible to livestock or in any position where access may be impeded by livestock.	WR-705.537.2

7.3.1.9 Luminaires and lighting installations

Luminaires shall comply with the BS EN 60598 series and be selected regarding their degree of protection against the ingress of dust, solid objects and moisture (e.g. 1P54), and suitability for mounting on:	WR-705.559

- a normally flammable surface (e.g. ▽F); and
- limited temperature of luminaire surface (e.g. ▽D).

7.3.1.10 Protection against electric shock

The protective measures of obstacles and placing out of reach are **not** permitted.	WR-705.410.3.5
The protective measures of non-conducting location and earth-free local equipotential bonding are **not** permitted.	WR-705.410.3.6

7.3.1.11 Protection against fire

Fire is a particular hazard in agricultural premises as normally there are large quantities of straw and other flammable material stored in these locations.

Electrical heating appliances used for the breeding and rearing of livestock shall comply with BS EN 60335-2-71 and shall be fixed so as to maintain an appropriate distance from livestock and combustible material, to minimise any risks of burns to livestock and of fire.	WR-705.422.6

 For radiant heaters the clearance shall be not less than 0.5 m or such other clearance as recommended by the manufacturer.

For fire protection purposes, RCDs shall be installed with a rated residual operating current not exceeding 300 mA.	WR-705.422.7
RCDs shall disconnect all live conductors.	WR-705.422.7
Where improved continuity of service is required, RCDs not protecting socket outlets shall be of the S type or have a time delay.	WR-705.422.7
In locations where a fire risk exists conductors of circuits supplied at extra-low voltage shall be protected:	WR-705.422.8

- either by barriers or enclosures affording a degree of protection of IPXXD or IP4X; or
- in addition to their basic insulation, by an enclosure of insulating material.

7.3.1.12 Safety services

For high-density livestock rearing, systems operating for the life support of livestock shall be taken into account as follows:
 	
• Where the supply of food, water, air and/or lighting to livestock is not ensured in the event of power supply failure, a secure source of supply shall be provided (e.g. an alternative or back-up supply) and separate final circuits for ventilation and lighting units shall be provided.
• Where electrically powered ventilation is necessary in an installation one of the following shall be provided:
  ○ a standby electrical source ensuring sufficient supply for ventilation equipment; or
  ○ temperature and supply voltage monitoring.
	WR-705.560.6

 Note: A notice should be placed adjacent to the standby electrical source, indicating that it should be tested periodically according to the manufacturer's instructions.

7.3.1.13 Selection and erection of wiring systems in relation to external influences

In locations accessible to, and enclosing, livestock, wiring systems shall be erected so that they are inaccessible to livestock or suitably protected against mechanical damage.	WR-705.522
Overhead lines shall be insulated.	WR-705.522
In areas of agricultural premises where vehicles and mobile agricultural machines are operated, the following methods of installation shall be applied: • Cables shall be buried in the ground at a depth of at least 0.6 m with added mechanical protection. • Cables in arable or cultivated ground shall be buried at a depth of at least 1 m. • Self-supporting suspension cables shall be installed at a height of at least 6 m.	WR-705.522

Special attention shall be given to the presence of different kinds of fauna (e.g. rodents).

7.3.1.14 Selection and erection of equipment:

Only electrical heating appliances with visual indication of the operating position shall be used.	WR-705.53

7.3.1.15 Socket outlets

Socket outlets of agricultural and horticultural premises shall comply with: • BS EN 60309-1; or • BS EN 60309-2 (when interchangeability is required); or • BS 1363, BS 546 or BS 196 (provided the rated current does not exceed 20 A).	WR-705.553.1

7.3.1.16 Supplementary bonding conductors

Protective bonding conductors shall be protected against mechanical damage and corrosion, and shall be selected to avoid electrolytic effects.	WR-705.544.2

7.3.1.17 Supplementary equipotential bonding

In locations intended for livestock, supplementary bonding shall connect all exposed conductive parts and extraneous conductive parts that can be touched by livestock.	WR-705.415.2.1
Where a metal grid is laid in the floor, it shall be included within the supplementary bonding of the location (Figure 705 on p. 187 of BS 7671:2008 shows an example of this).	WR-705.415.2.1
Extraneous conductive parts in, or on, the floor (e.g. concrete reinforcement in general or reinforcement of cellars for liquid manure) shall be connected to the supplementary equipotential bonding.	WR-705.415.2.1

Note: It is recommended that:

- spaced floors made of prefabricated concrete elements are part of the supplementary equipotential bonding;
- the supplementary equipotential bonding and the metal grid (if any) shall be erected so that it is durably protected against mechanical stresses and corrosion.

Where a metal grid is not laid in the floor, a TN-C-S supply is **not** recommended.

7.3.1.18 TN system

 A TN-C system shall **not** be used in the installation. WR-705.411.4

7.3.2 Conducting locations with restricted movement

A restrictive conductive location is one in which the surroundings consist mainly of metallic or conductive parts such as a large metal container or boiler. People employed inside these locations (e.g. a person working inside the boiler whilst using an electric drill or grinder) would have their freedom of movement physically restrained and a large proportion of their body would be in contact with the sides of that location and, therefore, prone to shock hazards.

The particular requirements of this section apply to:

- fixed equipment in conducting locations where movement of persons is restricted by the location; and
- supplies for mobile equipment for use in such locations.

7.3.2.1 Automatic disconnection of supply

If a functional earth is required for certain equipment (for example measuring and control equipment) equipotential bonding shall be provided between all exposed conductive parts and extraneous conductive parts inside the conducting location with restricted movement and the functional earth.   WR-706.411.1.2

7.3.2.2 Electrical separation

As a protective measure, the unearthed source shall have simple separation and shall be situated outside the conducting location with restricted movement.   WR-706.413.1.2

7.3.2.3 Extra-low voltage provided by SELV or PELV

A source for SELV or PELV shall be situated   WR-706.414.3(ii)
outside the conducting location with restricted
movement.

7.3.2.4 Protection against electric shock

The protective measures of obstacles and placing   WR-706.410.3.5
out of reach are **not** permitted.

In a conducting location with restricted movement   WR-706.410.3.10
the following protective measures apply to circuits
supplying the following current-using equipment:

- for the supply to a hand-held tool or an item
  of mobile equipment:
  - electrical separation
  - SELV.
- for the supply to handlamps:
  - SELV.
- for the supply to fixed equipment:
  - automatic disconnection of the supply with
    supplementary equipotential bonding; or
  - electrical separation; or
  - SELV; or
  - PELV.

7.3.2.5 Requirements for SELV and PELV circuits

Where SELV or PELV is used, whatever the nominal   WR-706.414.4.5
voltage, basic protection shall be provided by:

- basic insulation; or
- barriers or enclosures.

7.3.3 Construction and demolition site installations

Electrical installations at construction sites are there primarily to provide lighting
and power to enable work to proceed. As workmen will probably be working

ankle deep in wet, muddy conditions and using a selection of portable tools such as drills and grinders, they will be particularly susceptible to electric shock.

There are six levels of voltage normally associated with construction sites. These are:

25V single-line	for portable hand-lamps in damp and confined situations
50V single line centre-point earthed	for hand lamps in damp and confined situations
400V three line	for use with fixed or transportable equipment with a load of more than 3750 watts
230V single line	for site buildings and fixed lighting
110V three line	for transportable equipment with a load up to 3750 watts
110V single line	for transportable tools and equipment, such as floodlighting

Equipment used must be suitable for the particular supply to which it is connected and for the application it will meet on site. Where more than one voltage is in use, plugs and sockets must be non-interchangeable to prevent misconnection

Supplies will normally be obtained from the electrical supply company but remote sites could need an IT supply (such as a generator) and care must be taken in complying with the safety requirements for this particular source.

The following requirements apply to temporary installations for construction and demolition sites during the period of the construction or demolition work, including, for example:

- construction work of new buildings;
- repair, alteration, extension, demolition of existing buildings or parts of existing buildings;
- engineering works;
- earthworks;
- work of similar nature;

and are applicable to:

- the main switchgear and protective devices;
- installations of mobile and transportable electrical equipment;
- the interface between the supply system and the construction site installations.

They do **not** apply to:

- installations covered by the IEC 60621 series 2, where equipment of a similar nature to that used in surface mining applications is involved;
- installations in administrative locations of construction sites (e.g. offices, cloakrooms, meeting rooms, canteens, restaurants, dormitories, toilets).

7.3.3.1 Automatic disconnection of supply

A TN-C-S system shall **not** be used for the supply to a construction site, except for the supply to a fixed building of the construction site.

WR-704.411.3.1

7.3.3.2 Automatic disconnection in the case of a fault

For any circuit supplying one or more socket outlets with a rated current exceeding 32 A, an RCD having a rated residual operating current not exceeding 500 mA shall be provided.

WR-704.411.3.2.1

7.3.3.3 Extra-low voltage provided by SELV or PELV

Where SELV or PELV is used, whatever the nominal voltage, basic protection shall be provided by:

WR-704.414.4.5

- basic insulation; or
- barriers or enclosures.

7.3.3.4 Isolation devices

Each Assembly for Construction Sites (ACS) shall incorporate suitable devices for the switching and isolation of the incoming supply.

WR-704.537.2.2

A device for isolating the incoming supply shall be suitable for securing in the OFF position (e.g. by padlock or location of the device inside a lockable enclosure).

WR-704.537.2.2

Current-using equipment shall be supplied by ACSs comprising:

WR-704.537.2.2

- overcurrent protective devices; and
- devices affording fault protection; and
- socket outlets, if required.

Safety and standby supplies shall be connected by means of devices arranged to prevent interconnection of the different supplies.

WR-704.537.2.2

7.3.3.5 Protection against electric shock

 The protective measures of obstacles and placing out of reach are **not** permitted.

WR-704.410.3.5

A circuit supplying a socket outlet with a rated current up to and including 32 A and any other circuit supplying hand-held electrical equipment with rated current up to and including 32 A shall be protected by:

WR-704.410.3.10

- reduced low voltage; or
- automatic disconnection of supply with additional protection provided by an RCD; or
- electrical separation of circuits (where each socket outlet and item of hand-held electrical equipment is supplied by an individual transformer or by a separate winding of a transformer); or
- SELV or PELV.

7.3.3.6 Selection and erection of equipment

All assemblies on construction and demolition sites for the distribution of electricity shall be in compliance with the requirements of BS EN 60439-4.

WR-704.511.1

A plug or socket outlet with a rated current equal to or greater than 16 A shall comply with the requirements of BS EN 60309.

WR-704.511.1

7.3.3.7 Supplies

Equipment shall be identified with (and be compatible with) the particular supply from which it is energised and shall contain only components connected to one and the same installation, **except** for control or signalling circuits and inputs from standby supplies.

WR-704.313.3

7.3.3.8 Wiring systems

Cable shall **not** be installed across a site road or a walkway unless adequate protection of the cable against mechanical damage is provided.	WR-704.522.8.10
For reduced low-voltage systems, low temperature 300/500 V thermoplastic (BS 7919) or equivalent flexible cables shall be used.	WR-704.522.8.11
For applications exceeding reduced low voltage, flexible cable shall be HO7RN-F (BS 7919) type or equivalent having 450/750 V rating and resistant to abrasion and water.	WR-704.522.8.11

7.3.4 Electrical installations in caravan/camping parks and similar locations

Note: In order not to mix regulations on different subjects, such as those for electrical installation of caravan parks with those for electrical installation inside caravans, two sections have been created:

- Section 721, concerning electrical installations in caravans and motor caravans; and
- Section 708, concerning electrical installations in caravan parks, camping parks and similar locations.

The requirements of the latter section, therefore, apply to that portion of the electrical installation in caravan/camping parks and similar locations providing facilities for supplying leisure accommodation vehicles (including caravans) or tents. They do **not** apply to the internal electrical installations of leisure accommodation vehicles or mobile or transportable units.

The requirements of the Electricity Supply Regulations do not allow the supply neutral to be connected to any metalwork in a caravan (which means that only TT or TN-S systems may be used) and in general, the supply of electrical energy to caravan and tent sites must ensure that:

- wherever possible the supply is via underground cables;
- overhead supplies use insulated as opposed to bare cables;
- cables are installed outside the area of the caravan pitch and are at least 3.5 m above ground level (increased to 6 m where vehicle movements are possible);

- each socket must have its own individual overcurrent protection in the form of a fuse or circuit-breaker;
- cables that are run below caravan pitches must be provided with additional protection as shown in Figure 7.2;

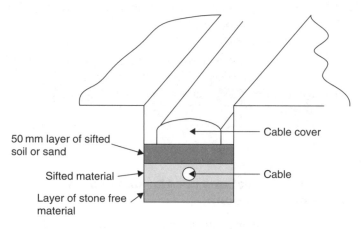

50 mm layer of sifted soil or sand

Sifted material

Layer of stone free material

Cable cover

Cable

Figure 7.2 Cable covers

- all sockets must be protected by an RCD complying with BS 4293, BS EN 61008-1 or BS EN 61009-1 with a 30 mA rating, either individually or in groups not exceeding three sockets.

The actual requirements (which only apply to installations supplying electricity to leisure accommodation vehicles in caravan/camping parks and similar locations) are as follows.

7.3.4.1 External influences

Electrical equipment installed outside in caravan parks shall comply at least with the following external influences: WR-708.512.2

- presence of water: AD4 (splashes), IPX4 in accordance with BS EN 60529
- presence of foreign solid bodies: AE2 (small objects), IP3X in accordance with BS EN 60529
- mechanical stress: AG3 (high severity), IK08 in accordance with BS EN 62262.

7.3.4.2 Overhead distribution circuits

All overhead conductors shall be insulated.	WR-708.521.1.2
Poles and other supports for overhead wiring shall be located or protected so that they are unlikely to be damaged by any foreseeable vehicle movement.	WR-708.521.1.2
Overhead conductors shall be at a height above ground of not less than 6 m in all areas subject to vehicle movement and 3.5 m in all other areas.	WR-708.521.1.2

7.3.4.3 Plug and socket outlets

Each socket outlet and its enclosure forming part of the caravan pitch electrical supply equipment shall comply with BS EN 60309-2 and meet the degree of protection of at least IP44 in accordance with BS EN 60529.	WR-708.553.1.8
The socket outlets shall be placed at a height of 0.5 m to 1.5 m from the ground to the lowest part of the socket outlet. In special cases, due to environmental conditions such as risk of flooding or heavy snowfall, the maximum height is permitted to exceed 1.5 m.	WR-708.553.1.9
The current rating of socket outlets shall be not less than 16 A.	WR-708.553.1.10
At least one socket outlet shall be provided for each caravan pitch.	WR-708.553.1.11
Each socket outlet shall be provided with individual overcurrent protection.	WR-708.553.1.12
Each socket outlet shall be protected individually by an RCD.	WR-708.553.1.13
Socket outlet protective conductors shall not be connected to any PEN conductor of the electricity supply.	WR-708.553A.1.14

7.3.4.4 Protection against electric shock

	The protective measures of obstacles and placing out of reach are **not** permitted.	WR-708.410.3.5

The protective measures of non-conducting location and earth-free local equipotential bonding are **not** permitted. WR-708.410.3.6

7.3.4.5 Supply voltage

The nominal supply voltage of the installation for the supply of leisure accommodation vehicles shall be 230 V a.c. single-line or 400 V a.c. three-line. WR-708.313.1.2

7.3.4.6 Switchgear and controlgear

Caravan pitch electrical supply equipment shall be located adjacent to the pitch and not more than 20 m from the connection facility on the leisure accommodation vehicle or tent when on its pitch. WR-708.530.3

No more than four socket outlets should be grouped in one location, in order to avoid the supply cable crossing a pitch other than the one intended to be supplied.

7.3.4.7 TN system

In the UK where the installation is supplied from a TN system, only a TN-S installation shall be installed. WR-708.411.4

In the UK the ESQCR (Electricity Safety, Quality and Continuity Regulations 2002) prohibit the use of a TN-C-S system for the supply to a caravan or similar construction.

7.3.4.8 Underground distribution circuits

Underground cables shall be buried at a depth of at least 0.6 m and, unless having additional mechanical protection, be placed outside any caravan pitch or away from any surface where tent pegs or ground anchors are expected to be present. WR-708.521.1.1

Unless mechanically protected, all underground cables **must** be installed outside of the caravan pitch and areas where tent pegs or ground anchors may be driven.

The following wiring systems are suitable for distribution circuits feeding caravan or tent pitch electrical supply equipment: • underground distribution circuits • overhead distribution circuits.	WR-708.521.1

 Note: The preferred method of supply is by means of underground distribution circuits

7.3.5 Electrical installations in caravans and motor caravans

Caravans and motor caravans are designed as leisure accommodation vehicles which are either towed (e.g. by a car) or self-propelled to a caravan site. They will often contain a bath or a shower and special requirements for such installations will apply. In addition to the normal dangers associated with fixed electrical installations, there is the potential hazard of totally unskilled people moving the caravan/motor caravan, connecting and disconnecting the mains supply, and ensuring that it is correctly earthed.

 Caravans that are used as mobile workshops will **also** be subject to the requirements of the Electricity at Work Regulations 1989, and locations containing baths or showers for medical treatment, or for disabled persons, may have additional special requirements.

The Regulations include a number of special requirements for caravans and mobile homes (as listed below) but these do not deal with:

• electrical circuits and equipment covered by the Road Vehicles Lighting Regulations 1989;
• installations covered by BS EN 1648-1; for 12 V d.c. extra-low-voltage installations in leisure accommodation vehicles.-1 and BS EN 1648-2.

 Note: A mobile home is defined as a 'transportable leisure accommodation vehicle that does not meet the requirements for use as a road vehicle' and is usually a permanent fixture on a caravan park. It normally has recognised power supplies and earthing and the internal electrical installation is outside the scope of the Regulations. Nevertheless, potential safety hazards still have to be considered.

7.3.5.1 Accessories

Every low-voltage socket outlet, other than those supplied by an individual winding of an isolating transformer, shall incorporate an earth contact.	WR-721.55.2.1
Every socket outlet supplied at extra-low voltage shall have its voltage visibly marked.	WR-721.55.2.2
Where an accessory is located in a position in which it is exposed to the effects of moisture it shall be constructed or enclosed so as to provide a degree of protection not less than IP44.	WR-721.55.2.3
Each luminaire in a caravan shall preferably be fixed directly to the structure or lining of the caravan.	WR-721.55.2.4
Where a pendant luminaire is installed in a caravan, provision shall be made for securing the luminaire to prevent damage when the caravan is in motion.	WR-721.55.2.4
Accessories for the suspension of pendant luminaires shall be suitable for the mass suspended and the forces associated with vehicle movement.	WR-721.55.2.4
A luminaire intended for dual voltage operation shall comply with the appropriate standard.	WR-721.55.2.5
The means of connection to the caravan pitch socket outlet shall be supplied with the caravan and shall comprise the following:	WR-721.55.2.6
a plug complying with BS EN 60309-2; and	
a flexible cord or cable of 25 m (± 2 m) length incorporating a protective conductor.	

 Annex A to BS 7671:2008 provides guidance for extra-low-voltage d.c. installations.

7.3.5.2 Cross-sectional areas of conductors of cables

The cross-sectional area of every conductor shall be not less than 1.5 mm².	WR-721.524.1

7.3.5.3 Identification

Instructions for use shall be provided with the caravan so that the caravan can be used safely. The instructions shall comprise: • a description of the installation • a description of the function of the RCD(s) and the use of the test button(s) • a description of the function of the main isolating switch.	WR-721.514.1

Note: The text for these instructions can be found in Figure 721 on p. 211 of BS 7671:2008.

7.3.5.4 Electrical inlets

Any a.c. electrical inlet on the caravan shall be an appliance inlet complying with BS EN 60309.	WR-721.55.1.1
The inlet shall be installed: • not more than 1.8 m above ground level; and • in a readily accessible position; and • such that it shall have a minimum protection of IP44 with or without a connector engaged; and • such that it shall not protrude significantly beyond the body of the caravan.	WR-721.55.1.2

7.3.5.5 External influences – vibration (AH)

As the wiring will be subjected to vibration, all wiring shall be protected against mechanical damage either by location or by enhanced mechanical protection.	WR-721.522.7.1
Wiring passing through metalwork shall be protected by means of suitable bushes or grommets, securely fixed in position.	WR-721.522.7.1
Precautions shall be taken to avoid mechanical damage due to sharp edges or abrasive parts.	WR-721.522.7.1

7.3.5.6 Isolation

Each installation shall be provided with a main disconnector which shall disconnect all live conductors and which shall be suitably placed for ready operation within the caravan.	WR-721.537.2.1.1
In an installation consisting of only one final circuit, the isolating switch may be the overcurrent protective device fulfilling the requirements for isolation.	WR-721.537.2.1.1
A notice shall be permanently fixed near the main isolating switch inside the caravan, bearing the text shown in Figure 721 on p. 211of BS 7671:2008.	WR-721.537.2.1.1.1

7.3.5.7 Mechanical stresses (AJ)

All cables, unless enclosed in rigid conduit and all flexible conduit, shall be supported at intervals not exceeding 0.4 m for vertical runs and 0.25 m for horizontal runs.	WR-721.522.8.1.3

7.3.5.8 Protection against electric shock

The protective measure electrical separation is **not** permitted (with the exception of shaver socket outlets).	WR-721.410.3.3.2
The protective measures of obstacles and placing out of reach are **not** permitted.	WR-721.410.3.5
The protective measures of non-conducting location and earth-free local equipotential bonding are **not** permitted.	WR-721.410.3.6

7.3.5.9 Protection against overcurrent – final circuits

Each final circuit shall be protected by an overcurrent protective device which disconnects all live conductors of that circuit.	WR-721.43.1

7.3.5.10 Protective conductors

All protective conductors shall be incorporated in a multicore cable or in a conduit together with the live conductors.

WR-721.543.2.1

7.3.5.11 Protective equipotential bonding

Structural metallic parts which are accessible from within the caravan shall be connected through main protective bonding conductors to the main earthing terminal within the caravan.

WR-721.411.3.1.2

7.3.5.12 Protective measure – automatic disconnection of supply

Where protection by automatic disconnection of supply is used, an RCD shall be used and the wiring system shall include a circuit protective conductor which shall be connected to:

WR-721.411.1

- the protective contact of the inlet; and
- the exposed conductive parts of the electrical equipment; and
- the protective contacts of the socket outlets.

7.3.5.13 Proximity of wiring systems to other services

Cables of low-voltage systems shall be run separately from the cables of extra-low-voltage systems, so that there is no risk of physical contact between the two wiring systems.

WR-721.528.1

No electrical equipment including wiring systems, except ELV equipment for gas supply control, shall be installed in any gas cylinder storage compartment.

WR-721.528.3.5

Where cables have to run through such a compartment, they shall be run at a height of less than 500 mm above the base of the cylinder(s), and such cables shall be protected against mechanical damage by installation within a continuous gas-tight conduit or duct passing through the compartment.

WR-721.528.3.5

Where installed, this conduit or duct shall be able to withstand an impact equivalent to AG3 without visible physical damage.	WR-721.528.3.5

7.3.5.14 Selection and erection of equipment

Where there is more than one electrically independent installation, each independent installation shall be supplied by a separate connecting device and shall be segregated in accordance with the relevant requirements of the Regulations.	WR-721.510.3

7.3.5.15 Supplies

The nominal supply system voltage shall be chosen from IEC 60038.	WR-721.313.1.2
The nominal a.c. supply voltage of the installation of the caravan shall not exceed 230 V single-line, or 400 V three-line.	WR-721.313.1.2
The nominal d.c. supply voltage of the installation of the caravan shall not exceed 48 V.	WR-721.313.1.2

7.3.5.16 TN system

Note: In the UK the ESQCR prohibit the use of a TN-C-S system for the supply to a caravan.

7.3.5.17 Types of wiring system

The wiring systems shall be installed using one or more of the following: • insulated single-core cables, with flexible Class 5 conductors, in non-metallic conduit • insulated single-core cables, with stranded Class 2 conductors (minimum of 7 strands), in non-metallic conduit • sheathed flexible cables.	WR-721.521.2
All cables shall, as a minimum, meet the requirements of BS EN 60332-1-2.	WR-721.521.2

Non-metallic conduits shall comply with BS EN   WR-721.521.2
61386-21.

Cable management systems shall comply with BS   WR-721.521.2
EN 61386.

7.3.6 Exhibitions, shows and stands

The particular requirements of this section apply to the temporary electrical installations in exhibitions, shows and stands (including mobile and portable displays and equipment) to protect users.

7.3.6.1 Classification of external influences

The external-influence conditions of the particular   WR-711.32
location where the temporary electrical installation
is erected (e.g. the presence of water or mechanical
stresses) shall be taken into account.

7.3.6.2 Electrical connections

Joints shall not be made in cables except where   WR-711.526.1
necessary as a connection into a circuit. Where
joints **are** made, these shall either use connectors in
accordance with relevant standards or be in enclosures
with a degree of protection of at least IP4X or IPXXD.

Where strain can be transmitted to terminals the   WR-711.526.1
connection shall include (i.e. incorporate) suitable
cable anchorage(s).

7.3.6.3 Electric discharge lamp installations

Installations of any luminous tube, sign or lamp as   WR-711.559.4.4
an illuminated unit on a stand, or as an exhibit, with
nominal power supply voltage higher than 230/400 V
a.c., shall comply with the following:

Location – the sign or lamp shall be installed out of   WR-711.559.4.5
arm's reach or shall be adequately protected to reduce
the risk of injury to persons.

Installation – the fascia or stand fitting material behind luminous tubes, signs or lamps shall be non-ignitable. — WR-711.559.4.6

Emergency switching devices – a separate circuit shall be used to supply signs, lamps or exhibits, which shall be controlled by an emergency switch. — WR-711.559.4.7

The switch shall be easily visible, accessible and clearly marked.

7.3.6.4 Electric motors – isolation

Where an electric motor might give rise to a hazard, the motor shall be provided with an effective means of isolation on all poles and such means shall be adjacent to the motor which it controls (see BS EN 60204-1). — WR-711.55.4

7.3.6.5 ELV transformers and electronic converters

ELV (extra-low voltage) transformers shall be mounted out of arm's reach of the public and shall have adequate ventilation. — WR-711.55.6

A manual reset protective device shall protect the secondary circuit of each transformer or electronic convertor. — WR-711.55.6

Access by a competent person for testing and by a skilled person competent in such work for maintenance shall be provided. — WR-711.55.6

Electronic converters shall conform with BS EN 61347-1. — WR-711.55.6

7.3.6.6 Extra-low voltage provided by SELV or PELV

Where SELV or PELV is used, whatever the nominal voltage, basic protection shall be provided by: — WR-711.414.4.5

- basic insulation; or
- by barriers or enclosures affording a degree of protection of at least IP4X or IPXXD.

7.3.6.7 Isolation

Every separate temporary structure, such as a
vehicle, stand or unit, intended to be occupied
by one specific user, and each distribution circuit
supplying outdoor installations, shall be provided
with its own readily accessible and properly
identifiable means of isolation.

WR-711.537.2.3

7.3.6.8 Lampholders

 Insulation piercing lampholders shall **not**
be used unless the cables and lampholders
are compatible, and provided the
lampholders are non-removable once
fitted to the cable.

WR-711.559.4.3

7.3.6.9 Luminaires and lighting installations

ELV lighting systems for filament lamps shall
comply with BS EN 60598-2-23.

WR-711.559.4.2

7.3.6.10 Other equipment

All such equipment shall be so fixed and protected that a focusing or concentration of heat is not likely to cause ignition of any material.

7.3.6.11 Protection against electric shock

A cable intended to supply temporary structures
shall be protected at its origin by an RCD whose
rated residual operating current does not exceed
300 mA.

WR-711.410.3.4

The protective measures of obstacles and placing
out of reach are **not** permitted.

WR-711.410.3.5

The protective measures of non-conducting location
and earth-free local equipotential bonding are **not**
permitted.

WR-711.410.3.6

7.3.6.12 Protection against fire and heat generation

Lighting equipment such as incandescent lamps, spotlights and small projectors and other equipment or appliances with high-temperature surfaces, shall be suitably guarded, and installed and located in accordance with the relevant standard.	WR-711.422.4.2
Showcases and signs shall be constructed of material having an adequate heat-resistance, mechanical strength, electrical insulation and ventilation, taking into account the combustibility of exhibits in relation to the heat generation.	WR-711.422.4.2
Stand installations containing a concentration of electrical equipment, luminaires or lamps liable to generate excessive heat shall not be installed unless adequate ventilation provisions are made, e.g. well-ventilated ceiling constructed of incombustible material.	WR-711.422.4.2

7.3.6.13 Protection against thermal effects

Luminaires mounted below 2.5 m (arm's reach) from floor level or otherwise accessible to accidental contact shall be firmly and adequately fixed, and so sited or guarded as to prevent risk of injury to persons or ignition of materials.	WR-711.559.5

7.3.6.14 Protection by RCDs

Each socket outlet circuit not exceeding 32 A and all final circuits other than for emergency lighting shall be protected by an RCD.	WR-711.411.3.3

7.3.6.15 Protective equipotential bonding

Structural metallic parts which are accessible from within the stand, vehicle, wagon, caravan or container shall be connected through the main protective bonding conductors to the main earthing terminal within the unit.	WR-711.411.3.1.2

7.3.6.16 Selection and erection of equipment

Switchgear and controlgear shall be placed in closed cabinets which can only be opened by the use of a key or a tool, except for those parts designed and intended to be operated by ordinary persons.	WR-711.51

7.3.6.17 Socket outlets and plugs

An adequate number of socket outlets shall be installed to allow user requirements to be met safely.	WR-711.55.7
Where a floor-mounted socket outlet is installed, it shall be adequately protected from accidental ingress of water and have sufficient strength to be able to withstand the expected traffic load.	WR-711.55.7

7.3.6.18 Supplies

The nominal supply voltage of a temporary electrical installation in an exhibition, show or stand shall not exceed 230/400 V a.c. or 500 V d.c.	WR-711.313

7.3.6.19 TN system

In the UK where the type of system earthing is TN, the installation shall be TN-S.	WR-711.411.4

7.3.6.20 Wiring systems

	Flexible cords shall **not** be laid in areas accessible to the public unless they are protected against mechanical damage.	WR-711.52

Armoured cables or cables protected against mechanical damage shall be used wherever there is a risk of mechanical damage.	WR-711.52

Wiring cables shall be copper, have a minimum
cross-sectional area of $1.5 \, \text{mm}^2$, and shall comply
with an appropriate British Standard for either
thermoplastic or thermosetting insulated electric
cables.

WR-711.52

Types of wiring system

Where no fire alarm system is installed in a building
used for exhibitions etc., cable systems shall be
either:

- flame retardant to BS EN 60332-1-2 and low
  smoke to BS EN 61034-2; or
- single-core or multicore unarmoured cables
  enclosed in metallic or non-metallic conduit or
  trunking, providing a degree of fire protection
  of at least IP4X.

WR-711.52

7.3.7 Floor and ceiling heating systems

This section applies to the installation of electric floor and ceiling heating systems which are erected as either thermal storage heating systems or direct heating systems. It does not apply to the installation of wall heating systems.

7.3.7.1 Ambient temperature (AA)

For cold tails (circuit wiring) and control leads
installed in the zone of heated surfaces, the increase
of ambient temperature shall be taken into account.

WR-753.522.1.3

7.3.7.2 Automatic disconnection of supply

RCDs with a rated residual operating current not
exceeding 30 mA shall be used as disconnecting
devices.

WR-753.411.3.2

7.3.7.3 Compliance with standards

Flexible sheet heating elements shall comply with   WR-753.511
the requirements of BS EN 60335-2-96.

Heating cables shall comply with BS 6351   WR-753.511
series.

7.3.7.4 Electrical separation

 The protective measure of electrical separation   WR-753.413.1.2
is **not** permitted.

7.3.7.5 External influences

Heating units for installation in ceilings shall   WR-753.512.2.5
have a degree of protection of not less than
IPX1.

Heating units for installation in a floor of concrete   WR-753.512.2.5
or similar material shall have a degree of ingress
protection not less than IPX7 and shall have the
appropriate mechanical properties.

7.3.7.6 Heating-free areas

For the necessary attachment of room fittings,   WR-753.520.4
heating-free areas shall be provided in such a way
that the heat emission is not prevented by such
fittings.

7.3.7.7 Heating units

To avoid the overheating of floor or ceiling heating   WR-753.424.1.1
systems in buildings, one or more of the following
measures shall be applied within the zone where

heating units are installed to limit the temperature
to a maximum of 80°C:

- appropriate design of the heating system
- appropriate installation of the heating system
  in accordance with the manufacturer's
  instructions
- use of protective devices.

Heating units shall be connected to the electrical installation via cold tails or suitable terminals.	WR-753.424.1.1
Heating units shall be inseparably connected to cold tails, for example, by a crimped connection.	WR-753.424.1.1
As the heating unit may cause higher temperatures or arcs under fault conditions, special measures should be taken when the heating unit is installed close to easily ignitable building structures, such as placing on a metal sheet, in metal conduit or at a distance of at least 10 mm in air from the ignitable structure.	WR-753.424.1.2

7.3.7.8 Identification and notices

The designer of the installation/heating system or   WR-753.514
installer shall provide a plan for each heating
system, containing the following details:

- manufacturer and type of heating
- number of heating units installed
- length/area of heating units
- rated power
- surface power density
- layout of the heating units in the form of a
  sketch, a drawing, or a picture
- position/depth of heating units
- position of junction boxes
- conductors, shields and the like
- heated area
- rated voltage
- rated resistance (cold) of heating units
- rated current of overcurrent protective device
- rated residual operating current of RCD

- insulation resistance of the heating installation and test voltage used
- leakage capacitance.

This plan shall be fixed to, or adjacent to, the distribution board of the heating system.   WR-753.514

7.3.7.9 Operational conditions

Precautions shall be taken not to stress the heating unit mechanically (for example, the material by which it is to be protected in the finished installation shall cover the heating unit as soon as possible).   WR-753.512.1.6

7.3.7.10 Presence of solid foreign bodies (AE)

Where heating units are installed there shall be heating-free areas where drilling and fixing by screws, nails and the like are permitted.   WR-753.522.4.3

The installer shall inform other contractors that no penetrating means, such as screws for door stoppers, shall be used in the area where floor or ceiling heating units are installed.   WR-753.522.4.3

7.3.7.11 Prevention of mutual detrimental influences

Heating units shall not cross expansion joints of the building or structure.   WR-753.515.4

7.3.7.12 Protection against burns

In floor areas where contact with skin or footwear is possible, the surface temperature of the floor shall be limited (for example, 35°C).   WR-753.423

7.3.7.13 Protection against electric shock

The protective measures of obstacles and placing out of reach are **not** permitted.	WR-753.410.3.5
The protective measures of non-conducting location and earth-free local equipotential bonding are **not** permitted.	WR-753.410.3.6

7.3.7.14 Protection by RCDs

A circuit supplying heating equipment of Class II construction or equivalent insulation shall be provided with additional protection by the use of an RCD.	WR-753.415.1

7.3.8 Locations containing a bath or shower

All locations containing a bath or shower shall be inspected to confirm that they comply with Part 701 of the Regulations.

When people use bathrooms and showers, most of the time they are naturally unclothed and wet and thus very vulnerable to electric shock due to their reduced body resistance (i.e. absence of shoes means less protection from shock, whilst water on their skin will tend to short-circuit its natural protection). Special measures are, therefore, required to ensure that the possibility of direct and/or indirect contact is reduced.

The following requirements apply to locations containing baths, showers and cabinets (containing a shower and/or bath) and the surrounding zones. They do **not** apply to emergency facilities in industrial areas and laboratories.

The requirements are based on three zones which take into account the limitations of walls, doors, fixed partitions, ceilings and floors. The three zones are as shown in Figure 7.3 and Table 7.1

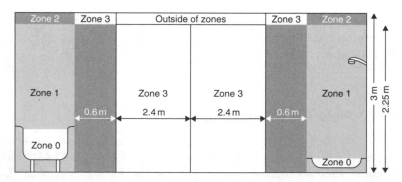

Figure 7.3 Zone dimensions

Table 7.1 Limitations of zones

Zone 0	Is the interior of the bath tub or shower basin
	Note: for showers without a basin, the height of zone 0 is 0.10 m and its surface extent has the same horizontal extent as zone 1
Zone 1	Zone 1 does not include zone 0 and is limited by:

- the finished floor level and the horizontal plane corresponding to the highest fixed shower head or water outlet or the horizontal plane lying 2.25 m above the finished floor level (whichever is higher)
- the vertical surface:
  circumscribing the bath tub or shower basin
  at a distance of 1.20 m from the centre point of the fixed water outlet on the wall or ceiling for showers without a basin.

 Note: The space under the bath tub or shower basin is considered to be zone 1. However, if the space under the bath tub or shower basin is only accessible with a tool, it is considered to be outside the zones.

Zone 2   Zone 2 is limited by:

- the finished floor level and the horizontal plane corresponding to the highest fixed shower head or water outlet or the horizontal plane lying 2.25 m above the finished floor level (whichever is higher)
- the vertical surface at the boundary of zone 1 and the parallel vertical surface at a distance of 0.60 m from the zone 1 border.

 Note: For showers without a basin, there is no zone 2 but an increased zone 1 is provided by the horizontal dimension of 1.20 m.

 For locations containing a bath or shower for medical treatment, special requirements may be necessary.

7.3.8.1 General characteristics

> Horizontal or inclined ceilings, walls with or without   WR-701.3
> windows, doors, floors and fixed partitions may be taken
> into account where these effectively limit the extent of
> locations containing a bath or shower as well as their zones.

7.3.8.2 Current-using equipment

In zone 0, current-using equipment shall **only** be installed provided that all the following requirements are met:

- The equipment complies with the relevant standard and is suitable for use in that zone according to the manufacturer's instructions for use and mounting.

- The equipment is fixed and permanently connected.
- The equipment is protected by SELV at a nominal voltage not exceeding 12 V a.c. rms or 30 V ripple-free d.c., the safety source being installed outside zones 0, 1 and 2.

In zone 1, the following fixed and permanently connected current-using equipment shall **only** be installed, provided it is suitable for installation in zone 1 according to the manufacturer's instructions:

- whirlpool units;
- electric showers;
- shower pumps;
- equipment protected by SELV or PELV at a nominal voltage not exceeding 25 V a.c. rms or 60 V ripple-free d.c., the safety source being installed outside zones 0, 1 and 2;
- ventilation equipment;
- towel rails;
- water heating appliances;
- luminaires.

7.3.8.3 Electric floor heating systems

For electric floor heating systems the protective measure 'protection by electrical separation' is **not** permitted.	WR-701.753

For electric floor heating systems, only heating cables or thin sheet flexible heating elements shall be installed **provided** that they have either a metal sheath or a metal enclosure or a fine mesh metallic grid.	WR-701.753
The fine mesh metallic grid, metal sheath or metal enclosure shall be connected to the protective conductor of the supply circuit.	WR-701.753

Compliance with the latter requirement is not required if the protective measure SELV is provided for the floor heating system.

7.3.8.4 Erection of switchgear, controlgear and accessories according to external influences

Note: The following requirements do not apply to switches and controls which are incorporated in fixed current-using equipment suitable for use in that zone or to insulating pull cords of cord operated switches.

In zone 0:

- switchgear or accessories shall **not** be installed.

In zone 1:

- only switches of SELV circuits supplied at a nominal voltage not exceeding 12 V a.c. rms or 30 V ripple-free d.c. shall be installed, the safety source being installed outside zones 0, 1 and 2.

In zone 2:

- switchgear, accessories incorporating switches or socket outlets shall not be installed with the exception of:
  - switches and socket outlets of SELV circuits, the safety source being installed outside zones 0, 1 and 2; and
  - shaver supply units complying with BS EN 61558-2-5.

Except for SELV socket outlets and shaver supply units complying with BS EN 61558-25, socket outlets are **prohibited** within a distance of 3 m horizontally from the boundary of zone 1.   WR-701.512.3

7.3.8.5 Extra-low voltage provided by SELV or PELV

Where SELV or PELV is used, whatever the nominal voltage, basic protection for equipment in zones 0, 1 and 2 shall be provided by: WR-701.414.4.5

- basic insulation; or
- barriers or enclosures affording a degree of protection of at least IPXXB or IP2X.

7.3.8.6 Protection by RCDs

Additional protection shall be provided to all circuits is the location, by the use of one or more RCDs.   WR-701.411.3.3

7.3.8.7 Protection for safety: protection against electric shock

 The protective measures of obstacles and placing out of reach are **not** permitted.   WR-701.410.3.5

The protective measures of non-conducting location and earth-free local equipotential bonding are **not** permitted.

WR-701.410.3.6

7.3.8.8 Protective measure – electrical separation

Protection by electrical separation shall **only** be used for:

WR-701.413

- circuits supplying one item of current-using equipment; or
- one single socket outlet.

7.3.8.9 Selection and erection of equipment

Installed electrical equipment shall have at least the following degrees of protection:

WR-701.512.2

- in zone 0: IPX7
- in zones 1 and 2: IPX4.

This requirement does not apply to shaver supply units complying with BS EN 61558-2-5 installed in zone 2 and located where direct spray from showers is unlikely.

WR-701.512.2

 Electrical equipment exposed to water jets (for example, for cleaning purposes) shall have a degree of protection of at least IPX5.

7.3.8.10 Supplementary equipotential bonding

Local supplementary equipotential bonding shall be established connecting together the terminals of the protective conductor of each circuit supplying Class I and Class II equipment to the accessible extraneous conductive parts, within a room containing a bath or shower, including the following:

WR-701.415.2

- metallic pipes supplying services and metallic waste pipes (e.g. water, gas)
- metallic central heating pipes and air conditioning systems

- accessible metallic structural parts of the building (metallic door architraves, window frames and similar parts are not considered to be extraneous conductive parts unless they are connected to metallic structural parts of the building).

Supplementary equipotential bonding may be installed outside or inside rooms containing a bath or shower, preferably close to the point of entry of extraneous conductive parts into such rooms.

WR-701.415.2

Where the location containing a bath or shower is in a building with a protective equipotential bonding system, supplementary equipotential bonding may be omitted where all of the following conditions are met:

WR-701.415.2

- all final circuits of the location comply with the requirements for automatic disconnection
- all final circuits of the location have additional protection by means of an RCD
- all extraneous conductive parts of the location are effectively connected to the protective equipotential bonding.

 Special requirements exist for locations containing baths and showers for medical use or for disabled persons (see Building Regulations Part M).

7.3.9 Marinas and similar locations

The particular requirements of this section are applicable only to circuits intended to supply pleasure craft or houseboats in marinas and similar locations. They do not apply to the supply to houseboats if they are directly supplied from the public network or to the internal electrical installations of pleasure craft or houseboats.

7.3.9.1 External influences

For marinas, particular attention shall be given to the likelihood of corrosive elements, movement of structures, mechanical damage, presence of flammable fuel and the increased risk of electric shock due to:

WR-709.512.2

- presence of water
- reduction in body resistance
- contact of the body with earth potential.

7.3.9.2 Impact (AG)

Equipment installed on or above a jetty, wharf, pier or pontoon shall be protected against mechanical damage (impact of medium severity AG2). Protection shall be afforded by one or more of the following:	WR-709.512.2.1.4

- the position or location selected to avoid damage by any reasonably foreseeable impact
- the provision of local or general mechanical protection
- installing equipment complying with a minimum degree of protection for external mechanical impact IK08 (see BS EN 62262).

7.3.9.3 Isolation

At least one means of isolation (with a maximum of four outlet sockets) shall be installed in each distribution cabinet.	WR-709.537.2.1.1
This switching device shall disconnect all live conductors including the neutral conductor.	WR-709.537.2.1.1

7.3.9.4 Overhead cables or overhead insulated conductors

All overhead conductors shall be insulated.	WR-709.521.1.8
Poles and other supports for overhead wiring shall be located or protected so that they are unlikely to be damaged by any foreseeable vehicle movement.	WR-709.521.1.8
Overhead conductors shall be at a height above ground of not less than 6 m in all areas subjected to vehicle movement and 3.5 m in all other areas.	WR-709.521.1.8

7.3.9.5 Plug and socket outlets

Socket outlets shall comply with BS EN 60309-1 above 63 A and BS EN 60309-2 up to 63 A.	WR-709.553.1.8
Every socket outlet shall meet the degree of protection of IP44 or such protection shall be provided by an enclosure.	WR-709.553.1.8
Every socket outlet shall be located as close as practicable to the berth to be supplied. Socket outlets shall be installed in the distribution board or in separate enclosures.	WR-709.553.1.9
A maximum of four socket outlets shall be grouped together in one enclosure.	WR-709.553.1.10
One socket outlet shall supply only one pleasure craft or houseboat.	WR-709.553.1.11
In general, single-line socket outlets with rated voltage 200–250 V and rated current 16 A shall be provided.	WR-709.553.1.12
Socket outlets shall be placed at a height of not less than 1 m above the highest water level. In the case of floating pontoons or walkways only, this height may be reduced to 300 mm above the highest water level provided that appropriate additional measures are taken to protect against the effects of splashing.	WR-709.553.1.13

7.3.9.6 Presence of corrosive or polluting substances (AF)

Equipment installed on or above a jetty, wharf, pier or pontoon shall be suitable for use in the presence of atmospheric corrosive or polluting substances (AF2).	WR-709.512.2.1.3

 Note: If hydrocarbons are present, AF3 is applicable.

7.3.9.7 Presence of water (AD)

In marinas, equipment installed on or above a jetty, wharf, pier or pontoon shall be selected as	WR-709.512.2.1.1

follows, according to the external influences which
may be present:

- water splashes (AD4): IPX4
- water jets (AD5): IPX5
- water waves (AD6): IPX6.

7.3.9.8 Presence of solid foreign bodies (AE)

Equipment installed on or above a jetty, wharf, pier or pontoon shall be selected with a degree of protection of at least IP3X in order to protect against the ingress of small objects (AE2).	WR-709.512.2.1.2

7.3.9.9 Protection against overcurrent

Each socket outlet shall be protected by an individual overcurrent protective device.	WR-709.533
A fixed connection for supply to each houseboat shall be protected individually by an overcurrent protective device.	WR-709.533

7.3.9.10 Protection against electric shock

	The protective measures of obstacles and out of reach are **not** permitted.	WR-709.410.3.5
	The protective measures of non-conducting location and earth-free local equipotential bonding are **not** permitted.	WR-709.410.3.6

7.3.9.11 Protection by RCDs

An RCD shall be used to protect socket outlets, individually, for fault protection by automatic disconnection of supply.	WR-709.531.2
Final circuits intended for fixed connection for the supply to houseboats shall be protected individually by an RCD.	

7.3.9.12 Supplies

The nominal supply voltage of the installation for the supply to pleasure craft or houseboats shall be 230 V a.c. single-line, or 400 V a.c. three-line.	WR-709.313.1.2

7.3.9.13 TN system

In the UK for a TN system, the final circuits for the supply to pleasure craft or houseboats shall **not** include a PEN conductor.   WR-709.411.4

Note: In the UK the ESQCR prohibit the use of a TN-C-S system for the supply to a boat or similar construction.

7.3.9.14 Wiring systems of marinas

The following wiring systems are suitable for distribution circuits of marinas: WR-709.521.1.4

- underground cables
- overhead cables or overhead insulated conductors
- cables with copper conductors and thermoplastic or elastomeric insulation and sheath installed within an appropriate cable management system taking into account external influences such as movement, impact, corrosion and ambient temperature
- mineral-insulated cables with a PVC protective covering
- cables with armouring and serving of thermoplastic or elastomeric material
- other cables and materials that are no less suitable than those listed above.

The following wiring systems shall **not** be used on or above a jetty, wharf, pier or pontoon: WR-709.521.1.5

- cables in free air suspended from or incorporating a support wire
- non-sheathed cables in conduit, trunking etc.
- cables with aluminium conductors
- mineral insulated cables.

Cables shall be selected and installed so that mechanical damage due to tidal and other movement of floating structures is prevented.	WR-709.521.1.6
Cable management systems shall be installed to allow the drainage of water by drainage holes and/or installation of the equipment on an incline.	WR-709.521.1.6

7.3.9.15 Underground cables

Underground distribution cables shall, unless provided with additional mechanical protection, be buried at a sufficient depth to avoid being damaged, e.g. by heavy vehicle movement.	WR-709.521.1.7

 Note: A depth of 0.5 m is generally considered as a minimum depth to fulfil this requirement.

7.3.10 Mobile and transportable units

For the purposes of this section, the term 'unit' is intended to mean a vehicle and/or mobile (self-propelled or towed) or transportable structure (such as a container or cabin) in which all or part of an electrical installation is contained and which is provided with a temporary supply by means of, for example, a plug and socket outlet.

7.3.10.1 IT system

The protective measure of non-conducting location is **not** permitted.	WR-717.418
The protective measure of earth-free local equipotential bonding is **not** recommended.	WR-717.418

An IT system can be provided by: • an isolating transformer or a low-voltage generating set, with an insulation monitoring device installed; or	WR-717.411.6.2

- a transformer providing simple separation and providing:
  - automatic disconnection of the supply in the case of a first fault between live parts and the frame of the unit; or
  - an RCD.

Each item of equipment used outside the unit shall be protected by a separate RCD.

WR-717.411.6.2

Additional protection by an RCD having the characteristics shall be provided for every socket outlet intended to supply current-using equipment outside the unit, with the exception of socket outlets which are supplied from circuits with protection by:

WR-717.415

- SELV; or
- PELV; or
- electrical separation.

The protective measures of obstacles and placing out of reach are **not** permitted.

WR-717.417

7.3.10.2 TN system

In the UK a TN-C-S system shall **not** be used to supply a mobile or transportable unit except:

WR-717.411.4

- where the installation is continuously under the supervision of a skilled or instructed person; and
- the suitability and effectiveness of the means of earthing has been confirmed before the connection is made.

7.3.10.3 Plugs and connectors

Plugs and connectors used to connect the unit to the supply shall comply with BS EN 60309-2 and shall also meet the following requirements:

WR-717.55.1

- Plugs shall have an enclosure of insulating material.
- Plugs and socket outlets shall afford a degree of protection of not less than IP44, if located outside.

- Appliance inlets with their enclosures shall
  provide a degree of protection of at least IP44.

The plug part shall be situated on the unit.

7.3.10.4 Protection against electric shock

Automatic disconnection of the supply shall be provided by means of an RCD.	WR-717.411.1

7.3.10.5 Protective equipotential bonding

Accessible conductive parts of the unit, such as the chassis, shall be connected through the main protective bonding conductors to the main earthing terminal within the unit.	WR-717.411.3.1.2
The main protective bonding conductors shall be finely stranded.	WR-717.411.3.1.2

7.3.10.6 Proximity to non-electrical services

No electrical equipment (including wiring systems), except ELV equipment for gas supply control, shall be installed in any gas cylinder storage compartment.	WR-717.528.3.5
Where cables have to run through such a compartment, they shall be run at a height of less than 500 mm above the base of the cylinder(s), and such cables shall be protected against mechanical damage by installation within a continuous gas-tight conduit or duct passing through the compartment.	WR-717.528.3.5
Where installed, this conduit or duct shall be able to withstand an impact equivalent to AG3 without visible physical damage.	WR-717.528.3.5

7.3.10.7 Selection and erection of equipment

Identification – a permanent notice shall be fixed to the unit in a prominent position, preferably adjacent	WR-717.514

to the supply inlet connector. The notice should state in clear and unambiguous terms the following:

- the type of supply which may be connected to the unit
- the voltage rating of the unit
- the number of lines and their configuration
- the on-board earthing arrangement
- the maximum power requirement of the unit.

7.3.10.8 Socket outlets

Socket outlets located outside the unit shall be provided with an enclosure affording a degree of protection not less than IP44.	WR-717.55.2

7.3.10.9 Wiring systems

Flexible cables (for connecting the unit to the supply), or cables of equivalent design, having a minimum cross-sectional area of 2.5 mm^2 copper, shall be used.	WR-717.52.1
The flexible cable shall enter the unit by an insulating inlet in such a way as to minimise the possibility of any insulation damage or fault which might energise the exposed conductive parts of the unit.	WR-717.52.1
The following or other equivalent cable types are permitted for the internal wiring of the unit: - thermoplastic or thermosetting insulated only cable (BS 6004, BS 7211, BS 7919) installed in conduits in accordance with BS EN 61386-1 - thermoplastic or thermosetting insulated and sheathed cable (BS 6004, BS 7211, BS 7919), if precautionary measures are taken to prevent mechanical damage due to any sharp-edged parts or abrasion.	WR-717.52.2

7.3.11 Rooms and cabins containing saunas

Saunas, similar to bathrooms and showers, are primarily used by people who are unclothed and wet and thus very vulnerable to electric shock due to their reduced body resistance (i.e. absence of shoes means less protection from shock, whilst water on their skin will tend to short circuit its natural protection). Special measures are, therefore, needed to ensure that the possibility of direct and/or indirect contact is reduced.

The Regulations requirements for hot air saunas are based on four zones which take into account the limitations of walls, doors, fixed partitions, ceilings and floors and the electric heater itself. The four zones are shown in Figure 7.4.

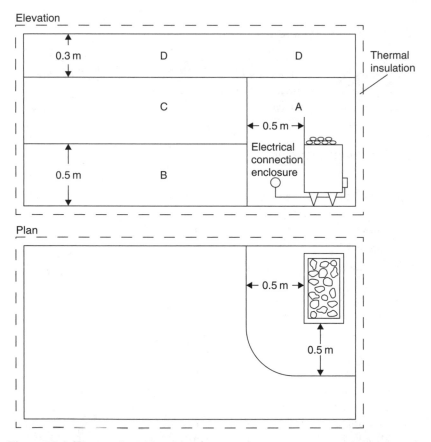

Figure 7.4 Zones of ambient temperature

Zone 1 is the volume containing the sauna heater, limited by the floor, the cold side of the thermal insulation of the ceiling and a vertical surface circumscribing the sauna heater at a distance 0.5 m from the surface of the heater. If the sauna heater is located closer than 0.5 m to a wall, then zone 1 is limited by the cold side of the thermal insulation of that wall. In zone 1, only the sauna heater and equipment belonging to the sauna heater shall be installed.

Zone 2 is the volume outside zone 1, limited by the floor, the cold side of the thermal insulation of the walls and a horizontal surface located 1.0 m above the floor. In zone 2, there is no special requirement concerning heat resistance of equipment.

Zone 3 is the volume outside zone 1, limited by the cold side of the thermal insulation of the ceiling and walls and a horizontal surface located 1.0 m above the floor. In zone 3, the equipment shall withstand a minimum temperature of 125°C and the insulation and sheaths of cables shall withstand a minimum temperature of 170°C.

External influences

The equipment shall have a degree of protection of at least IPX4.	WR-703.512.2
If cleaning by use of water jets may be reasonably expected, electrical equipment shall have a degree of protection of at least IPX5.	WR-703.512.2

7.3.11.1 Extra-low voltage provided by SELV or PELV

Where SELV or PELV is used, whatever the nominal voltage, basic protection shall be provided by: • basic insulation; or • barriers or enclosures affording a degree of protection of at least IPXXB or IP2X.	WR-703.414.4.5

7.3.11.2 Isolation, switching, control and accessories

Switchgear and controlgear which form part of the sauna heater equipment or of other fixed equipment installed in zone 2, may be installed within the sauna room or cabin.	WR-703.537.5
Other switchgear and controlgear (e.g. for lighting) shall be placed outside the sauna room or cabin.	WR-703.537.5
Socket outlets shall **not** be installed within the location containing the sauna heater.	WR-703.537.5

7.3.11.3 Protection against electric shock

The protective measures of obstacles and placing out of reach are **not** permitted.	WR-703.410.3.5
The protective measures of non-conducting location earth-free local equipotential bondings are **not** permitted.	WR-703.410.3.6

7.3.11.4 Protection by RCDs

Additional protection shall be provided for all circuits of the sauna, by the use of one or more RCDs.	WR-703.411.3.3

7.3.11.5 Standards

Sauna heating appliances shall comply with BS EN 60335-2-53 and be installed in accordance with the manufacturer's instructions.	WR-703.55

7.3.11.6 Wiring systems

The wiring system should be preferably installed outside the zones, i.e. on the cold side of the thermal insulation.	WR-703.52
If the wiring system is installed on the warm side of the thermal insulation in zones 1 or 3, it shall be heat-resisting.	WR-703.52
Metallic sheaths and metallic conduits shall **not** be accessible in normal use.	WR-703.52

7.3.12 Solar, pholtovoltaic (PV) power supply systems

Photovoltaics (PV) is a technology that converts light directly into electricity by using photons from sunlight to knock electrons into a higher state of energy, thereby creating electricity. Solar cells are packaged in photovoltaic modules (often electrically connected in multiples as solar photovoltaic arrays) and convert energy from the sun into electricity.

7.3.12.1 Accessibility

The selection and erection of equipment shall enable safe maintenance and shall not adversely affect provisions made by the manufacturer of the PV equipment to enable maintenance or service work to be carried out safely.	WR-712.513.1

7.3.12.2 Automatic disconnection of supply

On the a.c. side, the PV supply cable shall be connected to the supply side of the protective device for automatic disconnection of circuits supplying current-using equipment.	WR-712.411.3.2.1.1
Where an electrical installation includes a PV power supply system without at least simple separation between the a.c. side and the d.c. side, an RCD installed to provide fault protection by automatic disconnection of supply shall be type B according to IEC 60755 Amendment 2.	WR-712.411.3.2.1.2
Where the PV convertor is, by construction, not able to feed d.c. fault currents into the electrical installation, an RCD of type B according to IEC 60755 Amendment 2 is not required.	WR-712.411.3.2.1.2

7.3.12.3 Compliance with standards

PV modules shall comply with the requirements of the relevant equipment standard, e.g. BS EN 61215 for crystalline PV modules. PV modules of Class II construction or with equivalent insulation are recommended if U_{oc} STC of the PV strings exceeds 120 V d.c.	WR-712.511.1

> The PV array junction box, PV generator junction box and switchgear assemblies shall be in compliance with BS EN 60439-1.
>
> WR-712.511.1

7.3.12.4 Devices for isolation

> In the selection and erection of devices for isolation and switching to be installed between the PV installation and the public supply, the public supply shall be considered the source and the PV installation shall be considered the load.
>
> WR-712.537.2.2.1
>
> A switch-disconnector shall be provided on the d.c. side of the PV converter.
>
> WR-712.537.2.2.5
>
> All junction boxes (PV generator and PV array boxes) shall carry a warning label indicating that parts inside the boxes may still be live after isolation from the PV converter.
>
> WR-712.537.2.2.5.1

7.3.12.5 Earthing arrangement

> Earthing of one of the live conductors of the d.c. side is permitted, if there is at least simple separation between the a.c. side and the d.c. side.
>
> WR-712.312.2

 Note: Any connections with earth on the d.c. side should be electrically connected so as to avoid corrosion (see BS 7361-1:1991).

7.3.12.6 Earthing arrangements and protective conductors

> Where protective bonding conductors are installed, they shall be parallel to (and in as close contact as possible with) d.c. cables and a.c. cables and accessories.
>
> WR-712.54

7.3.12.7 Isolation, switching and control

> To allow maintenance of the PV convertor, means of isolating the PV convertor from the d.c. side and the a.c. side shall be provided.
>
> WR-712.537.2.1.1

7.3.12.8 Operational conditions and external influences

Electrical equipment on the d.c. side shall be suitable for direct voltage and direct current.	WR-712.512.1.1
PV modules may be connected in series up to the maximum allowed operating voltage of the PV modules and the PV convertor, whichever is lower.	WR-712.512.1.1
If blocking diodes are used, they shall be connected in series with the PV strings and their reverse voltage shall be rated for $2 \times U_{oc}$ STC of the PV string.	WR-712.512.1.1
PV modules shall be installed in such a way that there is adequate heat dissipation under conditions of maximum solar radiation for the site.	WR-712.512.2.1

7.3.12.9 Protection against electric shock

PV equipment on the d.c. side shall be considered to be energised, even when the system is disconnected from the a.c. side.	WR-712.410.3

 The protective measures of non-conducting location and earth-free local equipotential bonding are **not** permitted on the d.c. side.

7.3.12.10 Protection against electromagnetic interference (EMI) in buildings

To minimise voltages induced by lightning, the area of all wiring loops shall be as small as possible.	WR-712.444.4.4

7.3.12.11 Protection against fault current

The PV supply cable on the a.c. side shall be protected against fault current by an overcurrent protective device installed at the connection to the a.c. mains.	WR-712.434.1

7.3.12.12 Protection against overload on the d.c. side

Overload protection may be omitted:

- for PV string and PV array cables when the continuous current-carrying capacity of the cable is equal to or greater than 1.25 times I_{sc} STC at any location   WR-712.433.1
  WR-712.433.2
- the PV main cable if the continuous current-carrying capacity is equal to or greater than 1.25 times I_{sc} STC of the PV generator.

7.3.12.13 Protective measure – double or reinforced insulation

Protection by the use of Class II or equivalent insulation shall preferably be adopted on the d.c. side.   WR-712.412

7.3.12.14 Protective measure – Extra-low voltage provided by SELV or PELV

For SELV and PELV systems, U_{oc} STC replaces U_o and shall not exceed 120 V d.c.   WR-712.414.1.1

7.3.12.15 Selection and erection of wiring systems

PV string cables, PV array cables and PV d.c. main cables shall be selected and erected so as to minimise the risk of earth faults and short-circuits.   WR-712.522.8.1

Wiring systems shall withstand the expected external influences such as wind, ice formation, temperature and solar radiation.   WR-712.522.8.3

7.3.13 Swimming pools and other basins

The particular requirements of this section apply to the basins of swimming pools, the basins of fountains and the basins of paddling pools and to the surrounding zones of these basins. Swimming pools are, by design, wet areas and people using them are normally wet, which will increase their vulnerability to electric shock. Special measures are, therefore, needed to ensure that all possibility of direct and/or indirect contact is reduced.

Note: Except for areas especially designed as swimming pools, the requirements of this section do not apply to natural waters, lakes in gravel pits, coastal areas and the like.

Special requirements may be necessary for swimming pools for medical purposes.

7.3.13.1 Classification of external influences

As shown in Figure 7.5, these requirements are based on the dimensions of three zones.

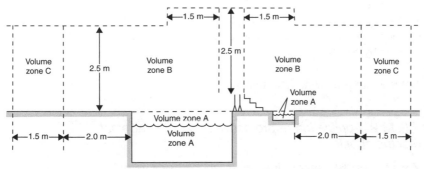

Note: The dimensions are measured taking account of walls and fixed partitions

Figure 7.5 Zone dimensions for swimming pools and paddling pools

Zones 1 and 2 may be limited by fixed partitions having a minimum height of 2.5 m.

Zone 0

This zone is the interior of the basin of the swimming pool or fountain including any recesses in its walls or floors, basins for foot cleaning and waterjets or waterfalls and the space below them.

Zone 1

This zone is limited by:

- zone 0;
- a vertical plane 2 m from the rim of the basin;

- the floor or surface expected to be occupied by persons;
- the horizontal plane 2.5 m above the floor or the surface expected to be occupied by persons.

Where the swimming pool or fountain contains diving boards, springboards, starting blocks, chutes or other components expected to be occupied by persons, zone 1 comprises the zone limited by:

- a vertical plane situated 1.5 m from the periphery of the diving boards, springboards, starting blocks, chutes and other components such as accessible sculptures, viewing bays and decorative basins;
- the horizontal plane 2.5 m above the highest surface expected to be occupied by persons.

Zone 2

This zone is limited by:

- the vertical plane external to zone 1 and a parallel plane 1.5 m from the former;
- the floor or surface expected to be occupied by persons;
- the horizontal plane 2.5 m above the floor or surface expected to be occupied by persons.

 Note: There is no zone 2 for fountains.

7.3.13.2 *Application of protective measures against electric shock*

Zones 0 and 1 (swimming pools and other basins)

 Except for fountains, in zone 0 and zone 1,   WR-702.410.3.4.1
only protection by SELV is permitted.

Equipment for use in the interior of basins which   WR-702.410.3.4.1
is only intended to be in operation when people
are not inside zone 0 shall be supplied by a circuit
protected by:

- SELV
- automatic disconnection of the supply using an RCD; or
- electrical separation.

| The socket outlet of a circuit supplying such equipment and the control device of such equipment shall have a notice in order to warn the user that this equipment shall be used only when the swimming pool is not occupied by persons. | WR-702.410.3.4.1 |

Zones 0 and 1 of fountains

| In zones 0 and 1, one or more of the following protective measures shall be employed:

• SELV
• automatic disconnection of supply using an RCD
• electrical separation. | WR-702.410.3.4.2 |

Zone 2 (swimming pools and other basins)

| One or more of the following protective measures shall be employed:

• SELV
• automatic disconnection of supply
• electrical separation. | WR-702.410.3.4.3 |

7.3.13.3 Current-using equipment of swimming pools

| In zones 0 and 1, it is **only** permitted to install fixed current-using equipment specifically designed for use in a swimming pool. | WR-702.55.1 |
| Equipment which is intended to be in operation only when people are outside zone 0 may be used in all zones provided that it is supplied by a protected circuit. | WR-702.55.1 |

It is permitted to install an electric heating unit embedded in the floor, provided that it:

WR-702.55.1

- is protected by SELV; or
- incorporates an earthed metallic sheath connected to the supplementary equipotential bonding and its supply circuit is additionally protected by an RCD; or
- is covered by an embedded earthed metallic grid connected to the supplementary equipotential bonding and its supply circuit is additionally protected by an RCD.

7.3.13.4 Erection according to the zones

In zones 0, 1 and 2, any metallic sheath or metallic covering of a wiring system shall be connected to the supplementary equipotential bonding.

WR-702.522.21

 Note: Cables should preferably be installed in conduits made of insulating material.

7.3.13.5 External influences

Electrical equipment shall have at least the following degree of protection according to BS EN 60529:

WR-702.512.2

- zone 0: IPX8
- zone 1: IPX4, IPX5 (where water jets are likely to occur for cleaning purposes)
- zone 2: IPX2 for indoor locations, IPX4 for outdoor locations, IPX5 where water jets are likely to occur for cleaning purposes.

7.3.13.6 Junction boxes

 A junction box:

- shall **not** be installed in zone 0
- shall **not** be installed in zone 1 unless it is a SELV circuit.

WR-702.522.24

WR-702.522.24

7.3.13.7 Limitation of wiring systems according to the zones

In zones 0 and 1, a wiring system shall be limited to that necessary to supply equipment situated in these zones.　　　　WR-702.522.22

7.3.13.8 Protective measures

The protective measures of obstacles and placing out of reach are **not** permitted.　　　　WR-702.410.3.5

The protective measures of non-conducting location and earth-free local equipotential bonding are **not** permitted.　　　　WR-702.410.3.6

All extraneous conductive parts in zones 0, 1 and 2 shall be connected by supplementary protective bonding conductors to the protective conductors of exposed conductive parts of equipment situated in these zones.　　　　WR-702.411.3.3

7.3.13.9 Requirements for SELV and PELV circuits

Where SELV is used, whatever the nominal voltage, basic protection shall be provided by:　　　　WR-702.414.4.5

- basic insulation; or
- barriers or enclosures and affording a degree of protection of at least IPXXB or IP2X

7.3.13.10 Special requirements for the installation of electrical equipment in zone 1 of swimming pools and other basins

Fixed equipment designed for use in swimming pools and other basins (e.g. filtration systems, jet stream pumps) and supplied at low voltage is permitted in zone 1, subject to all of the following requirements being met:

- The equipment shall be located inside an insulating enclosure providing at least Class II or equivalent insulation and providing protection against mechanical impact of medium severity (AG2).　　　　WR-702.55.4

Note: This regulation applies irrespective of the classification of the equipment.

- The equipment shall only be accessible via a WR-702.55.4
  hatch (or a door) by means of a key or a tool
  which shall disconnect all live conductors and
  the supply cable and the main disconnecting
  means shall be installed in a way which
  provides protection of Class II or equivalent
  insulation.
- The supply circuit of the equipment shall be
  protected by:
  - SELV; or
  - an RCD; or
  - electrical separation.

7.3.13.11 Switchgear and controlgear

In zones 0 or 1, switchgear or controlgear shall not be   WR-702.53
installed. In zones 0 or 1, a socket outlet shall **not** be
installed.

In zone 2, a socket outlet or a switch is permitted only   WR-702.53
if the supply circuit is protected by one of the following
protective measures:

- SELV
- automatic disconnection of supply using an RCD
- electrical separation.

 For a swimming pool where it is not possible to locate a socket outlet or switch outside zone 1, a socket outlet or switch, preferably having a non-conductive cover or coverplate, is permitted in zone 1 if it is installed outside (1.25 m) from the border of zone 0, is placed at least 0.3 m above the floor, and is protected by:

- SELV (Section 414), at a nominal voltage not exceeding 25 V a.c. rms or 60 V ripple-free d.c., the source for SELV being installed outside zones 0 and 1; or
- automatic disconnection of supply using an RCD; or
- electrical separation.

7.3.13.12 Underwater luminaires for swimming pools

A luminaire for use in the water or in contact with the water shall be fixed and shall comply with BS EN 60598-2-18.   WR-702.55.2

Underwater lighting located behind watertight portholes, and serviced from behind, shall comply with the appropriate part of BS EN 60598 and be installed in such a way that no intentional or unintentional conductive connection between any exposed conductive part of the underwater luminaires and any conductive parts of the portholes can occur.   WR-702.55.2

7.3.13.13 Wiring systems

The following regulations apply to surface wiring systems and to wiring systems embedded in the walls, ceilings or floors at a depth not exceeding 50 mm.

7.3.13.14 Additional requirements for the wiring of fountains

For a fountain, the following additional requirements shall be met: WR-702.522.23

- A cable for electrical equipment in zone 0 shall be installed as far outside the basin rim as is reasonably practicable and run to the electrical equipment inside zone 0 by the shortest practicable route;
- In zone 1, a cable shall be selected, installed and provided with mechanical protection to medium severity (AG2) and the relevant submersion in water depth (AD8).

7.3.13.15 Electrical equipment of fountains

Electrical equipment in zone 0 or 1 shall be provided with mechanical protection to medium severity (AG2), e.g. by use of mesh glass or by grids which can only be removed by the use of a tool.   WR-702.55.3

A luminaire installed in zone 0 or 1 shall be fixed and shall comply with BS EN 60598-2-18.	WR-702.55.3
An electric pump shall comply with the requirements of BS EN 60335-2-41.	WR-702.55.3

7.3.14 Temporary electrical installations for structures, amusement devices and booths at fairgrounds, amusement parks and circuses

This section specifies the minimum electrical installation requirements to facilitate the safe design, installation and operation of temporarily erected mobile or transportable electrical machines and structures which incorporate electrical equipment.

The machines and structures are intended to be installed repeatedly, without loss of safety, temporarily, at fairgrounds, amusement parks, circuses or similar places.

7.3.14.1 Automatic disconnection of supply

For supplies to a.c. motors, RCDs, where used, should be of the time-delayed type in accordance with BS EN 60947-2 or be of the S-type in accordance with BS EN 61008-1 or BS EN 61009-1 where necessary to prevent unwanted tripping.	WR-740.411

7.3.14.2 Cables and cable management systems

Conduit systems shall comply with BS EN 61386 series.	WR-740.521.1
Cable trunking systems and cable ducting systems shall comply with the relevant Part 2 of BS EN 50085.	WR-740.521.1
Tray and ladder systems shall comply with BS EN 61537.	WR-740.521.1
All cables shall meet the requirements of BS EN 60332-1-2.	WR-740.521.1

Cables shall have a minimum rated voltage of 450/750 V.	WR-740.521.1
The routes of cables buried in the ground shall be marked at suitable intervals.	WR-740.521.1
Buried cables shall be protected against mechanical damage.	WR-740.521.1
Armoured cables or cables protected against mechanical damage shall be used wherever there is a risk of mechanical damage due to external influence, e.g. >AG2.	WR-740.521.1
Mechanical protection shall be used in public areas and in areas where wiring systems are crossing roads or walkways.	WR-740.521.1
Where subjected to movement, wiring systems shall be of flexible construction.	WR-740.521.1

7.3.14.3 Electric dodgems

Electric dodgems shall only be operated at voltages not exceeding 50 V a.c. or 120 V d.c.	WR-740.55.9
The circuit shall be electrically separated from the supply mains by means of a transformer or a motor-generator set.	WR-740.55.9

7.3.14.4 Electrical connections

Joints shall not be made in cables except where necessary as a connection into a circuit. Where joints are made, these shall either use connectors in accordance with the relevant British Standard or the connection shall be made in an enclosure with a degree of protection of at least IP4X or IPXXD.	WR-740.526
Where strain can be transmitted to terminals the connection shall incorporate cable anchorage(s).	WR-740.526

7.3.14.5 Electrical supply

At each amusement device, there shall be a connection   WR-740.55.8
point readily accessible and permanently marked to
indicate the following essential characteristics:

- rated voltage
- rated current
- rated frequency.

7.3.14.6 Electric discharge lamp installations

The location of a luminous tube, sign or lamp shall be   WR-740.55.3.1
out of arm's reach or shall be adequately protected to
reduce the risk of injury to persons.

7.3.14.7 Emergency switching device

A separate circuit **shall** be used to supply luminous   WR-740.55.3.2
tubes, signs or lamps, which shall be controlled by
an emergency switch.

The switch **shall** be easily visible, accessible and   WR-740.55.3.2
marked in accordance with the requirements of the
local authority.

7.3.14.8 Floodlights

Where transportable floodlights are used, they shall be   WR-740.55.1.4
mounted so that the luminaire is inaccessible.

Supply cables shall be flexible and have adequate   WR-740.55.1.4
protection against mechanical damage.

Luminaires and floodlights shall be so fixed and protected that a focusing or
concentration of heat is not likely to cause ignition of any material.

7.3.14.9 Inspection and testing

The electrical installation between its origin and any   WR-740.6
electrical equipment shall be inspected and tested after
each assembly on site.

7.3.14.10 IT system

Where an alternative system is available, an IT system shall **not** be used.	WR-740.411.6

7.3.14.11 Lampholders

Insulation-piercing lampholders shall **not** be used unless the cables and lampholders are compatible and the lampholders are non-removable once fitted to the cable.	WR-740.55.1.2

7.3.14.12 Lamps in shooting galleries

All lamps in shooting galleries and other sideshows where projectiles are used **shall** be suitably protected against accidental damage.	WR-740.55.1.3

7.3.14.13 Low-voltage generating sets – generators

All generators shall be so located or protected as to prevent danger and injury to people through inadvertent contact with hot surfaces and dangerous parts.	WR-740.551.8
Electrical equipment associated with the generator shall be mounted securely and, if necessary, on anti-vibration mountings.	WR-740.551.8
Where a generator supplies a temporary installation, forming part of a TN, TT or IT system, care shall be taken to ensure that the earthing arrangements are in accordance with the Regulations.	WR-740.551.8
The neutral conductor of the star-point of the generator shall, except for an IT system, be connected to the exposed conductive parts of the generator.	WR-740.551.8

7.3.14.14 Luminaires

Every luminaire and decorative lighting chain shall: • have a suitable IP rating • be installed so as not to impair its ingress protection; and • be securely attached to the structure or support intended to carry it.	WR-740.55.1.1
Its weight shall not be carried by the supply cable, unless it has been selected and erected for this purpose.	WR-740.55.1.1
Luminaires and decorative lighting chains mounted less than 2.5 m (arm's reach) above floor level or otherwise accessible to accidental contact, shall be firmly fixed and so sited or guarded as to prevent risk of injury to persons or ignition of materials.	WR-740.55.1.1
Access to the fixed light source shall only be possible after removing a barrier or an enclosure which shall require the use of a tool.	WR-740.55.1.1
Lighting chains shall use HO5RN-F (BS 7919) cable or equivalent.	WR-740.55.1.1

7.3.14.15 RCDs

All final circuits for: • lighting • socket outlets rated up to 32 A; and • mobile equipment connected by means of a flexible cable or cord with a current-carrying capacity up to 32 A **shall** be protected by RCDs.	WR-740.415.1
The supply to a battery-operated emergency lighting circuit shall be connected to the same RCD protecting the lighting circuit.	WR-740.415.1
This requirement does not apply to: • circuits protected by SELV or PELV; or • circuits protected by electrical separation; or • lighting circuits placed out of arm's reach.	WR-740.415.1

7.3.14.16 Protection against electric shock

 The protective measure of obstacles is **not** permitted. | WR-740.410.3

 The protective measures of non-conducting location and earth-free local equipotential bonding are **not** permitted. | WR-740.410.3.6

Automatic disconnection of supply to the temporary electrical installation shall be provided at the origin of the installation by one or more RCDs with a rated residual operating current not exceeding 300 mA. | WR-740.410.3

Placing out of arm's reach is acceptable for electric dodgems. | WR-740.410.3

7.3.14.17 Protection against thermal effects

A motor which is automatically or remotely controlled and which is not continuously supervised shall be fitted with a manually reset protective device against excess temperature. | WR-740.422.3.7

7.3.14.18 Safety isolating transformers and electronic convertors

Safety isolating transformers shall comply with BS EN 61558-2-6 or provide an equivalent degree of safety. | WR-740.55.5

A manually reset protective device shall protect the secondary circuit of each transformer or electronic convertor. | WR-740.55.5

Safety isolating transformers shall be mounted out of arm's reach or be mounted in a location that provides equal protection, and shall have adequate ventilation. | WR-740.55.5

Access by competent persons for testing or by a skilled person competent in such work for protective device maintenance shall be provided. | WR-740.55.5

Electronic converters shall conform to BS EN 61347-2-2.	WR-740.55.5
Enclosures containing rectifiers and transformers shall be adequately ventilated and the vents shall not be obstructed when in use.	WR-740.55.5

7.3.14.19 Selection and erection of equipment

Switchgear and controlgear shall be placed in cabinets which can be opened only by the use of a key or a tool, except for those parts designed and intended to be operated by ordinary persons (see Appendix 5 to BS 7671:2008).	WR-740.51
Electrical equipment shall have a degree of protection of at least IP44.	WR-740.512.2

7.3.14.20 Socket outlets and plugs

An adequate number of socket outlets shall be installed to allow the user's requirements to be met safely.	WR-740.55.7

 Note: In booths and stands and for fixed installations, one socket outlet for each square metre or linear metre of wall is generally considered adequate.

7.3.14.21 Supplementary equipotential bonding

In locations intended for livestock, supplementary bonding shall connect all exposed conductive parts and extraneous conductive parts that can be touched by livestock.	WR-740.415.2.1
Where a metal grid is laid in the floor, it shall be included within the supplementary bonding of the location (see Figure 705 on p. 187 of BS 7671:2008).	WR-740.415.2.1
Extraneous conductive parts in, or on, the floor (such as concrete reinforcement in general or reinforcement of cellars for liquid manure) shall be connected to the supplementary equipotential bonding.	WR-740.415.2.1

It is recommended that spaced floors made of prefabricated concrete elements be part of the equipotential bonding.

WR-740.415.2.1

The supplementary equipotential bonding and the metal grid, if any, shall be erected so that it is durably protected against mechanical stresses and corrosion.

WR-740.415.2.1

7.3.14.22 Supply from the public network

Irrespective of the number of sources of supply, the line and neutral conductors from different sources shall not be interconnected downstream of the origin of the temporary electrical installation.

WR-740.313.3

7.3.14.23 Switchgear and controlgear – isolation

Every electrical installation of a booth, stand or amusement device shall have its own means of isolation, switching and overcurrent protection, which shall be readily accessible.

WR-740.537.1

Every separate temporary electrical installation for amusement devices and each distribution circuit supplying outdoor installations shall be provided with its own readily accessible and properly identified means of isolation.

WR-740.537.2.1.1

A device for isolation shall disconnect all live conductors (line and neutral conductors).

WR-740.537.2.2

7.3.14.24 TN system

Where the type of system earthing is TN, a PEN conductor shall not be used downstream of the origin of the temporary electrical installation.

WR-740.411.4

In the UK for a TN system, the final circuits for the supply to caravans or similar shall **not** include a PEN conductor.

WR-740.411.4

7.3.14.25 Voltage

The nominal supply voltage of temporary electrical installations in booths, stands and amusement devices shall not exceed 230/400 V a.c. or 440 V d.c.

WR-740.313.1.1

8

External influences

As promised in the previous edition of BS 7671, the chapter concerning external influences has now been further developed in accordance with the BS EN 60721 and BS EN 61000 series of standards on environmental conditions and the main requirements are as stated below.

512.2	External influences
512.2.1	Equipment shall be of a design appropriate to the situation in which it is to be used or its mode of installation shall take account of the conditions likely to be encountered.
512.2.2	If the equipment does not, by its construction, have the characteristics relevant to the external influences of its location, it may nevertheless be used on condition that it is provided with appropriate additional protection in the erection of the installation. Such protection shall not adversely affect the operation of the equipment thus protected.
512.2.3	Where different external influences occur simultaneously, they may have independent or mutual effects and the degree of protection shall be provided accordingly.
512.2.4	The selection of equipment according to external influences is necessary not only for proper functioning, but also to ensure the reliability of the measures of protection for safety complying with these Regulations generally. Measures of protection afforded by the construction of the equipment are valid only for the given conditions.

For the purpose of these Regulations, the following classes of external influence are conventionally regarded as normal (i.e. the requirement must generally satisfy applicable standards):

- AA Ambient temperature   AA4
- AB Atmospheric humidity   AB4
- Other environmental conditions XX1 of each parameter
  (AC to AS)

- Utilisation and construction of build-   XXI of each parameter,
  ings (B and C)   except
                                          XX2 for the parameter BC

Note: A list of external influences and their characteristics have been included as an appendix (i.e. Appendix 5) to BS 7671:2008 and the following notes concerning external influences are offered as guidance. Also included in this chapter are extracts from the current Regulations that have an impact on the environment.

8.1 Environmental factors and influences

The actual environment to which equipment is likely to be exposed is normally complex and comprises a number of environmental conditions. When defining the conditions for a certain application it is, therefore, necessary to consider all environmental influences which may be as a result of:

- conditions from the surrounding medium;
- conditions caused by the structure in which the equipment is situated or attached;
- influences from external sources or activities.

8.1.1 Combined environmental factors

Equipment may, of course, be simultaneously exposed to a large number of environmental factors and corresponding parameters. Some of the parameters are statistically dependent (e.g. low air velocity and low temperature; sun radiation and high temperature). Other parameters are statistically independent (e.g. vibration and temperature). The effect of a combination of environmental factors is, therefore, extremely important and has to be considered during manufacture and operation.

8.1.2 Sequences of environmental factors

Certain effects of exposing electrical equipment and electrical installations to environmental conditions are a direct result of two or more factors, or parameters, happening either simultaneously or after each other (e.g. thermal shock caused when exposing equipment to a high temperature immediately after exposing it to a low temperature). These possibilities must always be taken into account when designing and installing electrical equipment.

8.1.3 Environmental application

Although the conditions affecting electrical equipment mainly apply to the environment (ambient and created), consideration also has to be given to where the equipment will be operating and how it will be used.

For simplicity this can be broken down into two basic categories:

Conditions	The environmental conditions that have been identified as having an effect on equipment (Table 8.1).
Situations	The main uses of electronic equipment (Table 8.2).

8.1.4 Environmental conditions

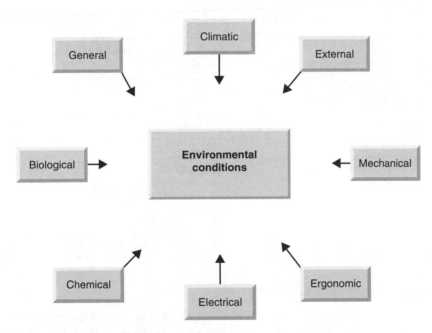

Figure 8.1 Environmental conditions

There are eight basic conditions that have a direct effect on electrical equipment and electrical installations (see Table 8.1).

Table 8.1 Basic conditions

Climatic	Externally generated influences
• Ambient temperature	• Temperature
• Solar radiation	• Precipitation (e.g. water spray)
• Condensation	• Pressure changes (e.g. tunnels)
• Relative atmospheric humidity	• Air movement
• Atmospheric pressure	• Dust
• Wind	
• Precipitation (i.e. rain, snow and hail)	
• Altitude	

(*continued*)

Table 8.1 (*continued*)

Mechanical	Ergonomic aspects
• Vibration • Shock (sinusoidal and random)	• Protecting the health of the engineer and end user • The comfort of the operator and end user • Achieving maximum task effectiveness

Electrical	Chemical
• Electromagnetic environment (EMC and EMI) • Susceptibility and generation • Transients (spikes and surges) • Power supplies • Earthing and bonding	• Pollution and contamination • Dangerous substances • Corrosion • Resistance to solvents

Biological	General
• Animals • Humans (vandalism) • Vegetation	• Safety • Reliability • Maintainability • Components • Waste • Earthquakes • Flammability and fire hazardous areas • Design of equipment

8.1.5 Equipment situations

Obviously not all equipment will be fully operational, all of the time, and so various equipment 'situations' also have to be considered (Table 8.2).

Table 8.2 Environmental conditions

Operational	Storage	Transportation
• When installed and operational • When installed and not in use	• When in storage	• When being transported

As a reference on the whole topic of environmental requirements, *Environmental Requirements for the Electromechanical and Electronic Equipment* (Tricker and Tricker) is recommended. This book has become known as the definitive reference for designers and manufacturers of electrical and electromechanical equipment worldwide. It has been written in the form of a reference book, and contains background guidance concerning the environment and environmental requirements. It also provides a case study of typical requirements, values and ranges currently being requested from industry in today's contracts. Test specifications aimed at proving conformance to these requirements are also listed and the most used national, European and international standards and specifications (e.g. BS, EN, CEN, CENELEC, IEC and ISO) are described in relative detail.

8.1.6 Requirements from the Regulations – General environmental conditions

Electrical equipment shall be selected so as to withstand safely the stresses, the environmental conditions and the characteristics of its location.	WR-133.3

 A coating of paint, varnish or similar product is generally **not** considered to comply with these requirements.

Equipment likely to be exposed to weather, corrosive atmospheres or other adverse conditions shall be so constructed or protected as may be necessary to prevent danger arising from such exposure.	WR-132.5.1
Equipment in surroundings susceptible to risk of fire or explosion shall be so constructed or protected, and such other special precautions shall be taken, as to prevent danger.	WR-132.5.2
The installation shall have an adequate level of immunity against electromagnetic disturbances so as to function correctly in the specified environment.	WR-131.6.4

8.1.7 Requirements from the Regulations – General external influences

Equipment installed shall be appropriate to the external influences foreseen.	WR-530.3

8.1.7.1 *Type of wiring and method of installation*

> The choice of the type of wiring system and the   WR-132.7
> method of installation shall include consideration of
> the following:
>
> - the nature of the location
> - the nature of the structure supporting the
>   wiring
> - accessibility of wiring to persons and
>   livestock
> - voltage
> - the electromechanical stresses likely to occur
>   due to short-circuit and earth fault currents
> - electromagnetic interference
> - other external influences (e.g. mechanical,
>   thermal and those associated with fire) to
>   which the wiring is likely to be exposed
>   during the erection of the electrical
>   installation or in service.

8.1.7.2 *Electrical connections*

> Where a connection is made in an enclosure the   WR-526.7
> enclosure shall provide adequate mechanical protection
> and protection against relevant external influences.

8.1.7.3 *Sealing of wiring system penetrations*

Any sealing arrangement shall resist external influences to the same degree as the wiring system with which it is used and, in addition, it shall meet all of the following requirements:

- It shall be resistant to the products of combustion to the same extent as the elements of building construction which have been penetrated.
- It shall provide the same degree of protection from water penetration as that required for the building construction element in which it has been installed.
- It shall be compatible with the material of the wiring system with which it is in contact.

- It shall permit thermal movement of the wiring system without reduction of the sealing quality.
- It shall be of adequate mechanical stability to withstand the stresses which may arise through damage to the support of the wiring system due to fire.

8.2 Ambient temperature

Temperature, humidity, rainfall, wind velocity and the duration of sunshine all affect the climate of an area. These elements are in turn the result of the interaction of a number of determining causes, such as latitude, altitude, wind direction, distance from the sea, relief, and vegetation. Elements and their determining causes are similarly inter-related, which also contributes to temperature changes; for example, the length of day is a factor which helps to determine temperature; however, the duration of actual sunshine is an element with far-reaching effects on plant and animal life.

Of all the elements that have an effect on man and equipment none is more vital to living organisms than temperature. Temperature has a large influence on where humans live and in the areas where they work. Protective housing or artificial heat sources may overcome low temperatures (and high altitudes); similarly, cooling devices and reflective coatings protect equipment from high temperatures. Temperature is therefore a particularly important aspect of the environment and its accurate measurement and definition requires careful consideration.

The ambient temperature at any given time is the temperature of the air measured under standardised conditions and with certain recognised precautions against errors introduced by radiation from the sun or other heated body. Temperature figures with respect to climate are generally 'shade' temperatures (i.e. the temperature of the air measured with due precautions taken to exclude the influence of the direct rays of the sun) and it is usual for the temperature to be much higher in the direct sunshine. Many mountain areas have air temperatures in the region of zero in winter but the presence of bright sunshine will produce a feeling of warmth and permits the wearing of light clothing.

Seasonal fluctuations in temperature do not pass below ground deeper than 60–80 ft. Below that depth, borings and mine-shafts show that the temperature increases (downwards) depending on the geographical position, location and depth.

On average, however, a rise of about 1°C may be taken for each 64 ft of descent. Assuming that this rate of increase is maintained, it stands to reason that the interior of the earth must be excessively hot and, therefore, it must warm the surface to some extent. It is not possible to determine the precise influence of this temperature increase but it has an effect on tunnels at a depth greater than 80 ft. As the heat from the core is virtually negligible on the surface of the earth (compared with that of the sun), it has not been considered in this book and the only source of heat that has been taken into account is the sun.

Figure 8.2 Enjoying the environment

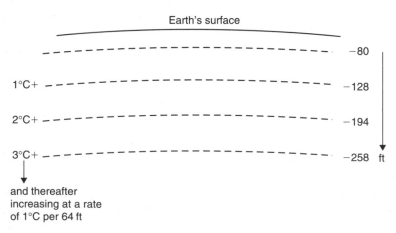

Figure 8.3 Temperature changes below the earth's crust

The difference between summer and winter temperatures for any locality is known as the 'annual range of temperature' or the 'absolute range of temperature' of that particular locality and it is the difference between the highest and lowest temperatures ever experienced at the place in question. The maximum and minimum temperatures are obviously not the same every year, and, should their average over a series of years be taken, it would be known as the 'mean annual extreme range'.

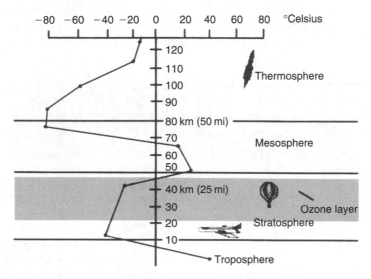

Figure 8.4 Temperature changes above the earth's surface

Although air near the surface (especially at night) may be cooler than the air just above it there is, generally speaking, a gradual falling off of temperature from the ground level up. Over thousands of feet this cooling averages 1°C per 300 ft and thus at approx. 25–55,000 ft (five to ten miles) above the ground the temperature will be down to 55–60°C below zero. Above this the air temperature ceases to fall off regularly; in fact it may even rise for a bit. Usually, however, it remains fairly constant and because of this it is sometimes referred to as the isothermal layer – but to meteorologists and airmen it is known as the stratosphere. The lower layers of the atmosphere (i.e. where temperature falls off with height) is known as the troposphere and the boundary layer between the two is called the tropopause.

8.2.1 Electrical installations

When equipment has been installed without any protection it can be expected to be exposed to more extreme air temperatures and more severe combinations of air temperatures and relative humidities than a similar piece of equipment that has been installed or housed in a temperature-controlled environment.

In addition to open air temperature, temperature stresses on equipment depend on a number of other environmental parameters (e.g. solar radiation, heating from adjacent equipment and air velocity) and these must be taken into account when designing, manufacturing and installing equipment.

The performance of equipment is also influenced and limited by the internal temperature of the equipment. Internal temperatures, in turn, depend on the external ambient conditions and the heat generated within the device itself. Indeed, whenever a temperature gradient exists within a system formed by a device and its surrounding environment, a process of heat transfer will follow.

Thus, in any generic (or system specific) specification or standard relating to ambient temperature it is necessary to consider the following.

Operating temperature range	The specified operating temperature for the equipment which must always be the lowest and the highest ambient temperature expected to be experienced by the equipment during its normal operation.
Storage temperature range	The specified storage temperature that is always the lowest and the highest ambient temperature that the equipment is expected to experience (with the power turned off) during storage or from exposure to climatic extremes.
	Note: The equipment is not normally expected to be capable of operating at these extreme temperatures, merely survive them without damage.

8.2.2 Typical requirements – ambient temperature

Table 8.3 lists the most common environmental requirements concerning ambient temperature.

Table 8.3 Typical requirements – ambient temperature

Temperature ranges	Equipment will need to be designed and manufactured to meet the full performance specification requirement for the selected temperature category.
Temperature increases	The design of equipment should always take into account temperature increases within cubicles and equipment cases so as to ensure that the components do not exceed their specified temperature ranges.
Temperature stresses	In addition to open air temperature, temperature stresses on equipment caused by other environmental parameters (e.g. solar radiation, air velocity and heating from adjacent equipment) will need to be considered.
Operational requirements	• The specified operating temperatures should be the lowest and the highest ambient temperatures expected to be experienced by equipment during normal operation.

(continued)

Table 8.3 (*continued*)

	• When equipment is turned on it should be expected to operate within the temperature ranges stipulated and be fully operational within a specified time after initial turn on – unless otherwise specified. • The permissible limit temperatures of the operating equipment are not allowed to be exceeded as a result of the temperature rise occurring in operation (including temporary acceleration).
Storage	The specified storage temperatures are normally the lowest and highest ambient temperatures that the sample is expected to experience (with the power turned off) during storage or exposure to climatic extremes. The equipment is not expected to be capable of operating at these extreme temperatures, but to survive them without damage.
Peripheral units	For peripheral units (measuring transducers etc.) or situations where equipment is in a decentralised configuration, ambient temperature ranges are frequently exceeded. In these cases the actual temperature occurring at the location of the equipment concerned needs to be considered during the design and installation.
Installation	When equipment is installed in a controlled climatic environment, provided the equipment is not required to operate outside of those conditions, the temperature range can normally be agreed between purchaser and supplier.

8.2.3 Requirements from the Regulations – ambient temperature

For the purpose of these Regulations, the AA2 and AA4 classes of ambient temperature (from $-40°C$ to $+40°C$) are generally recommended with AA4 conventionally being regarded as normal.

8.2.3.1 Mandatory safety requirements

Persons and livestock **shall** be protected against injury, and property shall be protected against damage, due to excessive temperatures or electromechanical stresses caused by any overcurrents likely to arise in live conductors.	WR-131.4
Where the temperature of an exposed part of electrical equipment is likely to cause injury to persons or livestock, that part **shall** be so located or guarded in order to prevent accidental contact.	WR-134.1.6

Note: Where necessary, suitable safety warning signs and/or notices shall be provided.

Persons, fixed equipment and fixed materials adjacent to electrical equipment **shall** be protected against harmful effects of heat or thermal radiation emitted by electrical equipment	WR-131.3.2

8.2.3.2 Electrical installations

Electrical installations shall be so arranged that: • the risk of ignition of flammable materials due to high temperature or electric arc is minimised • during normal operation of the electrical equipment, there shall be minimal risk of burns to persons or livestock.	WR-131.3.1

8.2.3.3 Conductors

Conductors other than live conductors, and any other parts intended to carry a fault current, shall be capable of carrying that current without attaining an excessive temperature.	WR-131.5
The cross-sectional area of conductors shall be determined for both normal operating conditions and, where appropriate, for fault conditions according to the admissible maximum temperature.	WR-132.6

8.2.3.4 Electrical equipment

Electrical equipment shall be so selected and erected that its normal temperature rise and foreseeable temperature rise during a fault cannot cause a fire.	WR-422.1.2

A temperature cut-out device shall have manual reset.

Electrical equipment shall be installed so that design temperatures are not exceeded.	WR-134.1.5
Electrical equipment that is likely to cause high temperatures or electric arcs shall be placed (or guarded) so as to minimise the risk of ignition of flammable materials.	WR-134.1.6

 Switchgear, protective devices, accessories and other types of equipment shall **not** be connected to conductors intended to operate at a temperature exceeding 70°C at the equipment in normal service.

8.2.3.5 Fixed electrical equipment

Fixed electrical equipment shall be selected and erected such that its temperature in normal operation will not cause a fire.	WR-421.2
Where fixed equipment may attain surface temperatures which could cause a fire hazard to adjacent materials, the equipment shall:	WR-421.2

- be mounted on a support which has low thermal conductance; or
- within an enclosure which will withstand such temperatures as may be generated; or
- be screened by materials of low thermal conductance which can withstand the heat emitted by the electrical equipment; or
- be mounted so as to allow safe dissipation of heat and at a sufficient distance from adjacent material on which such temperatures could have deleterious effects.

 Note: Any means of support shall be of low thermal conductance.

Where arcs, sparks or particles at high temperature may be emitted by fixed equipment in normal service, the equipment shall be:	WR-421.3

- totally enclosed in arc-resistant material; or

- screened, by arc-resistant material, from materials upon which the emissions could have harmful effects; or
- mounted so as to allow safe extinction of the emissions at a sufficient distance from materials upon which the emissions could have harmful effects.

Measures shall be taken to prevent an enclosure of electrical equipment such as a heater or resistor from exceeding the following temperatures: WR-422.3.2

- 90°C under normal conditions; and
- 115°C under fault conditions.

Electrical equipment such as installation boxes and distribution boards, that comply with the relevant standard for enclosure temperature rise.   WR-422.4.3

Accessible parts of fixed electrical equipment within arm's reach shall not attain a temperature in excess of the appropriate limit stated in Table 8.4.   WR-423.1

Each such part of the fixed installation likely to attain under normal load conditions, even for a short period, a temperature exceeding the appropriate limit in Table 8.4 shall be guarded so as to prevent accidental contact.   WR-423.1

Table 8.4 Temperature limit under normal load conditions for an accessible part of equipment within arm's reach (reproduced with permission of IET)

Accessible part	Material of accessible surfaces	Maximum temperature (°C)
A hand-held part	Metallic	55
	Non-metallic	65
A part intended to be touched but not hand-held	Metallic	70
	Non-metallic	80
A part which need not be touched for normal operation	Metallic	80
	Non-metallic	90

8.2.3.6 Accessories

Parts of a cable or flexible cord within an accessory, appliance or luminaire shall be suitable for the temperatures likely to be encountered – or shall be provided with additional insulation suitable for those temperatures.   WR-522.2.2

8.2.3.7 Cables

For groups containing non-sheathed or sheathed cables   WR-523.5
with different maximum operating temperatures, the
current-carrying capacity of all the non-sheathed or
sheathed cables in the group shall be based on the lowest
maximum operating temperature of any cable in the group
together with the appropriate group rating factor.

8.2.3.8 Conductors

Connections between conductors or between a conductor   WR-526.2
and other equipment shall take account of the temperature
attained at the terminals in normal service.

Where a soldered connection is used, the design shall take   WR-526.2
account of creep, mechanical stress and temperature rise
under fault conditions.

Where necessary, precautions shall be taken so that the   WR-526.4
temperature attained by a connection in normal service
shall not impair the effectiveness of the insulation of the
conductors connected to it or any insulating material used
to support the connection.

Where a cable is to be connected to a bare conductor   WR-526.4
or busbar its type of insulation and/or sheath shall be
suitable for the maximum operating temperature of the
bare conductor or busbar.

The current, including any harmonic current, to be carried   WR-523.1
by any conductor for sustained periods during normal
operation, shall be such that the appropriate temperature
limit specified in Table 8.5 is not exceeded.

Table 8.5 Maximum operating temperatures for types of cable insulation (reproduced with permission of IET)

Type of insulation	Temperature limit
Thermoplastic	70°C at the conductor
Thermosetting	90°C at the conductor
Mineral (thermoplastic covered or bare exposed to touch)	70°C at the sheath
Mineral (bare not exposed to touch and not in contact with combustible)	105°C at the sheath

Connections between conductors or between a conductor and other equipment shall take account of the temperature attained at the terminals in normal service.	WR-526.2
Where a cable is to be connected to a bare conductor or busbar its type of insulation and/or sheath shall be suitable for the maximum operating temperature of the bare conductor or busbar.	WR-526.4

8.2.3.9 Heating conductors and cables

The load of every floor-warming cable under operation shall be limited to a value such that the manufacturer's stated conductor temperature is not exceeded.	WR-554.4.4

8.2.3.10 Floor and ceiling heating systems

In floor areas where contact with skin or footwear is possible, the surface temperature of the floor shall be limited (for example, 35°C).	WR-753.423
For cold tails (circuit wiring) and control leads installed in the zone of heated surfaces, the increase of ambient temperature shall be taken into account.	WR-753.522.1.3

8.2.3.11 Immersed heating elements

Heaters for liquid or other substance shall have an automatic device to prevent a dangerous rise in temperature.	WR-554.2.1

8.2.3.12 Lighting equipment

Lighting equipment such as incandescent lamps, spotlights and small projectors and other equipment or appliances with high-temperature surfaces shall be suitably guarded, and installed and located in accordance with the relevant standard.	WR-711.422.4.2

8.2.3.13 Luminaires

Luminaires marked ▽ⅅ are designed to provide limited surface temperature. WR-422.3.2

Every luminaire shall have a limited surface temperature in accordance with BS EN 60598-2-24. WR-42WR-422.3.8

Bayonet lampholders B15 and B22 shall comply with BS EN 61184 and shall have the temperature rating T2 described in that standard. WR-559.6.1.7

Only independent lamp controlgear marked as suitable for independent use (according to the relevant Standard) shall be used external to a luminaire. WR-559.7

Only the following are permitted to be mounted on flammable surfaces: A 'class P' ▽P⁄ thermally protected ballast(s)/transformer(s), marked with the symbol. WR-559.8

A temperature declared thermally protected ballast(s)/transformer(s), marked with the symbol ▽ with a marked value equal to or below 130°C.

The generally recognised symbol of an independent ballast of EN 60417 is ⊟

8.2.3.14 Motors

A motor which is automatically or remotely controlled or which is not continuously supervised shall be protected against excessive temperature by a protective device with manual reset. WR-422.3.7

A motor with star-delta starting shall be protected against excessive temperature in both the star and delta configurations. WR-422.3.7

Where the motor is intended for intermittent duty and for frequent starting and stopping, account shall be taken of any cumulative effects of the starting or braking currents upon the temperature rise of the equipment of the circuit. WR-552.1.1

In temporary electrical installations for structures (such as amusement devices and booths at fairgrounds, amusement parks and circuses) a motor which is automatically or remotely controlled (and which is not continuously supervised) shall be fitted with a manually reset protective device against excess temperature.

WR-740.422.3.7

8.2.3.15 Type of wiring and method of installation

The choice of the type of wiring system and the method of installation shall include consideration of the following:

WR-132.7

- the nature of the location
- the nature of the structure supporting the wiring
- accessibility of wiring to persons and livestock
- voltage
- the electromechanical stresses likely to occur due to short-circuit and earth fault currents
- electromagnetic interference
- other external influences (e.g. mechanical, thermal and those associated with fire) to which the wiring is likely to be exposed during the erection of the electrical installation or in service.

8.2.3.16 Wiring system

A wiring system shall be selected and erected so as to be suitable for the highest and the lowest local ambient temperatures.

WR-522.1.1

In order to avoid the effects of heat from external sources, one or more of the following shall be used to protect a wiring system:

WR-522.2.1

- shielding
- placing sufficiently far from the source of heat
- selecting a system with due regard for the additional temperature rise which may occur
- local reinforcement or substitution of insulating material.

Heat from external sources may be radiated, conducted or convected, e.g.

- from hot water systems;
- from plant, appliances and luminaires;
- from a manufacturing process;
- through heat conducting materials;
- from solar gain of the wiring system or its surrounding medium.

In solar photovoltaic (PV) power supply systems, the wiring system shall withstand the expected external influences such as wind, ice formation, temperature and solar radiation.	WR-712.522.8.3

8.2.3.17 Wiring system components

Wiring system components, including cables and wiring accessories, shall only be installed or handled at temperatures within the limits stated in the relevant product specification or as given by the manufacturer.	WR-522.1.2

8.2.3.18 Agricultural and horticultural premises

For high-density livestock rearing systems operating for the life support of livestock (and where electrically powered ventilation is necessary) a system of temperature and supply voltage monitoring shall be used.	WR-705.560.6

8.2.3.19 Temporary electrical installations

For cold tails (circuit wiring) and control leads installed in the zone of heated surfaces of temporary electrical installations for structures (such as amusement devices and booths at fairgrounds, amusement parks and circuses) the increase of ambient temperature shall be taken into account.	WR-753.522.1.3

8.3 Solar radiation

Of all the factors that control the weather, the sun is by far the most power-ful, and practically everything that occurs on the earth is controlled, directly or indirectly, by it. The sun affects the places humans inhabit, in the kind of homes that are built, the work that is done, and the equipment that is used.

Less that one-millionth of the energy emitted from the sun's surface travels the ninety-odd million miles to reach this planet. The sun's energy crosses those miles in the form of short electromagnetic radio waves, identical in nature to those used in broadcasting, which pass through the atmosphere and are absorbed by the earth's surface. These waves warm the earth's surface and are then re-radiated back to space. The wavelength of the energy emitted by the earth is very much longer than that emitted by the sun (because the earth is much cooler than the sun) and these longer waves are not able to pass through the atmos-phere as freely as short waves. A large proportion of the energy emitted by the earth is absorbed by the water vapour and water droplets in the lower atmos-phere which in turn is re-radiated back to earth. Thus the earth plays the part of a receiving station absorbing short electromagnetic waves and converting them into longer electromagnetic waves, while the atmosphere acts as a trap contain-ing most of the longer electromagnetic waves before they are lost to space.

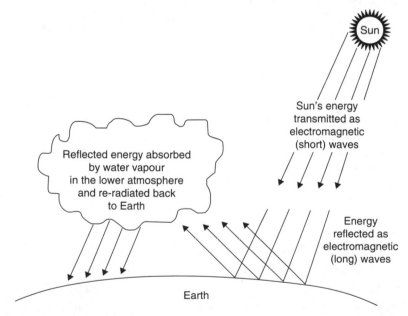

Figure 8.5 Solar radiation – energy

Radiation from the sun consists of rays of three differing wavelengths: heat rays, actinic rays and light rays. Heat rays and actinic rays are intercepted by solid bodies and produce peculiar effects in varying degrees according to the nature of the surface on which they fall. The light rays are responsible for daylight and both light rays and actinic rays are necessary for the life processes of plants. The heat rays' most important aspect is temperature, and the amount of sunshine (and therefore the temperature) will depend on latitude and the length of day.

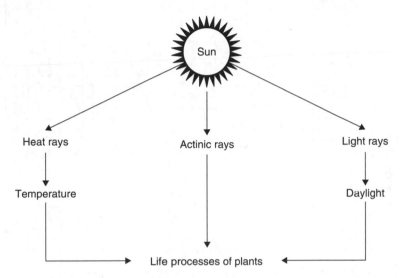

Figure 8.6 Sun's radiation

Radiant energy can be reflected from solid surfaces and intensified by that reflection. For example reflection from walls, which is frequently used for ripening peaches and pears. Reflection from bare ground can also assist in the ripening of melons and other creeping plants, while reflection from water surfaces enhances the 'climatic reputation' of waterside resorts. However, radiant energy can also cause damage to equipment as heat rays can warm the material or its environment to dangerous levels and photochemical degradation of materials can be caused by the ultraviolet content of solar radiation.

8.3.1 What are the effects of solar radiation?

On cloudless nights when atmospheric radiation is very low, objects exposed to the night sky will attain surface temperatures below that of the surrounding air temperature. For example (and by experiment), a horizontal disc thermally insulated from the ground and exposed to the night sky during a clear night can attain a temperature of $-14°C$ when the air temperature is $0°C$ and the relative humidity is close to 100%, and these values are of assistance when determining the 'under temperature' of components.

The sun's electromagnetic radiation consists of a broad spectrum of light ranging from ultraviolet to near infra-red. Owing to the distance of the sun

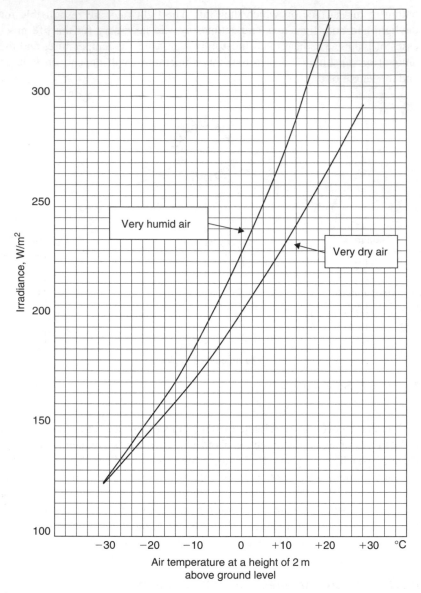

Figure 8.7 Lowest values of atmospheric radiation during clear nights (reproduced from BS 7527 Section 2.4; 1991 by kind permission of the BSI)

from the earth, solar radiation appears on the earth's surface as a parallel beam and the highest (maximum) level of radiation occurs at noon, on a cloudless day, at a surface perpendicular to the sun's rays.

Most of the sun's energy reaches the surface of the earth in the 0.3 to 0.4 μm range and the density of the solar radiated power (or irradiance – expressed in watts per square metre) is dependent on the content of aerosol particles, ozone and water vapour in the air. The actual amount of irradiance will vary considerably with geographical latitude and type of climate (temperature, humidity, air velocity etc.).

Having said that, an object subjected to solar radiation will obtain a temperature depending on the surrounding ambient air temperature, the intensity of radiation, the air velocity, the incidence angle of the radiation on the object, the duration of exposure, and the thermal properties of the object itself (e.g. surface reflectance, size, shape, thermal conductance and specific heat) together with other factors such as wind and heat conduction to mountings and surface absorbency.

8.3.2 Photochemical degradation of material

One of the biggest problems caused by solar radiation is the photochemical degradation of most organic materials, which in turn causes the elasticity and plasticity of certain rubber compounds and plastic materials to be affected and can, in exceptional cases, make optical glass opaque. Although solar radiation can bleach out colours in paints, textiles, paper etc. (a major consideration when trying to read the colour coding of components), by far the most important effect is the heating of materials.

The combined effect of solar radiation, atmospheric gases, temperature and humidity changes etc., is often termed 'weathering' and results in the 'ageing' and ultimate destruction of most organic materials, e.g. plastics, rubbers, paints, timber. Typical defects caused by weathering are:

- rapid deterioration and breakdown of paints;
- cracking and disintegration of cable sheathing;
- fading of pigments;
- bleached out colours in paints, textiles and paper.

8.3.3 Effects of irradiance

To guard against the effects of irradiance, the following guidelines should be considered when locating electrical equipment:

- The sun should be allowed to shine only on the smallest possible casing surfaces.
- Windows should be avoided on the sunny side of rooms housing electronic equipment.
- Heat-sensitive parts must be protected by heat shields made, for instance, of polished stainless steel or aluminium plate.
- Air conditioning plant and cooling fans (when used) in rooms housing electronic equipment should be efficient and reliable.
- Convection flow should sweep across the largest possible surfaces of materials with good conduction properties.

8.3.4 Heating effects

As previously shown, probably the most important effect of solar radiation –
heating – is mainly caused by the short term, high intensity radiation around
noon on cloudless days. Typical peak values of irradiance are shown in
Table 8.6 below.

Table 8.6 Typical peak values of irradiance in W/m² from a cloudless sky

Area	Large cities	Flat land	Mountainous areas
Subtropical climates and deserts	700	750	1180
Other areas	1050	1120	1180

As equipment (if fully exposed to solar radiation) in an ambient temperature (e.g. 35–40°C) can attain temperatures in excess of 60°C, one has to consider an item of equipment's outside surface. To a major extent the surface
reflectance of an object affects its temperature rise from solar heating and
changing the finish from a dark colour to a gloss white can cause a considerable reduction in temperature. On the downside, a pristine finish designed
specifically to reduce temperature can be expected to deteriorate in time and
result in an increase in temperature.

Another problem found in most of today's materials is that they are also
selective reflectors (i.e. their spectral reflectance factor changes with wavelength). For example paints are, in general, poor infra-red reflectors although
they may be very efficient as a visible warning. Care should, therefore, be
taken when selecting materials and finishes for equipment casings.

8.3.5 Typical requirements – solar radiation

Table 8.7 lists the most common environmental requirements concerning
solar radiation.

Table 8.7 Typical requirements – solar radiation

Survivability	Equipment that is exposed to the effect of solar radiation should remain unaffected.
Exposure	The sun should be allowed to shine only on the smallest possible casing surfaces and the convection flow should sweep across the largest possible surfaces of materials with good conduction properties.
Windows	Windows should be avoided on the sunny side of rooms housing electronic equipment.
Heat shields	Heat-sensitive parts shall be protected by heat shields made of (for instance) polished stainless steel or aluminium plate.
Air conditioning	Air conditioning plant and cooling fans (when used) in rooms housing electronic equipment should be efficient and reliable.

8.3.6 Requirements from the Regulations – solar radiation

The selected wiring system shall be selected and erected (or adequate shielding shall be provided) where and whenever significant solar radiation (AN2) or ultraviolet radiation is experienced or anticipated.	WR-522.11.1
Special precautions may need to be taken for equipment that is subject to ionising radiation.	WR-522.11.1
Solar photovoltaic (PV) modules shall be installed in such a way that there is adequate heat dissipation under conditions of maximum solar radiation for the site.	WR-712.512.2.1

 Note: See Section 7.3.12 for full details of the requirements for PV power supply systems.

8.4 Humidity

The atmosphere is normally described as 'a shallow skin or envelope of gases surrounding the surface of the earth which is made up of nitrogen, oxygen and a number of other gases which are present in very small quantities'. While the ratio of these components shows no appreciable variation either with latitude or altitude, the water vapour content of the atmosphere is subject to extremely wide fluctuations. The amount of water present in the air is referred to as humidity.

When air and water come into contact, they will exchange particles with each other (i.e. the air particles will pass into the water and water particles will pass into the air in the form of vapour). There is always a certain amount of water vapour present in the air and a certain amount of air present in water and there is always a constant movement between the two mediums. If there is only a small amount of water vapour present in the air, then more particles of water will pass from the water into the air than from the air into the water and so the water will gradually dry up or evaporate. Conversely, if the amount of water vapour in the air is large, then as many particles of vapour will pass from the air into the water, as from the water into the air – and the water will not evaporate. In such a case the air is said to be 'saturated' – or, to put it another way, it holds as much water vapour as it can possibly contain.

Water vapour is collected in the air above the oceans and is carried by the wind towards the land masses. The amount of water vapour in the air varies greatly depending on the place and season but in general, evaporation is most rapid at high temperature and slower at lower temperatures. We may, therefore, expect to find the greatest amount of water vapour over the oceans near the equator and the smallest amount over the land in a cold region such as North-East Asia in winter. However, even when the surface is covered with snow and ice, water evaporation may take place and occasionally, during a long frost, the snow will gradually disappear without melting.

Except for the water vapour being present, the composition of the atmosphere near the surface of the earth up to a height of some 2000 ft is practically uniform throughout the globe. However, at greater altitudes (i.e. above the atmospheric boundary in the tropopause), there is practically no water vapour, or water, in any form.

8.4.1 What is humidity?

Temperature and the relative humidity of air (in varying combinations) are climatic factors which act upon electrical equipment and installations during storage, transportation or operation. Humidity and the electrolytic damage resulting from moisture mostly affects plug points, soldered joints (in particular, dry joints), bare conductors, relay contacts and switches. Humidity also promotes metal corrosion (see Section 8.7 – Pollutants and contaminants) owing to its electrical conductivity

In many cases, however, environmental influences such as mechanical and thermal stresses are merely the forerunner of the impending destruction of components by humidity – especially as the majority of electronic component failures are caused by water!

 Note: 'Humidity' (in the context of this book) has been taken to cover relative humidity, absolute humidity, condensation, adsorption, absorption and diffusion and details of these 'sub sets' are provided in the following paragraphs.

8.4.2 Relative and absolute humidity and their effect on equipment performance

The performance of virtually all electrical equipment is influenced and limited by its internal temperature which, in turn, is dependent on the external ambient conditions and on the heat generated within the device itself. Fortunately, most electrical and electronic components (especially resistors) will normally remain dry when under load owing to the amount of internal/external heat dissipation. Indeed, many components either have to be de-rated in order to improve their reliability or, for reasons of circuit function, only energised intermittently.

8.4.2.1 Externally mounted equipment

Equipment and components that are mounted in external cabinets run the risk of coming into contact with water or water vapour (e.g. drifting snow, fog, dew, rain, spray water or water from hoses) and the equipment must, therefore, be adequately protected from such humidity in order to prevent the ingress of vapour into the system within the casing.

8.4.2.2 Housed equipment

In most locations (e.g. cabinets, equipment rooms, workshops and laboratories) although temperatures above 30°C may often occur, they are normally combined with a lower relative humidity than that found in the open air. In other rooms (e.g. offices), however, where several heat sources are present, temperatures and relative humidities can differ dramatically across the room.

The sun also plays its part because in certain circumstances (such as when equipment is placed in an unventilated enclosure) the intense heat caused through solar radiation can generate relative humidities in excess of 95% when combined with:

- high relative humidity caused by the release of moisture from hygroscopic materials;
- the breathing and perspiration of human beings;
- open vessels containing water or other sources of moisture.

8.4.3 Condensation

Condensation occurs when the surface temperature of an item is lower than that of the dew point (i.e. the temperature with a relative humidity of 100% at which condensation occurs) and can change electrical characteristics (e.g. decrease surface resistance, increase loss angle) between the absolute point at which atmospheric vapour condenses into droplets (i.e. the dew point), absolute humidity and vapour pressure.

For example, if a piece of equipment has a low thermal time constant, condensation (normally found on the surface of the equipment) will occur only if the temperature of the air increases very rapidly, or if the relative humidity is very close to 100%. Sudden changes in temperature may cause water to condense on parts of equipment and leakage currents can occur.

8.4.4 Adsorption

Adsorption is the amount of humidity that may adhere to the surface of a material and depends on the type of material, the surface structure and the vapour pressure. This layer of water (no matter how small) can cause electrical short circuits and material distortion etc.

8.4.5 Absorption

The quantity of water that can be absorbed by a material depends largely on the water content of the ambient air, and the speed of penetration of the water molecules generally increases with the temperature.

8.4.6 Diffusion

Water vapour can penetrate encapsulations of organic material (e.g. into a capacitor or semiconductor) by way of the sealing compound and into the casing. This factor is frequently overlooked and can become a problem, especially as the moisture absorbed by an insulating material can cause a variation in a number of electrical characteristics (e.g. reduced dielectric strength, reduced insulation resistance, increased loss angle, increased capacitance).

8.4.7 Protection

The effects of humidity mainly depend on temperature, temperature changes and impurities in the air. As shown in Table 8.8, there are three basic methods of protecting the active parts of equipment and components from humidity.

Table 8.8 Protective methods – humidity

Heating the surrounding air so that the relative humidity cannot reach high values	This method normally requires a separate heat source which (especially in the case of equipment mounted in external cabinets) usually means having to have a separate power supply. This method is disadvantaged by the reliability of the circuit, being dependent on the efficiency of the heating.
Hermetically sealing components or assemblies using hydroscopic materials	This is an extremely difficult process as the smallest crack or split can allow moisture to penetrate the component, particularly in the area of connecting wire entry points. Metal, glass and ceramic encapsulation do nevertheless produce some very satisfactory results.
Ventilation and the use of moisture-absorbing hygroscopic materials	Most water-retaining materials and paint etc. are suitable for the temporary absorption of excessively high air humidity in the casing, which, because of the risk of pollution and dust penetration, cannot be fully ventilated (i.e. air exchange with the outside temperature is not possible).

8.4.8 Typical requirements – humidity

Table 8.9 lists the most common environmental requirements concerning humidity.

Table 8.9 Typical requirements – solar radiation

Equipment interoperability	Equipment that is operated adjacent to the seashore (and, therefore, subject to extreme humidity) must be able to function equally well as the same equipment housed in the low humidity of (for example) the desert.

(continued)

Table 8.9 (*continued*)

External humidity levels	Equipment should be designed and manufactured to meet the following external humidity levels (limit values), over the complete range of ambient temperature values anticipated (see Table 8.10). **Note:** Meteorological measurements made over many years have shown that, within Europe, a relative humidity greater than 95% combined with a temperature above 30°C does not occur over long periods in free air conditions.
Condensation	Operationally caused infrequent and slight moisture condensation should not lead to malfunction or failure of the equipment.
Indoor installations	In all indoor installations, provision must be made for limiting the humidity of the ambient air to a maximum of 75% at −5°C.
Equipment in cubicles and cases	The design of equipment should take into account temperature rises within cubicles and equipment cases in order to ensure that the components do not exceed their specified temperature ranges.
Peripheral units	For peripheral units (e.g. measuring transducers) or equipment employed in a decentralised configuration (i.e. where ambient temperature ranges are exceeded) the actual temperature occurring at the location of the equipment concerned should be utilised when designing equipment.
Product configuration	All proposed and dated equipment, components or other articles must be tested in their production configuration without the use of any additional external devices that have been added expressly for the purpose of passing humidity testing.

Table 8.10 Humidity – external humidity levels

Duration	Limit value
Yearly average	75% relative humidity
On 30 days of the year, continuously	95% relative humidity
On the other days, occasionally	100% relative humidity
On the other days, occasionally	$30\,\text{g/m}^3$ occurring in tunnels

8.4.9 Requirements from the Regulations – humidity

For the purpose of these Regulations, the AB2 and AB4 classes of ambient temperature (between 5% and 100%) are generally recommended.

A wiring system shall be selected and erected so that no damage is caused by condensation or ingress of water during installation, use and maintenance.	WR-522.3.1

Special consideration needs to be given to wiring systems that are liable to frequent splashing, immersion or submersion.

Suitable means for the escape of condensation that might form in a wiring system or where water might collect shall be made.	WR-522.3.2
Wiring systems that could be subjected to waves (AD6) shall be protected from mechanical damage (such as impact, vibration and other mechanical stresses).	WR-522.3.3
Where corrosive or polluting substances (including water) could cause corrosion and/or deterioration, the parts of the wiring system likely to be affected shall be suitably protected (e.g. by protective tapes, paints or grease) and/or manufactured from a material resistant to such substances.	WR-522.5.1
If a wiring system is routed below services that are liable to cause condensation (such as water, steam or gas services), precautions shall be taken to protect the wiring system from harmful effects.	WR-528.3.2

8.5 Air Pressure and altitude

Air pressure, frequently referred to as atmospheric pressure, is 'the force exerted on a surface of a unit area caused by the earth's gravitational attraction on the air vertically above that area'. Air pressure varies with altitude

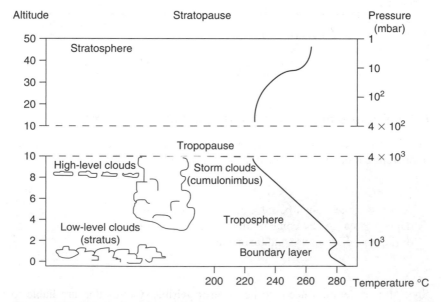

Figure 8.8 Atmospheric structure

(i.e. elevation above mean sea level) and location. For instance at the equator where the tradewinds of both hemispheres converge, there is a low pressure zone (known as the ITCZ or International Conveyance Zone), which is characterised by high humidity.

It is not widely appreciated that the location of equipment, especially with respect to its altitude above sea level, can effect the working of that equipment. But it is not just the height above sea level that has the most effect! Even air pressure variations at ground level have to be considered (see Table 8.11).

Table 8.11 Air pressure and altitude (reproduced with permission from ISO 2533)

Air pressure		Approximate altitude above sea level
(kPa)	(mbar)	(m)
1	10	31,200
2	20	26,600
4	40	22,100
8	80	17,600
15	150	13,600
25	250	10,400
40	400	7,200
55	550	4,850
70	700	3,000

8.5.1 Low air pressure

At altitudes above sea level, low air pressure can cause:

- leakage of gases or fluids from gasket sealed containers;
- ruptures of pressurised containers;
- change of physical or chemical properties;
- erratic breakdown or malfunction of equipment from arcing or corona;
- decreased efficiency of heat dissipation by convection and conduction in air which will effect equipment cooling (e.g. an air pressure decrease of 30% has been found to cause an increase of 12% in temperature);
- acceleration of effects due essentially to temperature (e.g. volatilisation of plasticisers, evaporation of lubricants).

8.5.2 Typical requirements – air pressure and altitude

Table 8.12 lists the most common environmental requirements concerning air pressure and altitude.

Table 8.12 Typical requirements – air pressure and altitude

High air pressure	High air pressure occurring in natural depressions and mines can have a mechanical effect on sealed containers and this should always be borne in mind when designing and installing electrical equipment.
Installations up to 2000 m above sea level	Electrical equipment must be capable of working to an altitude (h) from 120 to 2000 m above sea level – which corresponds to an air pressure range from 110.4 to 74.8 kPa.

8.6 Weather and precipitation

Water is one of the most remarkable substances on earth. It is the substance that we most often see in all its three states: liquid (water), solid (ice) and gas (steam). No living organism can exist without water and as much as half the weight of plants and animals is made up of water. Water in the oceans makes up approximately eleven times the volume of the solid part of the earth in addition, that is, to water frozen in ice floes, in lakes, rivers, within the ground and in living plants and animals.

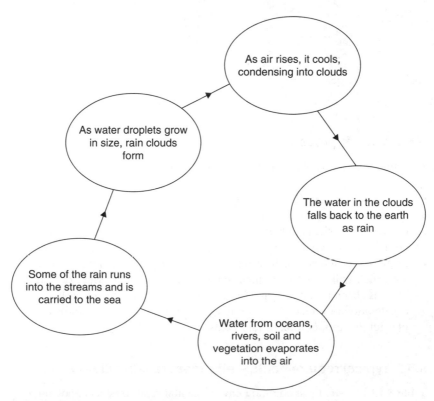

Figure 8.9 The hydrological cycle

Water is a constantly moving cycle. As the sun beats down, some of the surface water is evaporated; this water vapour rises as part of the air and is moved along by the wind. Should it pass over a land mass it may become a cloud and as more moisture is attracted to the cloud or the clouds pass over rising ground the water particles become larger and fall as rain, sleet or snow. See Figures 8.9 and 8.10.

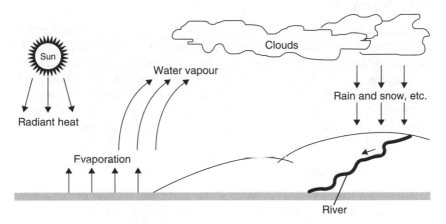

Figure 8.10 Simplified water cycle

As the rain comes into contact with the ground there are several avenues open to it: it may be re-vaporised and return to the atmosphere, be absorbed by the ground or remain on the surface of the ground and run downhill, forming streams and rivers, which run into the sea (or lakes); and the cycle begins once more.

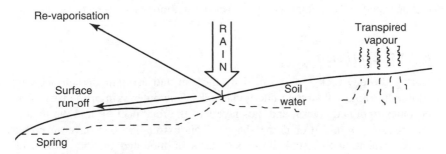

Figure 8.11 The effect of rain

8.6.1 Water

Water (in all of its three forms) is a major cause of failure in every application of electrical and electronic components. The humidity of the air and the possible formation of water particles must always be taken into consideration, especially as humidity possesses (almost without exception) a certain amount of electrical conductivity, which increases the possibility of corrosion of metals. Similarly, the ingress of water followed by freezing within electronic equipment can result in malfunction.

8.6.2 Salt water

Salt has an electrochemical effect on metallic materials (i.e. corrosion), which can damage and degrade the performance of equipment and/or parts that have been manufactured from metallic materials. Non-metallic materials can also be damaged by salt through a complex chemical reaction which is dependent on the supply of oxygenated salt solution to the surface of the material, its temperature and the temperature and humidity of the environment. This is particularly a problem in areas close to the sea or mountain ranges.

8.6.3 Ice and snow

Water in the form of ice can cause problems in the cooling of equipment or freezing and thawing, which will result in cracks occurring, breaking cases, etc.

Powdered snow can easily be blown through ventilation ducts and then melt down in equipment compartments and cubicles, which can cause damp problems in critical systems if not prevented in the original construction.

8.6.4 Weathering

As shown below, there are several types of 'weathering' (which is the collective term for the processes by which rock at or near the earth's surface is disintegrated and decomposed by the action of atmospheric agents, water, and living things).

8.6.4.1 Exfoliation

During the day rocks are warmed by the sun and at night the surface can cool more rapidly than the underlying rocks. The outer skin of the rock then becomes tight and cracks, and thus layers of the rock peel off and the mountain becomes rounded or dome shaped. This exfoliation can have an effect on equipment that is sunk into rock faces or mounted on the surface of equipment.

8.6.4.2 Freeze thaw

When water freezes it turns to ice, expanding by about one twelfth of its volume. If this water is in the joint or a crack in a casing then the space will become enlarged and the casing on either side will be forced apart. When the ice eventually thaws, more water will penetrate into the crack and the cycle repeats itself with the crack constantly enlarging.

8.6.4.3 Chemical weathering

Water can pick up quantities of sulphur dioxide from the atmosphere and will form a weak solution of acid. This acid can attack certain equipment housings and the process whereby the housing is worn away is known as chemical weathering.

8.6.4.4 Erosion

As the wind blows over dry ground it collects grit and 'throws' it vigorously against the surfaces nearby. This grit acts in a similar way to sandpaper and gradually wears away the surface with which it comes into contact.

8.6.4.5 Mass movement

Once solids such as sedimentary rocks have been broken up, there is often a downwards movement of the particles which have been broken off. This 'soil creep' can gather momentum and can, in certain circumstances, submerge equipment.

8.6.5 Typical requirements – weather and precipitation

Table 8.13 lists the most common environmental requirements concerning weather and precipitation.

Table 8.13 Typical requirements – solar radiation

Weather protection	Equipment that is operated adjacent to the seashore or on mountain ranges and therefore subject to water and precipitation must be able to function equally well as the same equipment housed in arid deserts.
Operation	All equipment should be capable of operating during rain, snow and hail and be unaffected by ice, salt and water.
Rain	All equipment should be capable of operating in rain and be capable of preventing the penetration of rainfall at a minimum rate of 13 cm/hour and an accompanying wind rate of 25 m/s.
Snow and hail	Consideration needs to be given to the effect of all forms of snow and/or hail. The maximum diameter of the hailstones is conventionally taken as 15 mm, but larger diameters can occur on occasions.
Salt water	Equipment should be capable of operating in (or be protected from) heavy salt spray, as would be experienced in seacoast areas and in the vicinity of salted roadways.

8.6.6 Requirements from the Regulations – weather and precipitation

8.6.6.1 Equipment

Equipment likely to be exposed to weather, corrosive atmospheres or other adverse conditions shall be so constructed or protected as may be necessary to prevent danger arising from such exposure.	WR-132.5.1

8.6.6.2 Water

A wiring system shall be selected and erected so that no damage is caused by condensation or ingress of water during installation, use and maintenance.	WR-522.3.1

 Special consideration needs to be given to wiring systems that are liable to frequent splashing, immersion or submersion.

Where water may collect or condensation may form in a wiring system, provision shall be made for its escape.	WR-522.3.2
Wiring systems that may be subjected to waves (AD6) shall be protected from mechanical damage (i.e. impact, vibration and other mechanical stresses).	WR-522.3.3
Where the presence of corrosive or polluting substances, including water, is likely to give rise to corrosion or deterioration, parts of the wiring system likely to be affected shall be suitably protected (e.g. by protective tapes, paints or grease) and/or manufactured from a material resistant to such substances.	WR-522.5.1
Where a wiring system passes through elements of building construction (such as floors, walls, roofs, ceilings, partitions or cavity barriers), the openings remaining after passage of the wiring system shall be sealed and provide the same degree of protection from water penetration as the building construction element in which it has been installed.	WR-527.2.1 WR-527.2.7
The seal and the wiring system shall be protected from dripping water.	WR-527.2.7

Where a wiring system is routed below
services liable to cause condensation (such
as water, steam or gas services), precautions
shall be taken to protect the wiring system
from harmful effects.

WR-528.3.2

8.6.6.3 Ice

Wiring systems shall withstand the expected
external influences such as **ice** formation.

WR-712.522.8.3

8.6.6.4 Marinas

In marinas, equipment installed on or above a
jetty, wharf, pier or pontoon shall be selected
as follows, according to the external influences
which may be present:

WR-709.512.2.1.1

- water splashes (AD4): IPX4
- water jets (AD5): IPX5
- water waves (AD6): IPX6.
- marinas: particular attention shall be given
  to the likelihood of corrosive elements,
  movement of structures, mechanical damage,
  presence of flammable fuel and the increased
  risk of electric shock due to the presence of
  water.

WR-709.512.2

 Note: See Chapter 7 for other requirements concerning weather and precipitation specific to special installations and locations.

8.6.6.5 Socket outlets

Socket outlets of wiring systems shall be
placed at a height of 0.5 m to 1.5 m from the
ground to the lowest part of the socket outlet.
In special cases, due to environmental
conditions such as risk of flooding or heavy
snowfall, the maximum height is permitted to
exceed 1.5 m.

WR-708.553.1.9

8.7 Pollutants and contaminants

Over the last twenty years environmental matters have become an area of widespread public concern, particularly those concerning the issues of pollution and contamination. Pollutants and contaminants come in many forms and can have an effect on the air, land or water courses and as pollutants move from one medium to another they may be deposited on equipment and equipment housing; they can cause extensive damage.

Pollution of the air can occur in both the troposphere and stratosphere, as shown in Figure 8.12 below. In the troposphere pollutants from chimneys (for example) are carried by the air and can be deposited over time and distance, thus having a limited life span before they are washed out or deposited on the ground. If pollutants are injected straight into the stratosphere (as with a volcanic eruption) they will remain there for some time and result in noticeable effects over the whole region. On the other hand, the roughness of the ground will produce air turbulence which will itself promote the mixing of pollutants and even low wind speeds will result in high pollutant concentrations.

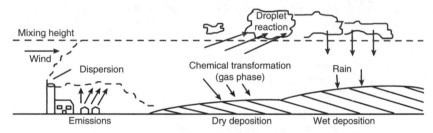

Figure 8.12 Processes involved between the emission of air pollutants and their deposition on the ground

Sources of natural pollutants include:

- sulphur – emitted by volcanoes and arising from biological processes;
- nitrogen – from biological processes in soil and lightning and biomass burning;
- hydrocarbons – methane from fermentation of rice paddies, fermentation of the digestive tract of ruminants (e.g. cows). Also released by insects, coal mining and gas extraction.

Sources of man-made pollutants include:

- carbon dioxide and carbon monoxide produced during the burning of fossil fuels;
- soot formation accompanied by carbon monoxide and generally due to inadequate or poor air supply;
- hydrocarbons. Most boilers and central heating units burning fossil fuels have very low emissions of gaseous hydrocarbons or oxygenated hydrocarbons such as aldehydes.

8.7.1 Pollutants

Although pollutant gases are normally only present in low concentrations, they can cause significant corrosion and a marked deterioration in the performance of contacts and connectors. The gases in operating environments which cause corrosion are oxygen, water vapour and the so-called pollutant gases which include sulphur dioxide, hydrogen sulphide, nitrogen oxides and chlorine compounds.

Silver and some of its alloys are particularly susceptible to tarnishing by the minute quantities of hydrogen sulphide that occur in many environments. The tarnish product is dark in colour, consists largely of b-silver sulphide and separable electrical connections using these materials may, therefore, suffer from increased resistance and contact noise as a result.

The amount of tarnishing (of a metal) is dependent upon the amount of humidity present. Less corrosion occurs below 70% relative humidity (RH) but above 80% RH the rate of tarnishing increases rapidly. Temperature also has an effect on the amount of tarnishing as the nature of the corrosion mechanism has a tendency to change at temperatures above 30°C.

8.7.1.1 Sulphur dioxide

Sulphur dioxide is the pollutant gas most commonly found in the atmosphere and is usually present in high concentrations in urban and industrial locations. In combination with other pollutants and moisture (e.g. humidity) it is responsible for the formation of high resistance, visible corrosion layers on all but the most noble metals (e.g. silver and gold) and alloys.

Although sulphur dioxide alone is less corrosive than other gases (such as sulphur trioxide, nitrogen oxides and chlorine compounds), the most extensive corrosion occurs when combustion products are present together with sulphur dioxide.

8.7.1.2 Hydrogen sulphide

Hydrogen sulphide is caused by bacterial reduction of sulphates in vegetation, soil, stagnant water and animal waste on a world-wide basis. In the atmosphere, hydrogen sulphide is oxidised to sulphur dioxide which, in turn, is brought to the ground by rain. In an aerobic soil, bacteria turns the sulphur dioxide to sulphates. Sulphate-reducing bacteria complete the cycle and turn the sulphates to hydrogen sulphide which is the principal natural sulphur input in the atmosphere and is, therefore, a widespread pollutant of air.

8.7.1.3 Nitrogen oxides

The production of nitrogen oxides is particularly significant because the rate of corrosion by sulphur dioxide is greatly accelerated in the presence of nitrogen dioxide, although the corrosion products have similar compositions.

8.7.2 Contaminants

Contaminants are composed of dust, sand, smoke and other particles that are contained within the air and these can have an effect on electrical equipment in various ways, especially:

- ingress of dust into enclosures and encapsulations;
- deterioration of electrical characteristics (e.g. faulty contact, change of contact resistance);
- seizure or disturbance in motion bearings, axles, shafts and other moving parts;
- surface abrasion (erosion and corrosion);
- reduction in thermal conductivity;
- clogging of ventilating openings, bushes, pipes, filters and apertures that are necessary for operation etc.

The presence of dust and sand in combination with other environmental factors such as water vapour can also cause corrosion and promote mould growth. Damp heat atmospheres cause corrosion in connection with chemically aggressive dust, and similar effects are caused by salt mist. Effects of ion-conducting and corrosive dusts (e.g. de-icing salts) need also to be considered.

8.7.2.1 Dust and sand

The concentration of dust and sand in the atmosphere varies widely with geographical locality, local climatic conditions and the type and degree of activity taking place. The amount of dust and sand found in the air is dependent on terrain, wind, temperature, humidity and precipitation. Under certain conditions enormous amounts of dust and sand may be temporarily released and this suspended dust will drift away with the wind (see Table 8.14) depending on its concentration and the size of the particles.

Table 8.14 Concentration of dust and sand (extracted with permission from a paper by Herne European Consultancy)

Region	Dust and sand concentration ($\mu g/m^3$)
Rural and suburban	40–110
Urban	100–450
Industrial	500–2000

Particles larger than $150\,\mu m$ are generally confined to the air layer in the first metre above ground and in this layer, about half of the sand grains move within the first $10\,mm$ above the surface.

The dust and sand appearing in enclosed and sheltered locations is generated by several sources (quartz, de-icing salts, fertilisers etc., penetrating into locations via ventilating ducts or badly fitting windows). The dust may also come from cloth or carpets in normal use in the working environment.

8.7.2.2 Dust

Dust may be defined as 'particulate matter of unspecified origin, composition and size ranging from $1\,\mu m$ to $150\,\mu m$ originating from quartz, flour, organic fibres etc'. Particles of less than $75\,\mu m$, because of their low terminal velocity, can remain suspended in the atmosphere for very long periods through the natural turbulence of the air. In sheltered and enclosed locations, the maximum grain size tends to be smaller (e.g. less than $100\,\mu m$) than in non weather-protected locations due to the filtering effect of the shelter.

The dust found in and around electrotechnical products may be generated by several different sources. The dust may be quartz, coal, de-icing salts, fertilisers, or small fibres from cotton or wool (real or artificial that have been generated from cloth or carpets by normal use in living rooms and offices), and penetration of dust into a piece of equipment can be:

- carried in by forced air circulation, for example for cooling purposes;
- carried in by the thermal motion of the air;
- pumped in by variations in the atmospheric pressure caused by temperature changes;
- blown in by wind.

Dust itself can act as a physical agent, chemical component, or both, and can cause one or more of the following harmful effects:

- seizure of moving parts;
- abrasion of moving parts;
- adding mass to moving parts thereby causing unbalance;
- deterioration of electric insulation;
- deterioration of dielectric properties;
- clogging of air filters;
- reduction of thermal conductivity;
- interference with optical characteristics;

and also

- corrosion and mould growth;
- overheating and fire hazard.

Dust adhering to the surface of materials may contain organic substances that provide a source of food for micro-organisms.

8.7.2.3 Sand

Sand is the term applied to 'segregated unconsolidated accumulation of detrital sediment, consisting mainly of tiny broken chips of crystalline quartz or other mineral, between 100 μm and 1000 μm in size'. Particles greater than 150 μm are unable to remain airborne unless continually subjected to strong winds, induced air-flows or turbulence. Sand is generally harder than most fused silica glass compositions and can, quite naturally, scratch the surface of most glass optical devices. Pressure applied over trapped grains of sand can also cause fractures to occur in equipment.

The electrostatic charges produced by friction of the particles in sand storms can interfere with the operation of equipment and sometimes be dangerous to personnel. The breakdown of insulators, transformers and lightning arresters and the failure of car ignition systems has also been known to occur as a result of such charges. The electrostatic voltages produced can be very large. Indeed, voltages as high as 150 kV have made telephone and telegraph communications inoperable during sand storms.

Quartz, because of its hardness, can result in rapid wear or damage to products, particularly moving parts. However, erosion of material requires that the presence of dust and sand is combined with a high velocity air stream over an extensive period of time.

Sand and the majority of dusts usually deposited on insulated surfaces are poor conductors in the absence of moisture. The presence of moisture, however, will result in the dissolving of the soluble particles and the formation of conducting electrolytes. For example, the leakage currents flowing over contaminated power line insulators can be of the order of one million times greater than those which flow through clean, dry insulators.

8.7.2.4 Smoke or fumes

Smoke or fumes are 'dispersive systems in the air consisting of particles below 1 μm'. As the particles are so small they do not usually affect equipment, provided that the equipment is properly designed.

8.7.2.5 Fauna and flora

With a few exceptions, fauna (rodents, insects, termites, birds etc.) and flora (plants, trees, seeds, fruit, blossom, mould, bacteria and fungi etc.) may be present at all locations where equipment is stored, transported or used. Whilst fauna may be the cause of damage inside buildings as well as open-air locations, damage by flora will predominantly occur in open-air conditions. Moulds and bacteria can, however, be present both inside buildings and in open-air conditions.

The frequency of this flora and fauna depends on temperature and humidity. In warm damp climates, fauna and flora, especially insects and micro-organisms such as mould and bacteria, will find favourable conditions of life. Humid or wet rooms in buildings (or rooms in which processes produce humidity) are suitable living spaces for rodents, insects and micro-organisms. The range of temperature in which moulds may grow is from 0°C to 40°C, and the most favourable temperatures for many cultures is between 20°C and 30°C.

If the surfaces of the products carry layers of organic substances (e.g. grease, oil, dust), or animal/vegetable deposits, the surfaces will become ideal locations for the growth of moulds and bacteria.

Effects of flora and fauna

The functioning of equipment and materials can be affected by physical attacks of fauna. Small animals and insects that feed from, gnaw at, eat into and chew at materials are particular problems as are termites cutting holes into material.

Materials, such as wood, paper, leather, textiles, plastics (including elastomers) and even some metals such as tin and lead are all susceptible to attack.

Larger animals can also cause damage by stroke, impact or thrust. These attacks can cause:

- physical breakdown of material, parts, units, devices;
- mechanical deformation or compression;
- surface deterioration;
- electrical failure caused by mechanical deterioration.

Deposits from fauna (especially insects, rodents, birds etc.) can be caused by the presence of the animal itself, nest-building, deposited feed stocks and metabolic products such as excrement and enzymes etc.

Deposits from flora may consist of detached parts of plants (leaves, blossom, seeds, fruits etc.) and growth layers of cultures of moulds or bacteria. These attacks can lead to:

- deterioration of material;
- metallic corrosion;
- mechanical failure of moving parts;
- electrical failure due to:
  - ○ increased conductivity of insulators;
  - ○ failure of insulation;
  - ○ increased contact resistance;
  - ○ electrolytic and ageing effects in the presence of humidity or chemical substances;
  - ○ moisture absorption and adsorption;
  - ○ decreased heat dissipation.

These in turn can cause interruption of electrical circuits, malfunctioning of mechanical parts and clouding of optical surfaces (including glass).

8.7.2.6 Mould

Surface contamination in the form of dusts, splashes, condensed volatile nutrients or grease may be deposited on equipment. When that equipment is exposed (in use, storage or transportation) to the atmosphere, and without proper protective covering, mould growth will occur and mould can cause unforeseen damage to equipment, whether constructed from mould-resistant materials or not!

Fungi grows in soil and in, or on, many types of common material. It propagates by producing spores which become detached from the main growth and later germinate to produce further growth. The spores are very small and easily carried by the wind (or moving air). They also adhere to dust particles carried in the air. Contamination can also occur due to handling. Spores may be deposited by the hands or in the film of moisture left by the hands.

Germination and growth

Moisture is essential in allowing the spores to germinate and when a layer of dust or other hydrophilic material (i.e. moisture retaining) is present on the surface, sufficient moisture may be abstracted by it, from the atmosphere. In addition to high humidity, spores require (on the surface of the specimen) a layer of material that will absorb the moisture. Mould growth is also encouraged by stagnant air spaces and lack of ventilation.

When the relative humidity is below 65%, no germination or growth will occur. The higher the relative humidity above this value, the more rapid the growth will be. Spores can survive long periods of very low humidity and even though the main growth may have died, they will germinate and start a new growth as soon as the relative humidity becomes favourable again (i.e. in excess of 65%). The optimum temperature of germination for the majority of moulds is between 20°C and 30°C.

Effects of mould growth

Moulds can live on most organic materials, but some of these materials are much more susceptible to attack than others. Growth normally occurs only on surfaces exposed to the air, and those which absorb or adsorb moisture will generally be more prone to attack.

Even where only a slightly harmful attack on a material occurs, the formation of an electrically conducting path across the surface due to a layer of wet mycelium (i.e. vegetative part of fungus) can drastically lower the insulation resistance between electrical conductors supported by an insulation material. When the wet mycelium grows in a position where it is within the electromagnetic field of a critically adjusted electronic circuit, it can cause a serious variation in the frequency-impedance characteristics of that circuit.

Among the materials that are very susceptible to attack are leather, wood, textiles, cellulose, silk and other natural resources. Most plastic materials, although less susceptible, are also prone to attack as they will probably contain oligomers (i.e. natural or synthetic compounds of, usually, high molecular weight that consist of millions of linked simple molecules), non-polymerised monomers (i.e. a molecule that can combine with others to form a polymer) and/or additives which may radiate to the surface and be a nutrient for fungi.

Some plastic materials depend, for a satisfactory life-span, on the presence of a plasticiser (i.e. substances added to plastics to make or keep them soft or pliable) which, if it is readily digested by fungi, will eventually give rise to failure of the main material.

Mould attack on materials usually results in a decrease of mechanical strength and/or changes in other physical properties and the growing mould on the surface of a material can yield acid products and other electrolytes which will cause a secondary attack on the material. This attack can lead to electrolytic or ageing effects, and even glass can lose its transparency due to this process. Oxidation or decomposition may be facilitated by the presence of catalysts secreted by the mould.

Prevention of mould growth

All insulating materials used should be chosen to give as great a resistance to mould growth as possible, thus maximising the time taken for mycelium to grow and minimising any damage to the material consequent upon such growth. The use of lubricants during assembly (varnishes, finishes etc.) is frequently necessary in order to obtain the required performance or durability of a product. Such materials should be chosen with regard to their ability to resist mould growth for even though it can be shown the lubricants do not support mould growth, they may collect dust which in turn will support mould growth.

Moisture traps which could possibly be formed during the assembly of equipment and in which mould can grow should be avoided. Examples of such less obvious traps are between unsealed mating plugs and sockets, or between printed circuit cards and edge connectors. Other preventatives of mould growth include:

- complete sealing of the equipment in (and with) a dry, clean atmosphere;
- continuous heating within an enclosure, which can ensure a sufficiently low humidity;
- operation of equipment within a suitably controlled environment;
- regularly replaced desiccants (e.g. silica beads);
- periodic and careful cleaning of enclosed equipment.

Where the material and functioning of the equipment allow such treatment, ultraviolet radiation or ozone may be used for sterilisation. Air currents flowing over the parts can retard the development of mould growth and can be used to control the action of acarids (i.e. mites and ticks).

8.7.3 Typical requirements – pollutants and contaminants

Table 8.15 lists the most common environmental requirements concerning pollutants and contaminants.

Table 8.15 Typical requirements – pollutants and contaminants

Pollutants	• Although the severity of pollution will depend upon the location of the equipment, the effects of pollution must be considered in the design of equipment and components. • Means need to be provided to reduce pollution by the effective use of protective devices. • The requirements of ISO 14001 regarding environmental protection and the prevention of pollution have to be met.
Contaminants	The following should be considered: • chemically active substances • biologically active substances • flora and fauna • dust • sand.
Mould	• In an assembled state, equipment needs to operate when exposed to airborne mould spores and within climates that will be conducive to the growth of moulds. • Insulating materials should be chosen to provide as much resistance to mould growth as possible and all materials used should be chosen with regard to their ability to resist mould growth.

8.7.4 Requirements from the Regulations – pollutants and contaminants

8.7.4.1 Wiring systems

In escape routes that are likely to be BD2 (difficult), BD3 (crowded) and BD4 (difficult and crowded), the wiring system selected shall have a limited rate of smoke production.	WR-422.2.1
A wiring system shall **not** be installed in the vicinity of services which produce heat, smoke or fumes likely to be detrimental to the wiring, **unless** it is protected from harmful effects by shielding arranged so as not to affect the dissipation of heat from the wiring.	WR-528.3.1
Where no fire alarm system is installed in a building used for exhibitions etc., cable systems shall be flame retardant to BS EN 60332-1-2 and low smoke to BS EN 61034-2.	WR-711.521

8.7.4.2 Safety services

The location of a safety source shall be properly and adequately ventilated so that exhaust gases, smoke or fumes from the safety source cannot penetrate areas occupied by persons.	WR-560.6.3

8.7.4.3 Dust

Precautions shall be taken to prevent the accumulation of dust or other substances which could adversely affect the heat dissipation from a wiring system.

Switchgear or controlgear shall be installed outside the location unless it is installed in an enclosure providing a degree of protection of at least IP4X or, in the presence of dust, IP5X.	WR-422.3.3
A heat storage appliance shall prevent the ignition of combustible dusts or fibres by the heat storing core.	WR-422.3.16
In a location where dust in significant quantity is present (AE4), additional precautions shall be taken to prevent the accumulation of dust or other substances in quantities which could adversely affect the heat dissipation from the wiring system.	WR-522.4.2
Luminaires shall be selected with regard to their degree of protection against the ingress of dust.	WR-705.559

8.7.4.4 Mould

Where previous or anticipated experience of mould constitute a hazard (AK2), the wiring system shall be selected accordingly and/or special protective measures shall be adopted.	WR-522.9.1

 Note: Possible preventive measures are closed types of installation (conduit or channel), maintaining distances to plants and regular cleaning of the relevant wiring system.

8.7.4.5 Fauna

Where previous or anticipated experience of fauna constitute a hazard (AL2), the wiring system shall be selected accordingly or special protective measures shall be adopted, for example, by:	WR-522.10.1

- the mechanical characteristics of the wiring system; or
- the location selected; or
- the provision of additional local or general protection against mechanical damage; or
- any combination of the above.

Where the conditions experienced or expected constitute a hazard (AK2), the wiring system shall be selected accordingly or special protective measures shall be adopted.	WR-522.9.1
In locations accessible to, and enclosing, livestock, special attention shall be given to the presence of different kinds of fauna (e.g. rodents).	WR-705.522

 Note: Possible preventive measures are closed types of installation (conduit or channel), maintaining distances to plants and regular cleaning of the relevant wiring system.

8.8 Mechanical

Mechanics is the branch of physics concerned with the motions of objects and their response to forces. Modern descriptions of such behaviour begin with a careful definition of such quantities as displacement (distance moved), time, velocity, acceleration, mass and force.

There is often a tendency to underestimate the effect that the mechanical environment can have on the reliability of equipment, especially the effects of vibration and shock. Mechanical stresses are normally attributed to a moving mass and there is frequently a tendency to underestimate the effect of the mechanical environment on the reliability of static installations. Experience suggests, however, that vibrations and shocks are a significant reliability, availability and maintainability (RAM) factor, not only from the point of view of vehicle mounted equipment, but also with respect to permanent installations.

If a spurious vibration acts on a printed circuit board (PCB), module or equipment, resonant oscillations will be induced in all components at their natural frequency. If, however, the frequency spectrum has some more or less distinctive frequency bands, the elements will perform forced oscillations at the cyclic frequency of the interference and generally depend both on the characteristics of the oscillator (i.e. the component) and on the interference.

8.8.1 Shock

'Shock' is generally defined as 'an impact shock characterised by a simple acceleration and free impact on a firm base' and is usually the result of a violent collision, or a heavy blow. Whilst it is difficult to design and install electrical equipment, components and systems to be completely immune to shock, precautions should, nevertheless, be taken to guard against potential problem areas.

8.8.2 Vibration

Components, equipment and other articles during transportation or in service may be subjected to conditions involving vibration of a harmonic pattern, generated primarily by rotating, pulsating or oscillating forces caused by machinery and seismic incidents.

8.8.3 Acceleration

Equipment, components and electrotechnical products that are likely to be installed in moving bodies (e.g. rotating machinery) will be subjected to forces caused by steady accelerations. In general the accelerations encountered in service will have different values along each of the major axes of the moving body, and, in addition, usually have different values in the opposite axes.

8.8.4 Protection

- The resonant frequency of components is greatly influenced by the length of their connecting wires and the actual length of these connecting wires may well be the decisive factor as to whether a component fails or remains functioning under given vibration and impact conditions.
- The amplitude of shocks on the equipment can be reduced by use of special mounting devices.
- Shock absorption is based on storing the impact and releasing it at a retarded rate. The peak acceleration is reduced and the high frequencies damped, thus providing protection for the components with their relatively higher natural frequencies (of some hundred hertz).
- Vibration dampers and shock absorbers are often used as a form of protection against mechanical stresses. The basic difference between vibration dampers and shock absorbers is that with the former the natural frequency lies below the interference frequency, whilst with the latter it is above.
 - Generally speaking, vibration dampers provide no protection whatsoever against shocks and, similarly, shock absorbers offer no protection against vibrations. Only in exceptional cases can vibration dampers, for high frequencies, be used as shock absorbers.
- Elastic suspension of equipment can cause a critical increase in amplitude at certain frequencies and, where translatory and rotary displacements greater than six degrees of freedom are possible, very complex motions may arise. Wherever possible, therefore, one should try to ensure that none of the (possible) resonant frequencies falls within the range of the induced displacements.
- Sheathing circuits by means of cast resins can, in most cases, be a very effective means of counteracting mechanical stresses combined with temperature humidity.

8.8.5 Typical requirements – mechanical

Table 8.16 lists the most common environmental requirements concerning mechanical factors.

Table 8.16 Typical requirements – mechanical

Vibrations and shocks	Any dampers or anti-vibration mountings must be integral with the equipment to prevent the unit being accidentally installed without them.
Mechanical shock	Equipment should be capable of withstanding shock pulses (e.g. a minimum of 20,000 shocks at a shock level of 20 g).
On or near the roadside	Equipment located on or near the roadside must be capable of withstanding vibrations and shocks.
Long-term exposure	Equipment must be capable of withstanding long-term exposure to shocks.
Encapsulated outdoor installations	Equipment contained in encapsulated outdoor installations must be capable of withstanding vibrations and shocks.
Closed rooms	Equipment located in closed-room installations must be capable of withstanding self-induced vibrations.
In service	Equipment should be capable of withstanding without deterioration or malfunction all mechanical stresses that occur in service.
Random vibration	Equipment should be capable of withstanding random vibration.

8.8.6 Requirements from the Regulations – mechanical

8.8.6.1 Mechanical stresses

A wiring system shall be selected and erected to avoid during installation, use or maintenance, damage to the sheath or insulation of cables and their terminations.	WR-522.8.1
A conduit system or cable ducting system (other than a pre-wired conduit assembly that has been specifically designed for the installation) that is going to be buried in the structure, shall be completely erected between access points before any cable is drawn in.	WR-522.8.2
The radius of every bend in a wiring system shall be such that conductors or cables do not suffer damage and terminals are not stressed.	WR-522.8.3
Where the conductors or cables are not supported continuously they shall be supported by suitable means at appropriate intervals in such a manner that the conductors or cables do not suffer damage by their own weight.	WR-522.8.4
Every cable or conductor shall be supported in such a way that it is not exposed to undue mechanical strain and so that there is no appreciable mechanical strain on the terminations of the conductors.	WR-522.8.5
A wiring system intended for the drawing in or out of conductors or cables shall have adequate means of access to allow this operation.	WR-522.8.6

A cable buried in the ground (that is not installed in a conduit or duct) shall incorporate an earthed armour or metal sheath or both, suitable for use as a protective conductor.	WR-522.8.10
The location of buried cables shall be marked by cable covers or a suitable marking tape.	WR-522.8.10
Buried conduits and ducts shall be suitably identified.	WR-522.8.10
Buried cables, conduits and ducts shall be at a sufficient depth to avoid being damaged by any reasonably foreseeable disturbance of the ground.	WR-522.8.10

 Note: See IEC 61386-24 for further details concerning underground conduits.

Cable supports and enclosures shall not have sharp edges liable to damage the wiring system.	WR-522.8.11
A cable or conductors shall not be damaged by the means of fixing.	WR-522.8.12
Cables, busbars and other electrical conductors which pass across expansion joints shall be so selected and/or erected that anticipated movement does not cause damage to the electrical equipment.	WR-522.8.13
No wiring system shall penetrate an element of building construction which is intended to be load-bearing unless the integrity of the load-bearing element can be assured after such penetration.	WR-522.8.14

8.8.6.2 Shock

For locations where the wiring system may be exposed to impact and mechanical shock due to vehicles and mobile agricultural machines etc., the external influences shall be classified AG3 and: • conduits shall provide a degree of protection against impact of 5 J according to BS EN 61386-2 • cable trunking and ducting systems shall provide a degree of protection against impact of 5 J according to BS EN 50085-2-1.	WR-705.522.16

8.8.6.3 Vibration

Stationary equipment which is moved temporarily for the purposes of connecting, cleaning etc. (e.g. a cooker or a flush-mounting unit for installations in a false floor) may be connected with a non-flexible cable however, if it is subject to vibration whilst in use it shall be connected by a flexible cable or cord.	WR-521.9.2
A wiring system (supported by or fixed to a structure or equipment) that is subject to vibration of medium severity (AH2) or high severity (AH3) shall be suitable for such conditions, particularly where cables and cable connections are concerned.	WR-522.7.1

 Where no vibration or movement can be expected, cable with non-flexible cores may be used.

8.8.6.4 Electrical connections

Connections between conductors or between a conductor and other equipment shall provide durable electrical continuity and adequate mechanical strength and protection.	WR-526.1
The selection of the means of connection shall take account of:	WR-526.2

- the material of the conductor and its insulation
- the number and shape of the wires forming the conductor
- the cross-sectional area of the conductor
- the number of conductors to be connected together
- the temperature attained at the terminals in normal service
- the provision of adequate locking arrangements in situations subject to vibration or thermal cycling.

8.8.6.5 Electrical connections in caravans and motor homes

In a caravan or motor caravan, the wiring will be subjected to vibration; all wiring shall be protected against mechanical damage either by location or by enhanced mechanical protection.	WR-721.522.7.1

Wiring passing through metalwork shall be
protected by means of suitable bushes or
grommets, securely fixed in position.

WR-721.522.7.1

Precautions shall be taken to avoid mechanical
damage due to sharp edges or abrasive parts.

WR-721.522.7.1

All cables, unless enclosed in rigid conduit and all
flexible conduit shall be supported at intervals not
exceeding 0.4 m for vertical runs and 0.25 m for
horizontal runs.

WR-721.522.8.1.3

8.8.6.6 Low-voltage generating sets – generators

Electrical equipment associated with the generator
shall be mounted securely and, if necessary, on anti-
vibration mountings.

WR-740.551.8

8.8.6.7 Protection against overcurrent

Persons and livestock shall be protected against
injury, and property shall be protected against
damage, due to electromechanical stresses caused by
any overcurrents likely to arise in live conductors.

WR-131.4

8.8.6.8 Protection against fault current

Electrical equipment, including conductors, shall
be provided with mechanical protection against
electromechanical stresses of fault currents as
necessary to prevent injury or damage to persons,
livestock or property.

WR-131.5

8.8.6.9 Cross-sectional area of conductors

The cross-sectional area of conductors shall be
determined for both normal operating conditions
and, where appropriate, for fault conditions
according to:

WR-132.6

- the electromechanical stresses likely to occur
  due to short-circuit and earth fault currents

- other mechanical stresses to which the conductors are likely to be exposed.

8.8.6.10 Waves

Wiring systems that may be subjected to waves (AD6)   WR-522.3.3
shall be protected from mechanical damage (i.e. impact, vibration and other mechanical stresses).

8.9 Electromagnetic compatibility

Most car owners normally accept that when they drive near electricity pylons, their listening pleasure may be interrupted by loud crackles and/or buzzing noises but with the increased use of electronic equipment, the problem of interference has become one of our prime concerns.

Although most forms of interference are usually tolerated as being 'one of those things that you cannot do much about', the design of modern sophisticated equipment has become so susceptible to electromagnetic interference that some form of regulation has had to be agreed.

Within Europe, this regulation is contained in the Electromagnetic Compatibility (EMC) Directive 2004/108/EC, which clearly states that all electronic equipment shall be constructed so that:

- the electromagnetic disturbance it generates does not exceed a certain level, so allowing radio and telecommunications equipment and other apparatus to operate as intended;
- the apparatus has an adequate level of intrinsic immunity to electromagnetic disturbance.

8.9.1 Typical requirements

Table 8.17 lists the most common environmental requirements concerning electromagnetic compatibility.

Table 8.17 Typical requirements – electromagnetic compatibility

Equipment	• The use of electronic equipment shall not interfere with the operation of other equipment. • All active electronic devices shall comply with the EMC Directive.
CE marking	Only CE (Conformity Europe) marked equipment may be offered for sale and all active equipment connected to an electrical installation shall carry a CE mark.

(*continued*)

Table 8.17 (*continued*)

Apparatus cases	Input/output connections from apparatus cases should always be of non-screened non-balanced signalling cable and are normally restricted in length.
Atmospheric disturbances	To counteract the affects of storms, it is generally recommended that all equipment should be capable of withstanding (as a minimum) the following overvoltages: Magnitude: 2000 V Rise time: $1.2\,\mu s$ Middle voltage time: $50\,\mu s$
Equipment immunity levels	• Equipment should be immune to induced common-mode voltages • Equipment should not experience a permanent loss of availability or suffer component damage for any induced common-mode voltage within this range.
Magnetic field	As low-frequency fields can influence cathode ray tubes, equipment should be capable of withstanding the following intensities:

Hz	A/m
5	0.8
50	3.0
250	1.5

|
| Power supply lines | Equipment should be immune to the following high-frequency bursts:
Initial peak-to-peak voltage   1 kV
Burst repetition rate   5 kHz |
| Transients | All electronic equipment should be capable of withstanding:
• transients (either directly induced or indirectly coupled) so that no damage or failure occurs during operation
• without damage or abnormal operation, transient non-repetitive surges. |

8.9.2 Requirements from the Regulations – electromagnetic compatibility

8.9.2.1 Electrical installations

Electrical installations **shall** be arranged so that they do not mutually interfere (including electromagnetic interference) with other electrical installations and non-electrical installations in a building.

> All fixed installations shall be in accordance with the current and relevant EMC regulations as shown in the EMC Directive 2004/108/EC.   WR-332.1

The previous EMC Directive (i.e. Directive 89/336/EC) was replaced by EMC Directive 2004/108/EC on 15 December 2004 and has been transposed into UK Law by the EMC Regulations 2006 (SI 2006/3418), which came into force on 20 July 2007. These new Regulations have introduced a new regime for fixed installations and require that **all** electrical and electronic apparatus marketed in the UK, including imports, satisfies the requirements of the EMC Directive.

Immunity levels of equipment shall be chosen taking into account:	WR-515.3.1
• the electromagnetic influences that can occur when the equipment is connected and erected as for normal use; and • the intended level of continuity of service necessary for the application. (See BS EN 50082.)	
Equipment shall be chosen with sufficiently low emission levels so that it cannot cause unacceptable electromagnetic interference with other electrical equipment. (See BS EN 50081.)	WR-515.3.2
Consideration shall be given by the planner and designer of the electrical installation to measures reducing the effect of induced voltage disturbances and electromagnetic interferences (EMI).	WR-332.2
To minimise voltages induced by lightning, the area of all wiring loops shall be as small as possible.	WR-712.444.4.4

8.9.2.2 Protection against voltage disturbances and measures against electromagnetic influences

Persons and livestock shall be protected against injury, and property shall be protected against any harmful effects, as a consequence of:	WR-131.6.1
• a fault between live parts of circuits supplied at different voltages	WR-131.6.2
• overvoltages such as those originating from atmospheric events or from switching	WR-131.6.3
• undervoltage and any subsequent voltage recovery.	
The installation shall have an adequate level of immunity against electromagnetic disturbances so as to function correctly in the specified environment.	WR-131.6.4
The installation design shall take into consideration the anticipated electromagnetic emissions generated by the installation or the installed equipment.	WR-131.6.4

8.9.2.3 Installation

The choice of the type of wiring system and the method of   WR-132.7
installation shall include consideration of the following:

- the nature of the location
- the nature of the structure supporting the wiring
- accessibility of wiring to persons and livestock
- voltage
- the electromechanical stresses likely to occur due to
  short-circuit and earth fault currents
- electromagnetic interference
- other external influences (e.g. mechanical, thermal
  and those associated with fire) to which the wiring
  is likely to be exposed during the erection of the
  electrical installation or in service.

Every installation shall be divided into circuits, as   WR-314.1
necessary, to:

- avoid hazards and minimise inconvenience in the
  event of a fault
- ensure safe inspection, testing and maintenance
- prevent the indirect energising of a circuit that is
  intended to be isolated
- reduce the possibility of unwanted tripping of RCDs
  due to excessive protective conductor currents
  produced by equipment
- reduce the effects of electromagnetic interferences
  (EMI)
- take account of any danger that may arise from the
  failure of a single circuit (such as a lighting circuit).

The electrical installation shall be arranged in such a way   WR-132.11
that no mutual detrimental influence will occur between
electrical installations and non-electrical installations.

Electromagnetic interference shall be taken into account.

8.9.2.4 Cables and conductors

Every conductor or cable shall have adequate strength   WR-521.5.1
and be so installed as to withstand the electromechanical
forces that may be caused by any current, including
fault current, it may have to carry in service.

Single-core cables armoured with steel wire or steel tape   WR-521.5.2
shall **not** be used for an a.c. circuit.

The conductors of an a.c. circuit installed in a
ferromagnetic enclosure shall be arranged so that all
line conductors, the neutral conductor (if any) and the
appropriate protective conductor are contained in the
same enclosure.

Where such conductors enter a ferrous enclosure, they
shall be arranged such that the conductors are only
collectively surrounded by ferrous material.

8.10 Fire

A fire will normally start when sufficient thermal energy from, for example
a burning cigarette or an electric short-circuit, is supplied to a combustible material. Following ignition, the fire will then produce its own thermal
energy, some of which will be used as feedback to maintain combustion and
some transferred via radiation and convection to other materials. These materials may also ignite and spread the fire.

The environmental conditions relating to the occurrence, development and
spread of fire within a building and its effect on electrotechnical products
exposed to fire are primarily covered by Section 8 (Fire Exposure) of IEC
721.2. This section provides background information for selecting the appropriate parameters and severities related to exposure of products to fire. More
detailed information on fire condition characteristics and fire hazard testing is
contained in specialist documentation.

The development of the fire generally consists of three processes:

- thermal
- aerodynamic
- chemical.

As a rule, radiation, convection and flame spread are the dominant physical
factors.

8.10.1 Fire growth

Once a fire has started in a space (e.g. a room) its growth and spread is determined by:

- site;
- volume;
- arrangement of the fuel or fire load, its distribution, continuity, porosity
  and combustion properties;

- aerodynamic conditions of the space;
- shape and size of the space;
- thermal properties of the space.

During the growth of a fire, a hot layer of gas builds up under the ceiling of the space. Under certain conditions, this gas layer can give rise to a rapid fire growth and flashover might occur.

8.10.2 Flashover

One normally defines flashover as the time when flames begin to emerge from the openings of the space, which correlates to a temperature of 500°C to 600°C in the upper gas layer.

Flashover marks the transition from the growing fire (pre-flashover) to the fully developed fire (post-flashover).

8.10.2.1 Pre-flashover

A pre-flashover fire primarily concerns the operation and function of products (e.g. detectors, alarm systems, associated cables and sprinklers) that are vital to maintaining the level of safety required for escape and/or the rescue of people caught in a fire.

Characteristics of pre-flashover fire

The ignitability properties of exposed material will depend on:

- the heat supplied;
- the exposure time;
- the presence or not of flames;
- the geometrical location;
- the thermal data.

together with time variations such as:

- rate of heat release;
- rate of flame spread;
- gas temperature.

Fire hazard of a pre-flashover

The fire hazard of a pre-flashover situation is normally considered in terms of a series of probabilities, which depends on:

- the presence of ignition sources;
- the presence of products;
- the product fire performance properties;
- the environmental factors;
- the presence of people;
- the presence/operation of detection and suppression devices;
- the availability of escape.

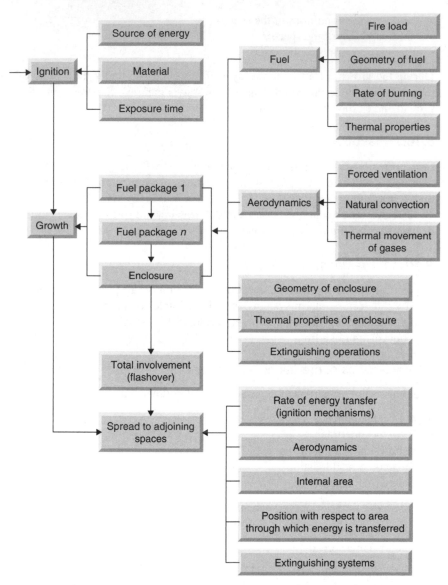

Figure 8.13 Factors affecting ignition, growth and spread of fire in a building

8.10.2.2 Post-flashover

Whilst most standards are normally concerned with conditions during the pre-flashover stage of a fire, conditions following flashover must also be considered. A post-flashover fire can seriously damage some of the structural and load-bearing elements of a building and the fire can then, quite easily, spread from one fire space to another via partitions and ventilation systems. This can, of course, seriously damage electrical equipment located in these voids. For example, in a large space it is quite possible that a fire, small in relation to that space, could be large enough to damage some of the structural elements in the post-flashover state. An important factor of the post-flashover fire, which is often overlooked, is the amount of smoke and toxic gases that can affect people in escape routes and remote safety areas in a building. Smoke and toxic gases can also significantly affect equipment.

Characteristics of post-flashover fire

The main characteristics of a post-flashover fire are:

- the rate of heat release;
- the gas temperature;
- the geometrical and thermal data for external flames;
- the smoke and its optical properties;
- the composition of the combustion products, particularly corrosive and toxic gases.

The possibility of a large external fire spreading from one storey to another in the same building (and eventually from one building to another) must also be considered. For these cases the first three characteristics – i.e. primarily gas temperature, geometrical and thermal data for the flames emerging from the window openings – are the most relevant.

8.10.3 Characteristics of smoke and gases as a fire product

Smoke is a mixture of heated gases, small liquid drops, and solid particles from the combustion. During a fire (pre- and post-flashover), smoke will be distributed within the building through the airflow between rooms and via ventilation ducts etc. In most circumstances this can have disastrous effects because smoke can not only damage and in some cases even destroy property; it can prevent the functioning of critical equipment. Most of the effects of smoke are of a chemical nature and the most prevalent is destruction or damage to electrotechnical products – in particular, corrosion caused by hydrogen chloride, which is a substance in smoke.

Metal surfaces, exposed to air under normal (non-fire) conditions, often have a chloride deposit up to $10\,mg/m^2$. Such an amount is normally not harmful. However, after exposure to smoke from a fire involving polyvinyl chloride (PVC), a surface contamination of up to thousands of milligrams per square metre can be found, often causing significant damage. Chloride contamination of electrotechnical equipment can be removed by, for instance, detergents, solvents, neutralising agents, ultrasonic vibrations, and clean air jets, but the procedures are not always effective, sometimes giving a temporary but not permanent cure.

Experiments, involving PVC-coated electrical wires and carried out on a scale large enough to be representative of real fires, are currently in hand.

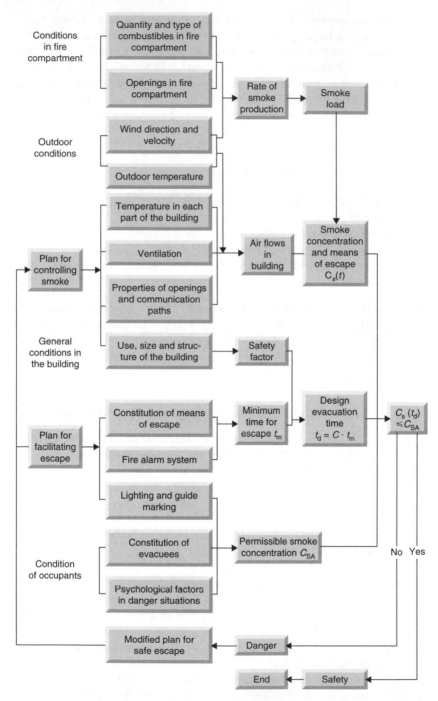

Figure 8.14 Flow diagram of a smoke control design system in a building

8.10.4 Building designs

In the design of buildings, the fire design of load-bearing structural elements and partitions is normally considered as a national problem and directly related to the results of standard national (and when available) international fire resistance tests. In such tests, the specimen is exposed, in a furnace, to a temperature rise, which is varied with time and within specified limits, according to the particular test being used.

Over the last decades, rapid progress has been made in the development of analytical and computational methods for determining the fire design of load-bearing and separating structures and structural elements. In the long term, it is foreseeable that this will develop into an analytical and/or computational design, directly based on a natural fire exposure. These will be specified with regard to the combustion characteristics of a fire load and the geometrical, ventilation and thermal properties of the fire space.

8.10.5 Test Standards

Fire tests on building materials, components and structures normally focus on the characteristics of pre-flashover fire. Simplified full-scale (i.e. room) tests for surface products against smoke and in particular toxic combustion products are already available, but considerable development work needs to be completed before a useful small-scale test is available.

If no mathematical model of a small-scale test is available, the test results should be statistically correlated directly to full-scale test data. If a validated mathematical model of a small-scale test exists, important material characteristics controlling the space fire growth can be given quantitative values which can then be used as input data in mathematical models of full-scale pre-flashover space fire for specified scenarios.

With a view to practical, long-term use the results of small-scale reaction to fire tests to predict fire hazard should be based on a fundamental and scientific approach.

Figure 5.15 outlines the structure of such an approach.

8.10.6 Other related standards and specifications

IEC 60695 Series	Fire hazard testing – guidance, tests and specifications for assessing fire hazard of electrotechnical products.
ISO 5657	Fire tests – reaction to fire – ignitability of building products.
ISO 5658	Reaction to fire tests – spread of flame on building products and vertical configuration.
ISO 5660	Fire tests – reaction to fire – rate of heat release from building products.

(continued)

ISO 9705	Fire tests – full-scale room test for surface products.
ISO TR 5924	Fire tests – reaction to fire – smoke generated by building products (dual-chamber test).
ISO TR 9112.1	Toxicity testing of fire effluents – general.

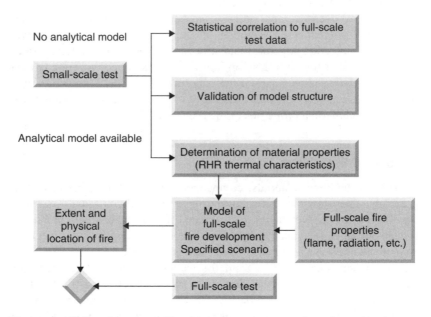

Figure 8.15 Combination of basic property tests and mathematical models for assessing the contribution of a tested material or product to the overall fire safety

8.10.7 Typical contract requirements – fire

In most contracts, reference is made to the IEC 60695 series of Standards which cover the assessment of electrotechnical products against a nominated fire hazard. CENELEC, on the other hand, show the requirement for equipment to operate in fire hazardous areas as three distinct clauses, as follows:

Class FO – no special fire hazard envisaged. This is considered a normal service condition and except for the characteristics inherent to the design of the equipment, no special measures need to be taken to limit flammability.

Class F1 – equipment subject to fire hazard. This is considered an abnormal condition and restricted flammability is required. Self-extinction of fire shall take place within a specified time period. Poor burning is permitted with negligible energy consumption.

The emission of toxic substances shall be minimised. Materials and products of combustion shall, as far as possible, be halogen free and shall contribute with a limited quantity of thermal energy to an external fire.

Class F2 – equipment subject to external fire. This is considered an abnormal condition and in addition to the requirements of Class F1, the equipment shall (by means of special provisions) be able to operate for a given time period when subjected to an external fire.

Materials are normally expected to conform to those requirements defined in EN 60721-3-3 and EN 60721-3-4.

8.10.8 Requirements from the Regulations – fire

BS 7671 has now been updated so as to maintain technical alignment with CENELEC harmonisation documents. One of the main changes concerns the requirements for safety services (e.g. emergency escape lighting, fire alarm systems, installations for fire pumps, fire rescue service lifts, smoke and heat extraction equipment) which now need to be observed.

Safety services have also been expanded in line with IEC standardisation.

8.10.8.1 Electrical installations

In electrical installations, risk of injury may result from excessive temperatures likely to cause burns, fires and other injurious effects. To guard against this happening:

Equipment in surroundings susceptible to risk of fire or explosion shall be so constructed and/or protected so as to prevent danger.	WR-132.5.2
The choice of the type of wiring system and the method of installation shall include consideration of the following:	WR-132.7

- the nature of the location
- the nature of the structure supporting the wiring
- accessibility of wiring to persons and livestock
- voltage
- the electromechanical stresses likely to occur due to short-circuit and earth fault currents
- electromagnetic interference
- other external influences (e.g. mechanical, thermal and those associated with fire) to which the wiring is likely to be exposed during the erection of the electrical installation or in service.

Note: In structures where the shape and dimensions are such as will facilitate the spread of fire, precautions shall be taken to ensure that the electrical installation cannot propagate a fire (e.g. chimney effect).

8.10.8.2 Selection and erection of installations in locations of national, commercial, industrial or public significance

The following measures may be considered:

- installation of mineral insulated cables according to BS EN 60702;
- installation of cables with improved fire-resisting characteristics in the case of a fire hazard;
- installation of cables in non-combustible solid walls, ceilings and floors;
- installation of cables in areas with constructional partitions having a fire-resisting capability for a time of 30 minutes or 90 minutes.

 Note: Where these measures are not practicable improved fire protection may be possible by the use of reactive fire protection systems.

8.10.8.3 Precautions where a particular risk of fire exists

Locations that include buildings or rooms with assets of significant value (such as national monuments, museums and other public buildings), or buildings such as railway stations and airports (that are generally considered to be of public significance), and, or buildings or facilities such as laboratories, computer centres and certain industrial and storage facilities, can be of commercial or industrial significance.

Electrical equipment shall be so selected and erected so that its normal temperature rise and foreseeable temperature rise during a fault cannot cause a fire.   WR-422.1.2

 A temperature cut-out device shall have manual reset.

Where BE2 conditions exist and where there is a risk of fire due to the manufacture, processing or storage of flammable materials such as:

- barns (due to the accumulation of dust and fibres);
- woodworking facilities;
- paper mills and textile factories (due to the storage and processing of combustible materials);

a fire risk will be present and the following precaution shall be observed:

A cable shall, as a minimum, satisfy the test under fire conditions specified in BS EN 60332-1-2.	WR-422.3.4
A cable not completely embedded in non-combustible material such as plaster or concrete or otherwise protected from fire shall meet the flame propagation characteristics as specified in BS EN 60332-1-2.	WR-422.3.4
A conduit system shall satisfy the test under fire conditions specified in BS EN 61386-1.	WR-422.3.4
A cable trunking system or cable ducting system shall satisfy the test under fire conditions specified in BS EN 50085.	WR-422.3.4
A cable tray system or cable ladder shall satisfy the test under fire conditions specified in BS EN 61537.	WR-422.3.4
Precautions shall be taken such that a cable or wiring system cannot propagate flame.	WR-422.3.4
Where the risk of flame propagation is high the cable shall meet the flame propagation characteristics specified in the appropriate part of the BS EN 50266 series.	WR-422.3.4
Conduit and trunking systems shall be in accordance with BS EN 61386-1 and BS EN 50085-1 respectively and shall meet the fire-resistance tests within these standards.	WR-422.4.6

8.10.8.4 Protection against the risk of fire

Where it is necessary to limit the consequence of fault currents in a wiring system from the point of view of fire risk:	WR-532.1

- the circuit shall be protected by an RCD for fault protection; and
- the RCD shall be installed at the origin of the circuit to be protected; and
- the RCD shall switch all live conductors; and
- the rated residual operating current of the RCD shall not exceed 300 mA; or
- the circuit will need to be continuously monitored by an insulation monitoring device which initiates an alarm on the occurrence of an insulation fault.

When selecting and erecting a luminaire, the thermal effects of radiant and convected energy on the surroundings shall be taken into account, including the fire-resistance of adjacent material:

- at the point of installation; and
- in the thermally affected areas.

A device for protection against fault current need not be provided where the wiring is installed in such a manner as to reduce to a minimum the risk of fire or danger to persons.	WR-434.3

The omission of devices for protection against overload is permitted for circuits supplying current-using equipment where unexpected disconnection of the circuit could cause danger or damage.

Examples of such circuits are:

- a circuit supplying a fire extinguishing device;
- a circuit supplying a safety service, such as a fire alarm or a gas alarm.

In such situations consideration should be given to the provision of an overload alarm.

8.10.8.5 Protection against thermal effects

Persons, livestock and property **shall** be protected against harmful effects of heat or fire which may be generated or propagated in electrical installations. These effects include:

- heat accumulation, heat radiation, hot components or equipment;
- failure of electrical equipment such as protective devices, switchgear, thermostats, temperature limiters, seals of cable penetrations and wiring systems;
- overcurrent;
- insulation faults or arcs, sparks and high temperature particles;
- harmonic currents;
- external influences such as lightning surge.

Note: The use of supplementary bonding does not exclude the need to disconnect the supply for other reasons, for example protection against fire, thermal stresses in equipment.

Electrical heating appliances used for the breeding and rearing of livestock shall comply with BS EN 60335-2-71 and shall be fixed so as to maintain an appropriate distance from livestock and combustible material, to minimise any risks of burns to livestock and of fire.	WR-705.422.6

For radiant heaters the clearance shall be not less than 0.5 m or such other clearance as recommended by the manufacturer.

For fire protection purposes, RCDs shall be installed with a rated residual operating current not exceeding 300 mA.	WR-705.422.7
RCDs shall disconnect all live conductors.	WR-705.422.7
Where improved continuity of service is required, RCDs not protecting socket outlets shall be of the S type or have a time delay.	WR-705.422.7
In locations where a fire risk exists conductors of circuits supplied at extra-low voltage shall be protected: • either by barriers or enclosures affording a degree of protection of IPXXD or IP4X; or • in addition to their basic insulation, by an enclosure of insulating material.	WR-705.422.8

8.10.8.6 Protection against fire heat generation

Lighting equipment such as incandescent lamps, spotlights and small projectors and other equipment or appliances with high-temperature surfaces, shall be suitably guarded, and installed and located in accordance with the relevant standard.	WR-711.422.4.2
Showcases and signs shall be constructed of material having an adequate heat-resistance, mechanical strength, electrical insulation and ventilation, taking into account the combustibility of exhibits in relation to the heat generation.	WR-711.422.4.2
Stand installations containing a concentration of electrical equipment, luminaires or lamps liable to generate excessive heat shall not be installed unless adequate ventilation provisions are made, e.g. well ventilated ceiling constructed of incombustible material.	WR-711.422.4.2

8.10.8.7 Protection against fire caused by electrical equipment

Electrical equipment shall not present a fire hazard to adjacent materials.	WR-421.1

Fixed electrical equipment shall be selected and erected such that its temperature in normal operation will not cause a fire.

WR-421.2

The heat generated by electrical equipment shall not cause danger or harmful effects to adjacent fixed material or to material.

WR-421.2

Where fixed equipment may attain surface temperatures which could cause a fire hazard to adjacent materials, the equipment shall:

WR-421.2

- be mounted on a support which has low thermal conductance; or
- within an enclosure which will withstand such temperatures as may be generated; or
- be screened by materials of low thermal conductance which can withstand the heat emitted by the electrical equipment; or
- be mounted so as to allow safe dissipation of heat and at a sufficient distance from adjacent material on which such temperatures could have deleterious effects.

 Note: Any means of support shall be of low thermal conductance.

Where arcs, sparks or particles at high temperature may be emitted by fixed equipment in normal service, the equipment shall be:

WR-421.3

- totally enclosed in arc-resistant material; or
- screened by arc-resistant material from materials upon which the emissions could have harmful effects; or
- mounted so as to allow safe extinction of the emissions at a sufficient distance from materials upon which the emissions could have harmful effects.

 **Note:** Arc-resistant material used for this protective measure shall be non-ignitable, of low thermal conductivity and of adequate thickness to provide mechanical stability.

Fixed equipment that could cause a concentration and focus of heat shall be at a sufficient distance from any fixed object or building element.

WR-421.4

Precautions shall be taken to prevent the spread of liquid,   WR-421.5
flame and other products of combustion for electrical
equipment containing a significantly high amount of
flammable liquid.

Materials used for the construction of enclosures for   WR-421.6
electrical equipment shall be capable of resisting heat and
fire in accordance with an appropriate product standard.

Every termination of a live conductor or connection or   WR-421.7
joint between live conductors shall be contained within an
enclosure.

8.10.8.8 Safety services

Safety services shall be regulated.   WR-351

Safety services may be required to operate at all material   WR-560.5.1
times that people or livestock are at risk including
during mains and local supply failure and through fire
conditions. To meet this requirement, specific sources,
equipment, circuits and wiring are necessary.

Some applications also have particular requirements.

Circuits of safety services shall not pass through   WR-560.7.3
locations exposed to fire risk (BE2) unless they are
fire-resistant.

For safety services required to operate in fire conditions: WR-560.5.2

- a safety source of supply shall be selected which
  will maintain a supply of adequate duration
- equipment shall be provided, either by
  construction or by erection, with protection
  ensuring fire-resistance of adequate duration.

Note: The safety source is generally additional to the normal source (i.e. the
public supply network) and examples include:

- emergency lighting;
- fire pumps;
- fire rescue service lifts;
- fire detection and alarm systems;

- CO detection and alarm systems;
- fire evacuation systems;
- smoke ventilation systems;
- fire services communication systems;
- essential medical systems;
- industrial safety systems.

> The device protecting a conductor against overload   WR-433.2.2
> may be installed so as to reduce to a minimum the risk
> of fire or danger to persons.

8.10.8.9 Protection against fault current

> A device for protection against fault current need not   WR-434.3
> be provided where the wiring is installed in such a
> manner as to reduce to a minimum the risk of fire or
> danger to persons.

The omission of devices for protection against overload is permitted for circuits supplying current-using equipment where unexpected disconnection of the circuit could cause danger or damage.

Examples of such circuits are:

- a circuit supplying a fire extinguishing device;
- a circuit supplying a safety service, such as a fire alarm or a gas alarm.

 In such situations consideration should be given to the provision of an overload alarm.

8.10.8.10 Wiring systems

> The risk of spread of fire shall be minimised by the   WR-527.1.1
> selection of appropriate materials and erection.
>
> A wiring system shall be installed so that the general   WR-527.1.2
> building structural performance and fire safety are not
> reduced.
>
> Cables not complying with the flame propagation   WR-527.1.4
> requirements of BS EN 60332-1-2 shall be limited
> to short lengths for connection of appliances to the
> permanent wiring system and shall not pass from one
> fire-segregated compartment to another.

Where no fire alarm system is installed in a building used for exhibitions etc., cable systems shall be either: WR-711.521

- flame retardant to BS EN 60332-1-2 and low smoke to BS EN 61034-2; or
- single-core or multicore unarmoured cables enclosed in metallic or non-metallic conduit or trunking, providing a degree of fire protection of at least IP4X.

Where both the live circuit conductors are uninsulated, either: WR-559.11.4.1

- they shall be provided with a protective device complying with the requirements of Regulation 559.11.4.2; or
- the system shall comply with BS EN 60598-2-23.

A device providing protection against the risk of fire shall meet the following requirements: WR-559.11.4.2

- The device shall continuously monitor the power demand of the luminaires.
- The device shall automatically disconnect the supply circuit within 0.3 s in the case of a short-circuit or failure which causes a power increase of more than 60 W.
- The device shall provide automatic disconnection while the supply circuit is operating with reduced power or if there is a failure which causes a power increase of more than 60 W.
- The device shall provide automatic disconnection upon connection of the supply circuit if there is a failure which causes a power increase of more than 60 W.
- The device shall be fail-safe.

Wiring systems on escape routes that are likely to be BD2 (difficult), BD3 (crowded) and BD4 (difficult and crowded), shall have a resistance to fire of at least 2 hours.   WR-422.2.1

Where a heating cable is required to pass through, or be in close proximity to, material which presents a fire hazard, the cable: WR-554.4.1

- shall be enclosed in material having the ignitability characteristic '13' as specified in BS 476-12; and
- shall be adequately protected from any mechanical damage reasonably foreseeable during installation and use.

8.10.8.11 Sealing of wiring system penetrations

Where a wiring system passes through elements of building construction (such as floors, walls, roofs, ceilings, partitions or cavity barriers) the openings remaining after passage of the wiring system shall be sealed according to the degree of fire-resistance for the respective element of building construction.	WR-527.2.1
A wiring system (such as a conduit system, cable ducting system, cable trunking system, busbar or busbar trunking system) which penetrates elements of building construction having specified fire-resistance shall be internally sealed to the degree of fire-resistance of the respective element before penetration as well as being externally sealed.	WR-527.2.4
In the event of a wiring system crossing (or being in the proximity of) underground telecommunication cables and underground power cables, a minimum clearance of 100 mm shall be maintained and a fire-retardant partition shall be provided between the cables.	WR-528.2

8.10.8.12 Firefighter's switches

A firefighter's switch shall be provided in the low voltage circuit supplying: • exterior electrical installations operating at a voltage exceeding low voltage; and • interior discharge lighting installations operating at a voltage exceeding low voltage.	WR-537.6.1
Every exterior installation in each single premises shall wherever practicable be controlled by a single firefighter's switch.	WR-537.6.2
Similarly, every internal installation in each single premises shall be controlled by a single firefighter's switch independent of the switch for any exterior installation.	WR-537.6.2
Every firefighter's switch shall comply with the following: • For an exterior installation, the switch shall be outside the building and adjacent to the equipment	WR-537.6.3

(or alternatively a notice indicating the position of the switch shall be placed adjacent to the equipment and a notice shall be fixed near the switch so as to render it clearly distinguishable).

- For an interior installation, the switch shall be in the main entrance to the building.
- The switch shall be placed in a conspicuous position, reasonably accessible to firefighters, and at not more than 2.75 m from the ground or the standing beneath the switch.
- Where more than one switch is installed on any one building, each switch shall be clearly marked to indicate the installation or part of the installation which it controls.

A firefighter's switch shall: WR-537.6.4

- be coloured red and have fixed on or near it a permanent nameplate marked with the words **'FIREFIGHTER'S SWITCH'**, the plate being of minimum size 150 mm by 100 mm, and having lettering easily legible from a distance appropriate to the site conditions but not less than 36 point; and
- have its ON and OFF positions clearly indicated by lettering legible to a person standing on the ground at the intended site, with the OFF position at the top; and
- be provided with a device to prevent the switch being inadvertently returned to the ON position; and
- be arranged to facilitate operation by a firefighter.

8.10.8.13 Inspection

 Note: Inspection shall precede testing and shall normally be done with that part of the installation under inspection disconnected from the supply.

The inspection shall be made to verify that the installed   WR-611.2
electrical equipment is:

- in compliance; and
- correctly selected and erected; and
- not visibly damaged or defective so as to impair safety.

The inspection shall include the checking (during   WR-611.3
erection) of presence of fire barriers, suitable seals and protection against thermal effects.

Where RCDs are also used for protection against fire, the conditions for protection by automatic disconnection of the supply shall be verified.

WR-612.8

8.10.8.14 Periodic inspection and testing

Periodic inspection comprising a detailed examination of the installation shall be carried out without dismantling, or with partial dismantling as required, supplemented by appropriate tests to show that the requirements for disconnection times for protective devices are complied with, to provide for:

WR-621.2

- safety of persons and livestock against the effects of electric shock and burns
- protection against damage to property by fire and heat arising from an installation defect
- confirmation that the installation is not damaged or deteriorated so as to impair safety
- the identification of installation defects and departures from the requirements of these Regulations that may give rise to danger.

8.10.8.15 Testing

In locations exposed to fire hazard, a measurement of the insulation resistance between the live conductors should be applied.

WR-612.3.2

Insulation resistance values are usually much higher than those of Table 8.18.

Table 8.18 Minimum values of insulation resistance

Circuit nominal voltage (V)	Test voltage d.c. (V)	Minimum insulation resistance
SELV and PELV	250	>0.5
Up to and including 500 V	500	>1.0
Above 500 V	1000	>1.0

9

Inspection and testing

Every installation (or alteration to an existing installation) shall, during erection and on completion before being put into service, be inspected and tested to verify, so far as is reasonably practicable, that the requirements of the Regulations have been met.

The verification shall be made by a competent person and on completion of the verification, a certificate shall be prepared.

To meet these requirements it is essential for any electrician engaged in inspection, testing and certification of electrical installations to have a **full** working knowledge of the IEE Wiring Regulations. The electrician must also have above average experience and knowledge of the type of installation under test in order to carry out **any** inspection and testing. Without this prerequisite, it could be quite dangerous – particularly concerning installations such as that shown in Figure 9.1!

Figure 9.1 Example of a non-conforming electrical installation! (courtesy StingRay)

9.1 What inspections and tests have to be completed and recorded?

Every installation must be inspected and tested during erection and on completion before being put into service:

- to verify that, so far as is reasonably practicable, that the requirements of BS 7671:2008 have been met;
- to verify that precautions have been taken to avoid danger to persons and damage to property and installed equipment during inspection and testing;
- to make an assessment of the frequency and type of maintenance (e.g. periodic inspection, testing, maintenance and repair) that an installation can reasonably be expected to receive during its intended life.

9.2 Inspections

An inspection is, most generally, an official examination or formal evaluation exercise involving measurements, tests, and gauges applied to certain characteristics with regard to an object or activity. The results of an inspection are usually compared with specified requirements (e.g. a British Standard) in order to determine whether an item or activity is in line with the relevant Standard(s) and achieves certain criteria and characteristics. Inspections are usually non-destructive.

9.2.1 General

Inspection shall precede testing and shall normally be done with that part of the installation under inspection disconnected from the supply.

Inspections shall be made to verify that the installed electrical equipment:

- complies with the requirements of the applicable British Standard, or Harmonised Standard appropriate to the intended use of the equipment;

Note: Equipment complying with a foreign national Standard may be used **only** if it provides the same degree of safety afforded by a British or Harmonised Standard.

- is correctly selected and erected in accordance with the Regulations;
- is not visibly damaged or defective so as to impair safety.

9.2.2. Inspection check list

In accordance with the requirements of BS 7671:2008 and for compliance with the Building Regulations, the inspection shall include the following items:

- access to switchgear and equipment;
- cable routing;

- choice and setting of protective and monitoring devices;
- connection of accessories and equipment;
- connection of conductors;
- connection of single-pole devices for protection or switching in line conductors;
- continuity of all protective conductors;
- continuity of all ring final circuit conductors;
- earth electrode resistance;
- earth fault loop impedance;
- erection methods;
- functional testing;
- identification of conductors;
- insulation of non-conducting floors and walls;
- insulation resistance;
- labelling of protective devices, switches and terminals;
- polarity;
- presence of danger notices and other warning signs;
- presence of diagrams, instructions and similar information;
- presence of fire barriers, suitable seals and protection against thermal effects;
- prevention of mutual (i.e. detrimental) influence;
- presence of undervoltage protective devices;
- prospective fault current;
- protection against electric shock:
  - capability of equipment to withstand mechanical, chemical, electrical and thermal influences and stresses normally encountered during service;
  - exposed conductive parts;
  - insulating enclosures;
  - insulation of operational electrical equipment;
  - verification of the quality of the insulation;
- protection against electric shock by direct and/or indirect contact:
  - SELV;
  - limitation of discharge of energy;
- protection against direct current:
  - barriers or an enclosure;
  - insulation of live parts;
  - obstacles;
  - PELV;
  - placing out of reach;
- protection against external influences;
- protection against indirect contact:
  - automatic disconnection of supply;
  - earth free local equipotential bonding;
  - earthed equipotential bonding;
  - earthing and protective conductors;
  - earthing arrangements for combined protective and functional purposes; non-conducting location (absence of protective conductors);

- o electrical separation;
- o main equipotential bonding conductors;
- o use of Class II equipment or equivalent insulation;
- o supplementary equipotential bonding conductors;
- selection of conductors for current-carrying capacity and voltage drop;
- selection of equipment appropriate to external influences;
- site applied insulation:
  - o protection against direct contact;
  - o protection against indirect contact;
  - o supplementary insulation.

 The Building Regulations **specifically state** that inspections **shall** include the design, construction, inspection and testing of any new electrical installation or new work associated with an alteration or addition to an existing installation.

In addition to the above list of mandatory inspections for compliance with the IEE Wiring Regulations and the Building Regulations, the following are some of the additional inspections that electricians usually complete during initial and periodic inspections and tests of electrical installations:

- cables and conductors (current carrying capacity, insulation and/or sheath);
- correct connection of accessories and equipment;
- electrical joints and connections (to ensure that they meet stipulated requirements concerning conductance, insulation, mechanical strength and protection);
- emergency switching;
- insulation;
- insulation monitoring devices (design, installation and security);
- inspection of an other electrical installations;
- isolation and switching devices (and their correct location);
- locations with risks of fire due to the nature of processed and/or stored materials;
- plug and socket outlets;
- protection against electric shock – special installations or locations;
- protection against earth insulation faults;
- protection against mechanical damage;
- protection against overcurrent;
- protection by extra-low-voltage systems (other than SELV);
- protection by non-conducting location;
- protection by residual current devices;
- protection by separation of circuits;
- supplies;
- supplies for safety services;
- wiring systems (selection and erection, temperature variations).

 Note: Details concerning tests to confirm compliance with the Regulations are contained in Section 11.5.

9.2.3 Requirements from the Regulations – inspection

Every electrical connection and joint shall be accessible   WR-526.3
for inspection, except for the following: WR-543.3.3

- a joint designed to be buried in the ground
- a compound-filled or encapsulated joint
- a connection between a cold tail and the heating element as in ceiling heating, floor heating or a trace heating system
- a joint made by welding, soldering, brazing or appropriate compression tool
- a joint forming part of the equipment complying with the appropriate product standard.

The inspection shall be made to verify that the installed   WR-611.2
electrical equipment is:

- in compliance; and
- correctly selected and erected; and
- not visibly damaged or defective so as to impair safety.

The inspection shall include at least the checking of the   WR-611.3
following items, where relevant to the installation and, where necessary, during erection:

- connection of conductors
- identification of conductors
- routing of cables in safe zones (or protection against mechanical damage)
- selection of conductors for current-carrying capacity and voltage drop, in accordance with the design
- connection of single-pole devices for protection or switching in line conductors only
- correct connection of accessories and equipment
- presence of fire barriers, suitable seals and protection against thermal effects
- methods of protection against electric shock
- prevention of mutual detrimental influence
- presence of appropriate devices for isolation and switching correctly located
- presence of undervoltage protective devices
- labelling of protective devices, switches and terminals

- selection of equipment and protective measures appropriate to external influences
- adequacy of access to switchgear and equipment
- absence of danger notices and other warning signs
- absence of diagrams, instructions and similar information
- erection methods.

9.3 Testing

Testing any electrical installation (even the simplest ones) can be very dangerous unless it is carried out safely – dangerous not just to the tester himself but to bystanders and other people.

As a minimum, the electrician must:

- have an above-average experience and knowledge of the type of installation under test;
- have a thorough understanding of the correct application and use of the relevant test instruments (and their associated leads, probes and accessories);
- ensure that the test equipment being used has recently been inspected, correctly maintained and (where necessary) calibrated either against a workshop standard or a national standard;
- observe the safety measures and procedures set out in HSE Guidance Note GS38 concerning the safe use of instruments and their accessories.

The following are précised details of the most important elements of the IEE Wiring Regulations that an electrician must test for in order to confirm that the electrical installation meets the fundamental design requirements of the IEE Wiring Regulations and is installed in conformance with the requirements of that British Standard.

9.3.1 Initial inspection and tests

In accordance with both the Wiring Regulations and the Building Regulations, all new installations (plus additions and/or alterations to existing circuits) need an initial verification to:

- ensure equipment and accessories meet the requirements of the relevant standard;
- comply with the requirements of BS 7671;
- comply with the requirements of the Building Regulations;
- ensure that the installation is not damaged thereby impairing safety.

The following tests **shall** be carried out (and in the following order) before the installation is energised:

- a continuity test of all protective conductors (including main and supplementary equipotential bonding);
- a continuity test of all ring final circuit conductors;
- a measurement of the insulation resistance between live conductors and between each live conductor and earth;
- a measurement of the insulation resistance of the main switchboard and each distribution circuit;
- confirmation that insulation for protection against direct and/or indirect contact meets requirements;
- verification that the separation of circuits is protected by SELV, PELV and/or electrical separation meets requirements;
- a measurement of the insulation resistance of live parts from those of other circuits and from earth;
- confirmation that functional extra-low-voltage circuits meet all the test requirements for low-voltage circuits;
- a test to ensure that the amount of protection against direct contact that is provided by a barrier or an enclosure (provided during erection) meets requirements;
- verification (by measurement) that the amount of protection against indirect contact provided by a non-conducting location meets requirements;
- a polarity test to verify that fuses, single-pole control and protective devices, lamp holders and wiring meet requirements.

The following tests shall be carried out when the installation is energised:

- a measurement of the electrode resistance to earth for earthing systems incorporating an earth electrode;
- a measurement of earth loop impedance;
- a measurement of prospective short-circuit and earth fault;
- functional tests to verify the effectiveness of a residual current devices and test assemblies (e.g. switchgear, controlgear, drives, controls and interlocks) to show that they are properly mounted, adjusted and installed in accordance with the Regulations.

9.3.2 Protective measures

BS 7671:2008 stipulates that a continuity test of all protective conductors (including main and supplementary equipotential bonding) shall be made to ensure that the correct degree of protection is being provided by the following protective measures by confirming, checking and testing:

Protection against overload current	That the protective device is capable of breaking any overload current flowing in

the circuit conductors before the current can damage the insulation of the conductors.

Protection by earth-free local equipotential bonding	That earth-free local equipotential bonding prevents the appearance of a dangerous voltage between simultaneously accessible parts in the event of failure of the basic insulation.
Protection by electrical separation	That equipment used as a fixed source of supply has been manufactured so that the output is separated from the input and from the enclosure by insulation for protection against indirect contact.

 Note: This form of protection is intended for an individual circuit and is aimed at preventing shock current through contact with exposed conductive parts, which might be energised by a fault in the basic insulation of that circuit.

Protection by extra-low-voltage systems (other than SELV)	That if an extra-low-voltage system complies with the requirements for SELV, it is not to be connected to a live part or a protective conductor forming part of another system and not connected to:

- earth;
- an exposed conductive part of another system;
- a protective conductor of any system; or
- an extraneous conductive part.

Protection against direct contact has been provided by either:

- insulation capable of withstanding 500 V a.c. rms for 60 seconds; or
- barriers or enclosures with a degree of protection of at least IP2X or IPXXB.

 This form of protection against direct contact is **not** required if the equipment is within a building in which main equipotential bonding is applied and the voltage does not exceed:

- 25 V a.c. rms or 60 V ripple-free d.c. when the equipment is normally only used in dry locations and large-area contact of live parts with the human body is not to be expected;

- 6 V a.c. rms or 15 V ripple-free d.c. in all other cases.

 Note: When an extra-low-voltage circuit is used to supply equipment whose insulation does not comply with the minimum test voltage required for the primary circuit, then the insulation of that equipment shall be reinforced to withstanding a voltage of 1500 V a.c. rms for 60 seconds.

Protection by insulation of live parts

That the insulation protection has been designed to prevent contact with live parts.

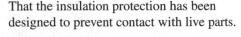 Whilst, generally speaking, this method is for protection against direct contact, it also provides a degree of protection against indirect contact.

Protection by non-conducting location

That this form of protection prevents simultaneous contact with parts which may be at different potentials through failure of the basic insulation of live parts.

 Whilst this protection is **not** recognised in the Regulations for general use it may be applied in special situations provided that they are under effective supervision.

 Protection by non-conducting location shall **not** be used in installations and locations subject to increased risk of shock such as agricultural and horticultural premises, caravans, swimming pools etc.

Protection by residual current devices

That:

- parts of an TT system that are protected by a single residual current device have been placed at the origin of the installation unless that part between the origin and the device complies with the requirements for protection by using Class II equipment or an equivalent insulation.

 Where there is more than one origin this requirement applies to each origin.

	• installations forming part of an IT system have been protected by a residual current device supplied by the circuit concerned or make use of an insulation monitoring device.
Protection by SELV	That circuit conductors for each SELV system have been physically separated from those of any other system. Where this proves impracticable, SELV circuit conductors have been:
	• insulated for the highest voltage present; • enclosed in an insulating sheath additional to their basic insulation.
Protection by the use of Class II equipment or equivalent insulation	That this form of protection prevents a fault in the basic insulation causing a dangerous voltage to appear on the exposed metalwork of electrical equipment.

A typical method for testing the continuity of protective conductors is illustrated in Figure 9.2 and basically this involves bridging the line conductor to the protective conductor at the distribution board (so as to include all of the circuit) and then testing between line and earth terminals at each point of the circuit.

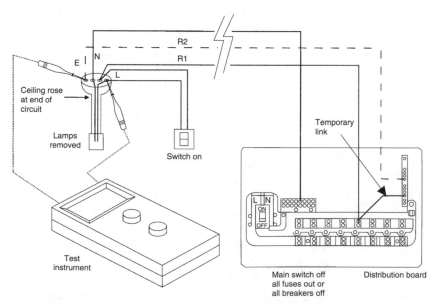

Figure 9.2 Connections for testing continuity of protective conductors (courtesy IET)

9.3.3 Design requirements

Check to confirm that the number and type of circuits required for lighting, heating, power, control, signalling, communication and information technology, etc. have taken consideration of:

- the location and points of power demand;
- the loads to be expected on the various circuits;
- the daily and yearly variation of demand;
- any special conditions;
- requirements for control, signalling, communication and information technology, etc.

9.3.4 Electricity distributor

Confirm that the electricity distributor has:

- evaluated and agreed proposals for new installations or significant alterations to existing ones;
- maintained the supply within defined tolerance limits;
- provided an earthing facility for all new connections;
- installed the cut-out and meter in a safe location;
- ensured that the cut-out and meter it is mechanically protected and can be safely maintained;
- provided certain technical and safety information to the consumer to enable them to design their installations;
- ensured that their equipment on consumers' premises:
  - ○ is suitable for its purpose;
  - ○ is safe in its particular environment;
  - ○ clearly shows the polarity of the conductors.

9.3.5 Site insulation

Confirm that:

- insulation applied on site to protect against direct contact is capable of withstanding, without breakdown or flashover, an applied test voltage as specified in the British Standard for similar type-tested equipment;
- supplementary insulation applied to equipment during erection to protect against indirect contact is tested to ensure that the insulating enclosure;
- protects to at least IP2X or IPXXB, and is capable of withstanding, without breakdown or flashover, an applied test voltage as specified in the British Standard for similar type-tested equipment.

9.3.6 Insulation resistance

Check to confirm that:

- the insulation resistance (measured with all its final circuits connected but with current-using equipment disconnected) between live conductors

and between each live conductor and earth, is not less than that shown in
Table 9.1;

Table 9.1 Minimum values of insulation resistance

Circuit nominal voltage (V)	Test voltage d.c. (V)	Minimum insulation resistance (MΩ)
SELV and PELV	250	0.25
Up to and including 500V (with the exception of the above systems)	500	0.5
Above 500V	1000	1.0

- the separation of live parts from those of other circuits and from earth is in
  accordance with the values show in Table 9.1, by measuring the insulation
  resistance.

Note: This test (more usually referred to as *meggering*) is, therefore, aimed
at ensuring that the insulation of conductors, accessories and equipment is
still capable of preventing dangerous leakage current between conductors and
between conductors and earth.

To determine the insulation resistance between live conductors, test between
the live (line and neutral) conductors at the distribution board (see Figure 9.3).

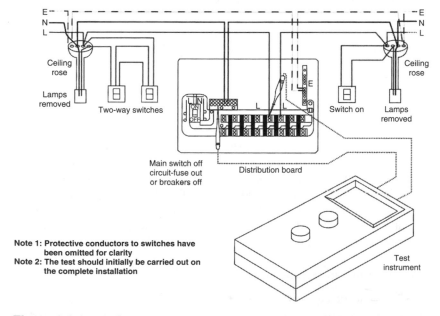

Figure 9.3 Insulation resistance tests between live conductors of a circuit
(courtesy IET)

The resistance readings obtained should be greater than the minimum values shown in Table 9.1.

To check the insulation resistance to earth:

- Single phase – test between the live conductors (line and neutral) and the circuit protective conductors at the distribution board as shown in Figure 9.4.

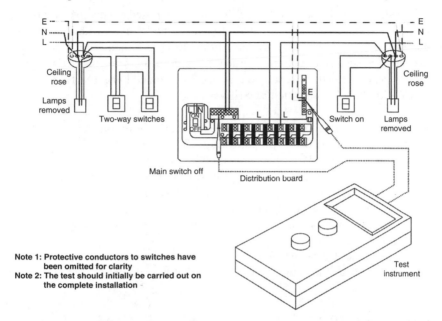

Note 1: Protective conductors to switches have
         been omitted for clarity
Note 2: The test should initially be carried out on
         the complete installation

Figure 9.4 Insulation resistance tests to earth (courtesy IET)

- Three phase – test to earth all live conductors (including the neutral) connected together. Resistance figures should be greater than the minimum figures shown in Table 9.1.

 Note:

- Measurements shall be carried out with direct current.
- When the circuit includes electronic devices, only a measurement to protective earth shall be made with the line and neutral connected together.

 Precautions may be necessary to avoid damage to electronic devices.

9.3.7 Protection against direct and indirect contact

The two methods for protecting against shock from both direct and indirect contact are:

- SELV (separated extra-low voltage) – i.e. where the system voltage does not exceed extra-low (e.g. 50 V a.c., 120 V ripple-free d.c.) and associated wiring etc. is separated from all other circuits of higher voltage.

- Limitation of discharge of energy – i.e. where equipment is arranged so that the current that can flow through the body (or livestock) is limited to a safe level (e.g. electric fences).

9.3.7.1 SELV

 The system is not be deemed to be SELV if any exposed conductive part of an extra-low-voltage system is capable of coming into contact with an exposed conductive part of any other system. In addition, a system which does **not** use a device such as an autotransformer, potentiometer, semiconductor device etc., to provide electrical separation, is also not deemed to be a SELV system.

Confirm that for protection by SELV:

- the nominal circuit voltage does **not** exceed extra-low voltage;
- the supply is from one of the following:
  - a safety isolating transformer complying with BS 3535;
  - a motor-generator with windings providing electrical separation equivalent to that of the safety isolating transformer specified above;
  - a battery or other form of electrochemical source;
  - a source independent of a higher-voltage circuit (e.g. an engine driven generator);
  - electronic devices which (even in the case of an internal fault) restrict the voltage at the output terminals so that they do not exceed extra-low voltage.

Confirm and test that:

- a mobile source for SELV has been selected and erected in accordance with the requirements for protection by the use of Class II equipment or by equivalent insulation;
- all live parts of a SELV system are:
  - electrically separated from that of any other higher voltage system;

 Note: this electrical separation shall be not less than that between the input and output of a safety isolating transformer
  - not connected to earth;
  - not connected to a live part or a protective conductor forming part of another system;
- circuit conductors for each SELV system are physically separated from those of any other system and where this proves impracticable, SELV circuit conductors are:
  - insulated for the highest voltage present; and (where this proves impracticable);
  - enclosed in an insulating sheath additional to their basic insulation;
- conductors of systems with a higher voltage than SELV are separated from the SELV conductors by an earthed metallic screen or an earthed metallic sheath;

- SELV circuit conductors that are contained in a multicore cable with other circuits having different voltages are insulated, individually or collectively, for the highest voltage present in the cable or grouping;
- electrical separation between live parts of a SELV system (including relays, contactors and auxiliary switches) and any other system is maintained;
- exposed conductive parts of a SELV system are **not** be connected to:
  - earth;
  - an exposed conductive part of another system;
  - a protective conductor of any system;
  - an extraneous conductive part;

Note: Except where that electrical equipment is mainly required to be connected to an extraneous conductive part (in which case, measures shall be incorporated so that the parts cannot attain a voltage exceeding extra-low voltage).

- if the nominal voltage of a SELV system exceeds 25 V a.c. rms or 60 V ripplefree d.c., protection against direct contact is provided by one or more of the following:
  - a barrier (or an enclosure) capable of providing protection to at least IP2X or IPXXB;
  - insulation capable of withstanding a type-test voltage of 500 V a.c. rms for 60 seconds;
- the socket outlet of a SELV system is:
  - incompatible with the plugs used for other systems in use in the same premises;
  - does not have a protective conductor contact;
- luminaire supporting couplers which have a protective conductor contact, are **not** be installed in a SELV system.

9.3.7.2 Protection – by limitation of discharge of energy

Protection against both direct and indirect contact shall be deemed to be provided when the equipment incorporates a means of limiting the amount of current which can pass through the body of a person or livestock to a value lower than that likely to cause danger.

9.3.8 Protection against direct contact

Electric shock caused by direct contact (i.e. when a body part directly touches live parts of equipment or systems that are intended to be live) is particularly dangerous, as the full voltage of the supply can be developed across the body. On the whole, however, if an electrical installation has been designed and installed correctly, then there shouldn't be too much risk from direct contact – but carelessness (such as changing an electric light bulb without switching the mains off first) or overconfidence (such as working on a circuit with the power on) are the prime causes of injuries and death from electric shock.

The main protective methods against direct contact causing an electric shock are:

- barriers or an enclosure;
- insulation of live parts;
- by obstacles and placing out of reach;
- PELV and FELV;
- placing out of reach.

 The use of a residual current device (RCD) cannot prevent direct contact, but may be used to supplement other protective means that are used.

9.3.8.1 Protection by a barrier or an enclosure provided during erection

Test to ensure that the degree of protection against direct contact provided by a barrier or an enclosure (provided during erection) is not less than IP2X or IPXXB or IP4X, as appropriate.

9.3.8.2 Protection by insulation of live parts

Complete a functional test to verify that protection by insulation of live parts does prevent contact with live parts.

 Note: Whilst, generally speaking, this basic form of insulation protection is for protection against direct contact, it also provides a degree of protection against indirect contact.

9.3.8.3 Protection by obstacles

Complete a functional test to verify that protection by obstacles prevents unintentional contact with a live part, but **not** intentional contact by deliberate circumvention of the obstacle.

 The application shall be limited to protection against direct contact and in an area accessible only to skilled persons.

 For some installations and locations where an increased risk of shock exists, this protective measure shall **not** be used.

9.3.8.4 PELV and FELV systems

Protective extra-low-voltage (PELV) and functional extra-low-voltage (FELV) systems shall provide protection against electric shock and meet the following requirements:

- use barriers or enclosures with a degree of protection of at least IP2X or IPXXB; or
- insulation capable of withstanding 500 V a.c. rms for 60 seconds.

This form of protection against direct contact is not required if the equipment is within a building in which main equipotential bonding is applied and the voltage does not exceed:

- 25 V a.c. rms or 60 V ripple-free d.c. when the equipment is normally only used in dry locations and large-area contact of live parts with the human body is not to be expected;
- 6 V a.c. rms or 15 V ripple-free d.c. in all other cases.

• For extra-low-voltage systems that do not comply with the requirements for SELV in some respect then protection against direct contact shall be provided by one or more of the following:

- barriers or enclosures;
- insulation corresponding to the minimum voltage required for the primary circuit.

• When an extra-low-voltage circuit is used to supply equipment whose insulation does not comply with the minimum test voltage required for the primary circuit, then the insulation of that equipment shall be reinforced to withstand a voltage of 1500 V a.c. rms for 60 seconds.

• If the primary circuit of the functional extra-low-voltage source is protected by automatic disconnection, then the exposed conductive parts of equipment in that functional extra-low-voltage system shall be connected to the protective conductor of the primary circuit.

• If the primary circuit of the functional extra-low-voltage source is protected by electrical separation, then the exposed conductive parts of equipment in that functional extra-low-voltage system shall be connected to the non-earthed protective conductor of the primary circuit.

• All socket outlets and luminaire supporting couplers in a functional extra-low-voltage system shall use a plug which is dimensionally different from those used for any other system in use in the same premises.

9.3.8.5 Protection by placing out of reach

Check to ensure that:

• bare (or insulated) overhead lines being used for distribution between buildings and structures are installed in accordance with the Electricity Safety, Quality and Continuity Regulations 2002;

• bare live parts (other than overhead lines) are not within arm's reach.

Note:

• If access to live equipment (from a normally occupied position) is restricted by an obstacle (i.e. such as a handrail, mesh or screen) with a degree of protection less than IP2X or IPXXB, the extent of arm's reach shall be measured from that obstacle.

• If a bulky or long conducting object is normally handled in these areas, the distances shall be increased accordingly.

- Bare live parts (other than an overhead line) are not to be within 2.5 m of:
  - an exposed conductive part;
  - an extraneous conductive part;
  - a bare live part of any other circuit.
- If a bulky or long conducting object is normally handled in these areas, the distances required shall be increased accordingly.

No additional protection against overvoltages of atmospheric origin is necessary for:

- installations that are supplied by low-voltage systems which do not contain overhead lines;
- installations that are supplied by low-voltage networks which contain overhead lines and their location is subject to less than 25 thunderstorm days per year;
- installations that contain overhead lines and their location is subject to less than 25 thunderstorm days per year;

provided that they meet the required minimum equipment impulse withstand voltages shown in Table 44.4 on p. 85 of BS 7671:2008.

Suspended cables having insulated conductors with earthed metallic coverings are considered to be an 'underground cable'.

- Check to ensure that installations that are supplied by (or include) low-voltage overhead lines, incorporate protection against overvoltages of atmospheric origin or (if the location is subject to more than 25 thunderstorm days per year) and that this protection against overvoltages of atmospheric origin shall be provided in the installation of the building by:
  - a surge protective device with a protection level not exceeding Category II; or
  - by other means providing an equivalent attenuation of overvoltages.
- Where protective measures against indirect contact only have been dispensed with, confirm that:
  - overhead line insulator brackets (and metal parts connected to them) are not within arm's reach;
  - the steel reinforcement of steel reinforced concrete poles in not accessible;
  - exposed conductive parts (including small isolated metal parts such as bolts, rivets, nameplates not exceeding 50 mm × 50 mm and cable clips) cannot be gripped or cannot be contacted by a major surface of the human body;
  - there is no risk of fixing screws used for non-metallic accessories coming into contact with live parts;
  - inaccessible lengths of metal conduit do not exceed 150 mm^2;
  - metal enclosures mechanically protecting equipment comply with the relevant British Standard;

○ unearthed street furniture that is supplied from an overhead line is inaccessible whilst in normal use.

9.3.9 Protection against indirect contact

Indirect contact (i.e. touching conductive parts which are not meant to be live, but which have become live due to a fault) is the other main cause of electric shock. Again this is particularly dangerous and the main protection against indirect contact is for the electrical installation to be correctly earthed and for the circuit to be fitted with some form of overcurrent cut-out device.

The main protective methods against indirect contact causing an electric shock are:

* use of Class II equipment;
* automatic disconnection of supply;
* earth-free local equipotential bonding;
* earthed equipotential bonding;
* earthing and protective conductors;
* earthing arrangements for combined protective and functional purposes;
* non-conducting location (absence of protective conductors);
* electrical separation;
* main equipotential bonding conductors;
* supplementary equipotential bonding conductors.

9.3.9.1 Use of Class II equipment

Class II equipment (often referred to as double insulated equipment) is typical of modern equipment intended to be connected to a fixed electrical installation (e.g. household appliances, portable tools and similar loads) and where all live parts are insulated so as to prevent a fault in the basic insulation causing a dangerous voltage to appear on the exposed metalwork of electrical equipment.

To verify compliance, confirm that:

* circuits supplying Class II equipment have a circuit protective conductor that is run to (and terminated at) each point in the wiring and at each accessory;

Except suspended lamp holders which have no exposed conductive parts.

* the metalwork of exposed Class II equipment is mounted so that it is not in electrical contact with any part of the installation that is connected to a protective conductor;
* when Class II equipment is used as the sole means of protection against indirect contact, the installation or circuit concerned is under effective supervision whilst in normal use.

This form of protection shall **not** be used for circuits that include socket outlets or where a user can change items of equipment without authorisation.

9.3.9.2 Automatic disconnection of supply

This intention of this form of protection is to prevent a dangerous voltage occurring between simultaneously accessible conductive parts. For installations and locations with increased risk of shock (e.g. those in Part 7 of the Regulations such as agricultural and horticultural buildings, saunas), additional measures may be required. For example:

- automatic disconnection of supply by means of a residual current device (RCD) with a rated residual operating current not exceeding 30 mA;
- supplementary equipotential bonding;
- reduction of maximum fault clearance time.

Confirm and test that:

- installations which are part of a TN system meet the requirements for earth fault loop impedance and for circuit protective conductor impedance as specified in BS 7671:2008 regarding specified times for automatic disconnection of supplies;
- for circuits supplying fixed equipment which are outside of the earthed equipotential zone and which have exposed conductive parts that could be touched by a person who has direct contact with earth, that the earth fault loop impedance ensures that disconnection occurs within the time stated in Table 9.2;

Table 9.2 Maximum disconnection times for TN systems

Installation nominal voltage U_0 (volts)	Maximum disconnection time t (seconds)
120	0.8
230	0.4
400	0.2
Greater than 400	0.1

- if the installation is part of a TT system, all socket outlet circuits are protected by a residual current device;
- automatic disconnection using a residual current device is not applied to a circuit incorporating a PEN conductor;
- installations that provide protection against indirect contact by automatically disconnecting the supply have a circuit protective conductor run to (and terminated at) each point in the wiring and at each accessory.

 Excepting suspended lamp holders which have no exposed conductive parts.

9.3.9.3 Earth-free local equipotential bonding

Earth-free local equipotential bonding is effectively a Faraday cage, where all metal is bonded together (but **not** to earth!) so as to prevent the appearance of a dangerous voltage occurring between simultaneously accessible parts in the event of failure of the basic insulation.

Confirm that:

- earth-free local equipotential bonding has **only** been used in special situations which are earth-free;
- a warning notice (warning that earth-free local equipotential bonding is being used) has been fixed in a prominent position adjacent to every point of access to the location concerned.

 For some installations and locations with an increased shock risk (e.g. agricultural and horticultural, saunas), earth-free local equipotential bonding shall to be used.

9.3.9.4 Earthing and protective conductors

A protective conductor may consist of one or more of the following:

- a single-core cable;
- a conductor in a cable;
- an insulated or bare conductor in a common enclosure with insulated live conductors;
- a fixed bare or insulated conductor;
- a metal covering (for example, the sheath, screen or armouring of a cable);
- a metal conduit or other enclosure or electrically continuous support system for conductors;
- an extraneous conductive part.

Verify and test that:

- the thickness of tape or strip conductors is capable of withstanding mechanical damage and corrosion (see BS 7430);
- the connection of earthing conductors to the earth electrode are:
  - soundly made;
  - electrically and mechanically satisfactory;
  - labelled in accordance with the Regulations;
  - suitably protected against corrosion;
- all installations have a main earthing terminal to connect the following to the earthing conductor:
  - the circuit protective conductors;
  - the main bonding conductors;
  - functional earthing conductors (if required);
  - lightning protection system bonding conductor (if any);
- earthing conductors are capable of being disconnected to enable the resistance of the earthing arrangement to be measured;
- any joint:
  - is capable of disconnection only by means of a tool;
  - is mechanically strong;
  - ensures the maintenance of electrical continuity;

Note: For convenience (and if required) this may be combined with the main earthing terminal or bar.

- unless a protective conductor forms part of a multicore cable (or cable trunking or a conduit is used as a protective conductor) that the cross-sectional area, up to and including $6\,\text{mm}^2$ has been protected, throughout, by a covering at least equivalent in insulation to that of a single-core non-sheathed cable having a voltage rating of at least 450/750 V.

Where PME (protective multiple earthing) conditions apply, verify and test that:

- the main equipotential bonding conductor has been selected in accordance with the neutral conductor of the supply and Table 9.3;

Table 9.3 Minimum cross-sectional area of the main equipotential bonding conductor in relation to the neutral

Copper equivalent cross-sectional area of the supply neutral conductor	Minimum copper equivalent cross-sectional area of the main equipotential bonding conductor
$35\,\text{mm}^2$ or less	$10\,\text{mm}^2$
over $35\,\text{mm}^2$ up to $50\,\text{mm}^2$	$16\,\text{mm}^2$
over $50\,\text{mm}^2$ up to $95\,\text{mm}^2$	$25\,\text{mm}^2$
over $95\,\text{mm}^2$ up to $150\,\text{mm}^2$	$35\,\text{mm}^2$
over $150\,\text{mm}^2$	$50\,\text{mm}^2$

Local distributor's network conditions may require a larger conductor.

- buried earthing conductors have a cross-sectional area not less than that stated in Table 9.4.

Table 9.4 Minimum cross-sectional areas of a buried earthing conductor

	Protected against mechanical damage	Not protected against mechanical damage
Protected against corrosion by a sheath		$16\,\text{mm}^2$ copper $16\,\text{mm}^2$ coated steel
Not protected against corrosion	$25\,\text{mm}^2$ copper $50\,\text{mm}^2$ steel	$25\,\text{mm}^2$ copper $50\,\text{mm}^2$ steel

All protective conductors – particularly main equipotential and supplementary bonding conductors – must be tested for continuity using a low resistance ohmmeter.

Verify and test that:

- the cross-sectional area of all protective conductors (less equipotential bonding conductors) is not less than $S = \sqrt{I^2t}/k$;

- if the protective conductor:
  - is not an integral part of a cable; or
  - is not formed by conduit, ducting or trunking; or
  - is not contained in an enclosure formed by a wiring system

then the cross-sectional area shall be not less than:

  - $2.5\,mm^2$ copper equivalent if protection against mechanical damage is provided; or
  - $4\,mm^2$ copper equivalent if mechanical protection is not provided;
- protective conductors buried in the ground shall have a cross-sectional area not less than that stated in Table 9.4.

9.3.9.5 Earthing arrangements for combined protective and functional purposes

The following may serve as a PEN conductor provided that the part of the installation concerned is not supplied through an RCD:

- a conductor of a cable not subject to flexing and with a cross-sectional area not less than $10\,mm^2$ (for copper) or $16\,mm^2$ for aluminium (this applies to a fixed installation);
- the outer conductor of a concentric cable where that conductor has a cross-sectional area not less than $4\,mm^2$.

Verify and test (where necessary) that PEN conductors have only been used if:

- authorisation to use a PEN conductor has been obtained by the distributor; or
- the installation is supplied by a privately owned transformer or converter and there is no metallic connection (except for the earthing connection) with the distributor's network; or
- the installation is supplied from a private generating plant;
- the outer conductor of a concentric cable is not common to more than one circuit;
- the conductance of the outer conductor of a concentric cable (and the terminal link or bar):
  - for a single-core cable, is not less than that of the internal conductor;
  - for a multicore cable in a multiphase or multipole circuit, is not less than that of one internal conductor;
  - for a multicore cable serving a number of points contained within one final circuit (or where the internal conductors are connected in parallel), is not less than that of the internal conductors;
- the continuity of all joints in the outer conductor of a concentric cable (and at a termination of that joint) is supplemented by an additional conductor, additional (that is) to any means used for sealing and clamping the outer conductor;
- isolation devices or switching have **not** been inserted in the outer conductor of a concentric cable;

- PEN conductors of all cables have been insulated or have an insulating covering suitable for the highest voltage to which it may be subjected;
- if neutral and protective functions are provided by separate conductors, those conductors are not then be re-connected together beyond that point;
- separate terminals (or bars) have been provided for the protective and neutral conductors at the point of separation;
- PEN conductors have been connected to the terminals or bar intended for the protective earthing conductor and the neutral conductor.

Note: Where earthing is required for protective as well as functional purposes, then the requirements for protective measures shall take precedence.

9.3.9.6 Non-conducting location

This form of protection (as the name implies) consists of an area in which the floor, walls and ceiling are all insulated and within which protective conductors and socket outlet do not have an earthing connection.

Test that the insulation of extraneous conductive parts is:

- not less than $0.5\,M\Omega$ (when tested at 500 V d.c.);
- is able to withstand a test voltage of at least 2 kV a.c. rms;
- does not pass a leakage current exceeding 1 mA in normal use.

Check that the:

- degree of protection against indirect contact provided by a non-conducting location is verified by measuring the resistance of the location's floors and walls to the installation's main protective conductor at not less than three points on each relevant surface;

Note: One of these measurements shall be not less than 1 m and not more than 1.2 m from any extraneous conductive part in the location. The other two measurements shall be made at greater distances.

- insulation of extraneous conductive parts (to satisfy the requirements for protection for a non-conducting location):
  - are not less than $0.5\,M\Omega$ when tested at 500 V d.c.; and
  - are able to withstand a test voltage of at least 2 kV a.c. rms; and
  - do not pass a leakage current exceeding 1 mA in normal use.

9.3.9.7 Electrical separation

This form of protection is intended for an individual circuit and is aimed at preventing shock current through contact with exposed conductive parts which might be energised by a fault in the basic insulation of that circuit.

Verify that:

- protection by electrical separation has been applied to the supply of individual items of equipment by means of a transformer complying with BS 3535 (the secondary of which is not earthed) or a source affording equivalent safety;
- protection by electrical separation has been used to supply several items of equipment from a single separated source(but **only** for special situations);
- equipment used as a fixed source of supply is either:
  - selected and/or installed with Class II or equivalent protection; or
  - manufactured so that the output is separated from the input and from the enclosure by insulation satisfying the conditions for Class II;
- the supply source to the circuit is either:
  - an isolating transformer complying with BS 3535; or
  - a motor-generator;
- mobile supply sources (fed from a fixed installation) are selected and/or installed with Class II or equivalent protection;
- source supplies are only supplying more than one item of equipment provided that:
  - all exposed conductive parts of the separated circuit are connected together by an insulated and non-earthed equipotential bonding conductor;
  - the non-earthed equipotential bonding conductor is not connected to a protective conductor, or to an exposed conductive part of any other circuit or to any extraneous conductive part;
  - all socket outlets are provided with a protective conductor contact (that is connected to the equipotential bonding conductor);
  - all flexible equipment cables (other than Class H equipment) have a protective conductor for use as an equipotential bonding conductor;
  - exposed conductive parts which are fed by conductors of different polarity (and which are liable to a double fault occurring) are fitted with an associated protective device;
  - any exposed conductive part of a separated circuit cannot come in contact with an exposed conductive part of the source;
- live parts of a separated circuit are not connected (at any point) to another circuit or to earth;
- live parts of a separate circuit are electrically separated from all other circuits;

Note: Live parts of relays, contactors etc. included in a separated circuit (and between a separated circuit and other live parts of other circuits) shall be similarly electrically separated.

- separated circuits, preferably, use a separate wiring system;

If this is not feasible, multicore cables (without a metallic sheath) or insulated conductors (in an insulating conduit) may be used.

- the voltage of an electrically separated circuit does not exceed 500 V;
- all parts of a flexible cable (or cord) that is liable to mechanical damage is visible throughout its length;
- for circuits supplying a single piece of equipment, no exposed conductive part of the separated circuit is connected:
  - ○ to the protective conductor of the source;
  - ○ to any exposed conductive part of any other circuit;
- a warning notice (warning that protection by electrical separation is being used) is fixed in a prominent position adjacent to every point of access to the location concerned.

9.3.9.8 Main equipotential bonding conductors

All main equipotential (and supplementary) bonding conductors must be tested for continuity.

The normal approach is to connect the leads from a low resistance ohmmeter to the ends of the bonding conductor as shown in Figure 9.5 – making sure that one end is disconnected from its bonding clamp, otherwise the measurement may include the resistance of parallel paths of other earthed metalwork.

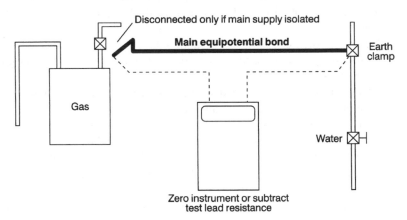

Figure 9.5 Continuity testing main and equipotential bonding conductors

Confirm that main equipotential bonding conductors have for each installation been connected to the main earthing terminal of that installation. This can include the following:

- water service pipes;
- gas installation pipes;
- other service pipes and ducting;
- central heating and air conditioning systems;
- exposed metallic structural parts of the building;
- the lightning protective system.

Note: Where an installation serves more than one building the above requirement shall be applied to each building.

9.3.9.9 Supplementary equipotential bonding

Using a low-resistance ohmmeter (similar to that employed for testing main equipotential bonding conductors described above), test all supplementary equipotential bonding conductors for continuity, particularly in locations intended for livestock, to confirm that:

- supplementary bonding connects all exposed and extraneous conductive parts which can be touched by livestock;
- metallic grids laid in the floor for supplementary bonding are connected to the protective conductors of the installation.

9.3.9.10 Locations with increased risk of shock

For installations and locations with increased risk of shock (e.g. saunas, bathrooms and agricultural/horticultural premises) certain additional measures may be required, such as:

- automatic disconnection of supply by means of a residual current device with a rated residual operating current not exceeding 30 mA;
- supplementary equipotential bonding;
- reduction of maximum fault clearance time.

In these cases, test to confirm that, for circuits supplying fixed equipment which are outside of the earthed equipotential zone (and which have exposed conductive parts that could be touched by a person who has direct contact with earth), the earth fault loop impedance ensures that disconnection occurs within the time stated in Table 9.5.

Table 9.5 Maximum disconnection times for TN systems

Installation nominal voltage U_o (volts)	Maximum disconnection time t (seconds)
120	0.8
230	0.4
400	0.2
Greater than 400	0.1

Test, measure and confirm that:

- if the installation is part of a TT system, all socket outlet circuits have been protected by a residual current device;
- automatic disconnection using a residual current device has **not** been applied to a circuit incorporating a PEN conductor;

- installations that provide protection against indirect contact by automatically disconnecting the supply have a circuit protective conductor run to (and terminated at) each point in the wiring and at each accessory;

Excepting suspended lamp holders which have no exposed conductive parts.

- one or more of the following types of protective device have been used:
  - an overcurrent protective device;
  - a residual current device;
- where a residual current device is used in a TN-C-S system, a PEN conductor has not been used on the load side;
- the protective conductor to the PEN conductor is on the source side of the residual current device;
- the maximum disconnection time to a circuit supplying socket outlets and to other final circuits which supply portable equipment intended for manual movement during use, or hand-held Class I equipment, does not exceed those values shown in Table 9.6.

Table 9.6 Maximum earth fault loop impedance (Z_s) for fuses, for 0.4 s disconnection time with U_o of 230V (see Regulation 413.02.10) (courtesy BSI)

General purpose (gG) fuses to BS 88-2.1 and BS 88-6

Rating (amperes)	6	10	16	20	25	32	40	50
Z_s (ohms)	8.89	5.33	2.82	1.85	1.5	1.09	0.86	0.63

Note: This requirement does not apply to a final circuit supplying an item of stationary equipment connected by means of a plug and socket outlet where precautions are already taken to prevent the use of the socket outlet for supplying hand-held equipment, nor to reduced low-voltage circuits.

- where a fuse is used to satisfy this disconnection requirement, maximum values of earth fault loop impedance (Z_s) corresponding to a disconnection time of 0.4 s are as stated in Table 9.6 for a nominal voltage to earth (U_o) of 230 V;
- for a distribution circuit and a final circuit supplying only stationary equipment, the maximum disconnection time of 5 s is not exceeded.

Note:

1. The circuit loop impedances given in Table 9.6 should not be exceeded when the conductors are at their normal operating temperature. If the conductors are at a different temperature when tested, then the reading should be adjusted accordingly.
2. See appropriate British Standard for types and rated currents of fuses other than those mentioned in Table 9.6.

9.3.9.11 Protection by separation of circuits

Ensure that:

- the separation of circuits shall be verified for protection by:
  - ○ SELV;
  - ○ PELV;
  - ○ electrical separation;
- the separation of live parts from those of other circuits and those from earth is verified by measuring that the insulation resistance is in accordance with the values shown in Table 9.7;

Table 9.7 Minimum values of insulation resistance

Circuit nominal voltage (V)	Test voltage d.c. (V)	Minimum insulation resistance (MΩ)
SELV and PELV	250	0.25
Up to and including 500 V (with the exception of the above systems)	500	0.5
Above 500 V	1000	1.0

- functional extra-low-voltage circuits meet all the test requirements for low-voltage circuits.

9.3.9.12 Polarity

Complete a polarity test (see Figure 9.6) to verify that:

- fuses and single-pole control and protective devices are only connected in the line conductor;

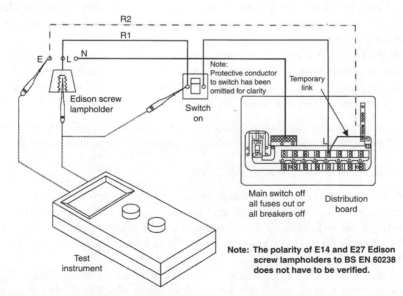

Figure 9.6 Polarity test on a lighting circuit (courtesy IET)

- circuits (other than BS EN 60238 E14 and E27 lamp holders) which have an earthed neutral conductor centre contact bayonet (and Edison screw lamp holders) have their outer or screwed contacts connected to the neutral conductor;
- wiring has been correctly connected to socket outlets and similar accessories.

9.3.10 Additional tests with the supply connected

Other than insulation tests the following tests are to be completed with the supply connected:

- re-check of polarity;
- earth electrode resistance;
- earth fault loop impedance;
- prospective fault current.

9.3.10.1 Polarity

Repeat the polarity test to verify that:

- fuses and single-pole control and protective devices are only connected in the line conductor;
- circuits (other than BS EN 60238 E14 and E27 lamp holders) which have an earthed neutral conductor centre contact bayonet (and Edison screw lamp holders) have their outer or screwed contacts connected to the neutral conductor;
- wiring has been correctly connected to socket outlets and similar accessories.

9.3.10.2 Earth electrode resistance

If the earthing system incorporates an earth electrode as part of the installation, measure the electrode resistance to earth (using test equipment similar to that shown in Appendix 9.1).

If the electrode under test is being used in conjunction with an RCD protecting an installation, test (prior to energising the remainder of the installation) between the line conductor at the origin of the installation and the earth electrode with the test link open.

 Note: The resulting impedance reading (i.e. the electrode resistance) should then be added to the resistance of the protective conductor for the protected circuits.

9.3.10.3 Earth fault loop impedance

This is an extremely important test to ensure that under earth fault conditions, overcurrent devices disconnect fast enough to reduce the risk of electric shock. The Regulations stipulate: *If the protective measures employed require a knowledge of earth fault loop impedance, then the relevant impedances must be measured.*

Where a fuse is used, maximum values of earth fault loop impedance (Z_s) corresponding to a disconnection time of 0.4 s are as stated in Table 9.8 for a nominal voltage to earth (U_o) of 230 V.

Table 9.8 British Standards for fuse links (courtesy Scaddon)

	Standard	Current rating	Voltages rating	Breaking capacity	Notes
1	BS 2950	Range 0.05–25A	Range 1000V (0.05 A) to 32V (25 A) a.c. and d.c.	Two or three times current rating	Cartridge fuse links for telecommunication and light electrical apparatus. Very low breaking capacity
2	BS 646	1, 2, 3 and 5 A	Up to 250V a.c. and d.c.	1000 A	Cartridge fuse intended for fused plugs and adapters to BS 546: 'round-pin' plugs
3	BS 1362 cartridge	1, 2, 3, 5, 7, 10 and 13 A	Up to 250V a.c.	6000 A	Cartridge fuse primarily intended for BS 1363: 'flat pin' plugs
4	BS 1361 HRC cut-out fuses	5, 15, 20, 30, 45 and 60 A	Up to 250V a.c.	16,500 A 33,000 A	Cartridge fuse intended for use in domestic consumer units. The dimensions prevent interchangeability of fuse links which are not of the same current rating
5	BS 88 motors	Four ranges, 2–1200 A	Up to 660V, but normally 250 or 415V a.c. and 250 or 500V d.c.	Ranges from 10,000 to 80,000 A in four a.c. and three d.c. categories	Part 1 of Standard gives performance and dimensions of cartridge fuse links, whilst Part 2 gives performance and requirements of fuse carriers and fuse bases designed to accommodate fuse links complying with Part 1
6	BS 2692	Main range from 5 to 200 A; 0.5 to 3 A for voltage transformer protective fuses	Range from 2.2 to 132kV	Ranges from 25 to 750 MVA (main range) 50 to 2500 MVA (VT fuses)	Fuses for a.c. power circuits above 660V
7	BS 3036 rewirable	5, 15, 20, 30, 45, 60, 100, 150 and 200 A	Up to 250V to earth	Ranges from 1000 to 12,000 A	Semi-enclosed fuses (the element is a replacement wire) for a.c. and d.c. circuits
8	BS 4265	500mA to 6.3 A 32 mA to 2 A	Up to 250V a.c.	1500 A (high breaking capacity); 35 A (low breaking capacity)	Miniature fuse links for protection of appliances of up to 250V (metric standard)

Note: See appropriate British Standard for types and rated currents of fuses other than those mentioned in Table 9.9.

Table 9.9 Maximum conductor operating temperatures for a floor-warming cable

Type of cable	Maximum conductor operating temperature (°C)
General-purpose PVC over conductor	70
Enamelled conductor, polychlorophene over enamel, PVC overall	70
Enamelled conductor PVC overall	70
Enamelled conductor, PVC over enamel, lead-alloy 'E' sheath overall	70
Heat-resisting PVC over conductor	85
Nylon over conductor, heat-resisting PVC overall	85
Synthetic rubber or equivalent elastomeric insulation over conductor	85
Mineral insulation over conductor, copper sheath overall	Temperature dependent on type of seal employed, outer covering etc.
Silicone-treated woven-glass sleeve over conductor	180

Using test equipment (similar to that listed in Appendix 9.1) complete the following tests:

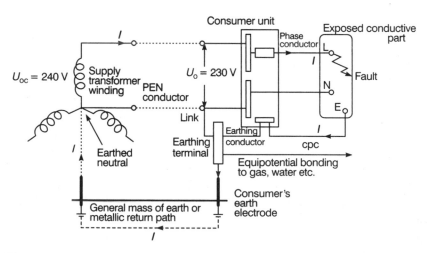

Figure 9.7 Testing earth fault loop impedance

Ensure that all main equipotential bonding is in place; connect the test equipment to the line, neutral and earth terminals at the remote end of the circuit under test. Press to test and record.

Note: The circuit loop impedances given in Table 9.8 should not be exceeded when the conductors are at their normal operating temperature. If the conductors are at a different temperature when tested, then the reading should be adjusted accordingly.

9.3.10.4 Prospective fault current

Verify, test and ensure that:

- prospective fault currents (under both short-circuit and earth fault conditions):
  - have been assessed for each supply source;
  - are calculated at every relevant point of the complete installation either by enquiry or by measurement;
  - are measured at the origin and at other relevant points in the installation;
- protection of wiring systems against overcurrent takes into account minimum and maximum fault current conditions;
- fault current protective devices are provided:
  - at the supply end of each parallel conductor where two conductors are in parallel;
  - at the supply and load ends of each parallel conductor where more than two conductors are in parallel;
- fault current protective devices are:
  - less than 3 m in length between the point where the value of current-carrying capacity is reduced and the position of the protective device;
  - installation so as to minimise the risk of fault current;
  - installation so as to minimise the risk of fire or danger to persons;
- fault current protective devices are placed (on the load side) at the point where the current-carrying capacity of the installation's conductors is likely to be lessened owing to:
  - the method of installation;
  - the cross-sectional area;
  - the type of cable or conductor used;
  - inherent environmental conditions;
- conductors are capable to carrying fault current without overheating.

Note: A single protective device may be used to protect conductors in parallel against the effects of fault current occurring.
Ensure that:

- fault current protection devices are capable of breaking;
- the breaking capacity rating of each device is not less than the prospective short-circuit current or earth fault current at the point at which the device is installed;

Note: A lower breaking capacity is permitted if another protective device is installed on the supply side.

- the characteristics of each device used for overload current and/or for fault current protection have been co-ordinated so that the energy let through (i.e. by the fault current protective device) does not exceed the overload current protective device's limiting values;
- devices providing protection against both overload current and fault current are capable of breaking;

Note: Overload current protection devices may have a breaking capacity below the value of the prospective fault current at the point where the device is installed.

- circuit breakers used as fault current protection devices:
  - ○ are capable of making any fault current up to and including the prospective fault current;
  - ○ break any fault current flowing before that current causes danger due to thermal or mechanical effects produced in circuit conductors or associated connections;
  - ○ break and make up any overcurrent up to and including the prospective fault current at the point where the device is installed;
- safety services with sources that are incapable of operating in parallel, are protected against electric shock and fault current.

9.3.11 Insulation tests

When an electrical installation fails an insulation test, the installation must be corrected and the test made again. If the failure influences any previous tests that were made, then those tests must also be repeated.

9.3.11.1 Locations with a risk of fire due to the nature of processed and/or stored materials

Test that wiring systems (less those using mineral insulated cables and busbar trunking arrangements) have been protected against earth insulation faults as follows:

- in TN and TT systems, by residual current devices having a rated residual operating current (I_{An}) not exceeding 300 mA;
- in IT systems, by insulation monitoring devices with audible and visible signals.

Note:

- Adequate supervision is required to facilitate manual disconnection as soon as appropriate.
- The disconnection time of the overcurrent protective device, in the event of a second fault, shall not exceed 5 s.

9.3.11.2 Protection against electric shock – insulation tests

Confirm, measure and test that:

- circuit conductors for each SELV system are physically separated from those of any other system or (i.e. where this proves impracticable) confirm that SELV circuit conductors are:
  - insulated for the highest voltage present;
  - enclosed in an insulating sheath additional to their basic insulation;
- equipment is capable of withstanding all mechanical, chemical, electrical and thermal stresses normally encountered during service;

Note: Paint, varnish, lacquer or similar products are **not** generally considered to provide adequate insulation for protection against direct contact in normal service.

- exposed conductive parts that might attain different potentials through failure of the basic insulation of live parts have been arranged so that a person will not come into simultaneous contact with two exposed conductive parts, or an exposed conductive part and any extraneous conductive part.

This may be achieved if the location has an insulating floor and insulating walls and one or more of the following arrangements apply:

- the distance between any separated exposed conductive parts (and between exposed conductive parts and extraneous conductive parts) is not less than 2.5 m (1.25 m for parts out of arm's reach);
- if protective obstacles (that are not connected to earth or to exposed conductive parts and which are made out of insulating material) are used between exposed conductive parts and extraneous conductive parts;
- the insulation is of acceptable electrical and mechanical strength;
- if the nominal voltage of a SELV system exceeds 25 V a.c. rms or 60 V ripple-free d.c., then protection against direct contact has been provided by one (or more) of the following:
  - insulation capable of withstanding a type-test voltage of 500 V a.c. rms for 60 seconds; or
  - a barrier (or an enclosure) capable of providing protection to at least IP2X or IPXXB;
- in IT systems, an insulation monitoring device has been provided so as to indicate the occurrence of a first fault from a live part to an exposed conductive part or to earth;
- insulating enclosures:
  - are not pierced by conductive parts (other than circuit conductors) likely to transmit a potential;
  - do not contain any screws of insulating material – the future replacement of which by metallic screws could impair the insulation provided by the enclosure;
  - do not adversely affect the operation of the equipment protected;

Note: Where the insulating enclosure has to be pierced by conductive parts (e.g. for operating handles of built-in equipment and for screws) protection against indirect contact shall not be impaired.

- live parts are completely covered with insulation which:
  - can only be removed by destruction;
  - is capable of durably withstanding electrical, mechanical, thermal and chemical stresses normally encountered during service;
- the basic insulation of operational electrical equipment is at least the degree of protection IP2X or IPXXB;
- where insulation has been applied during the erection of the installation, the quality of the insulation has been verified.

Where the risk of electric shock is increased by a reduction in body resistance and/or by contact with earth potential, confirm that protection has been provided by insulation of live parts, protection by obstacles, protection by barriers or enclosures or SELV.

9.3.11.3 Protection against electric shock – special installations or locations – verification tests

Where SELV or PELV is used (whatever the nominal voltage) in locations containing a bath, shower, hot air sauna and/or in a restrictive conductive location, confirm that protection against direct contact has been provided by:

- insulation capable of withstanding a type-test voltage of 500 V a.c. rms for 1 minute; or
- barriers and/or enclosures providing protection to at least IP2X or IPXXB.

The above requirements do not apply to a location in which freedom of movement is not physically constrained.

9.3.11.4 RCDs and RCBOs – verification tests

Test (on the load side of the RCD and as near as practicable to its point of installation and between the line conductor of the protected circuit and their associated circuit protective conductor) that:

- general purpose RCDs:
  - do not open with a leakage current flowing equivalency to 50% of the rated tripping current;
  - open in less that 200 ms with a leakage current flowing equivalency to 100% of the rated tripping current of the RCD;
- general purpose RCCDs to BS EN 61008 or RCBOs to BS EN 61009:
  - do not open with a leakage current flowing equivalency to 50% of the rated tripping current of the RCD;
  - open in less that 300 ms with a leakage current flowing equivalency to 100% of the rated tripping current of the RCD (unless it is a Type S (or

selective) device that incorporates an intentional time delay, in which case it should trip between 130 ms and 500 ms);

- RCD protected socket outlets to BS 7288;
  - o do not open with a leakage current flowing equivalency to 50% of the rated tripping current of the RCD;
  - o open in less that 200 ms with a leakage current flowing equivalency to 100% of the rated tripping current of the RCD.

9.3.11.5 Selection and erection of wiring systems

Test to confirm that:

- all electrical joints and connections meet stipulated requirements concerning conductance, insulation, mechanical strength and protection;
- cables that are run in a thermally insulated spaces are not covered by the thermal insulation;
- the current-carrying capacity of cables that are installed in thermally insulated walls or above a thermally insulated ceiling conforms with Appendix 4 to the Regulations;
- the current-carrying capacity of cables that are totally surrounded by thermal insulation for less than 0.5 m, has been reduced according to the size of cable, its length and the thermal properties of the insulation;
- the insulation and/or sheath of cables connected to a bare conductor or busbar is capable of withstanding the maximum operating temperature of the bare conductor or busbar;
- wiring systems are capable of withstanding the highest and lowest local ambient temperatures likely to be encountered – or are provided with additional insulation suitable for those temperatures;
- wiring systems have been selected and erected so as to minimise (i.e. during installation, use and maintenance) damage to the sheath and insulation of cables and insulated conductors and their terminations.

9.3.11.6 Site applied insulation

Test to confirm that:

- insulation applied on site to protect against direct contact is capable of withstanding, without breakdown or flashover, an applied test voltage as specified in the British Standard for similar type-tested equipment;
- supplementary insulation applied to equipment during erection (i.e. to protect against indirect contact):
  - o protects to at least IP2X or IPXXB; and
  - o is capable of withstanding, without breakdown or flashover, an applied test voltage as specified in the British Standard for similar type-tested equipment.

9.3.11.7 Supplies for safety services (IT systems)

In an IT system, confirm that continuous insulation monitoring has been provided to give audible and visible indications of a first fault.

9.3.11.8 Periodic Inspections and tests

Periodic inspection and testing of all electrical installations shall be carried out to confirm that the installation is in a satisfactory condition for continued service and this inspection shall consist of careful scrutiny of the installation (dismantled or otherwise) using appropriate tests.

The aim of periodic inspection and testing is to:

- confirm that the safety of the installation has not deteriorated or has been damaged;
- ensure the continued safety of persons and livestock against the effects of electric shock and burns;
- identify installation defects and non-compliance with the requirements of the Regulations, which may give rise to danger;
- protect property being damaged by fire and heat caused by a defective installation.

Precautions shall be taken to ensure that inspection and testing does not cause:

- danger to persons or livestock;
- damage to property and equipment (even if the circuit is defective).

The frequency of periodic inspection and testing of installations will depend on:

- the type of installation, its use and operation;
- the frequency and quality of maintenance; and
- the external influences to which it is subjected.

 Periodic inspection and testing of supervised installations may take the form of continuous monitoring and maintenance by skilled persons. Appropriate records shall be kept.

9.3.12 Verification tests

All completed installations (including additions and/or alteration to existing installations) shall be inspected and tested for conformance to the requirements of BS 7621 – as amended).

9.3.12.1 Accessibility of connections

Confirm that all connections and joints are accessible for inspection, testing and maintenance, unless:

- they are in a compound-filled or encapsulated joint;
- the connection is between a cold tail and a heating element;
- the joint is made by welding, soldering, brazing or compression tool.

9.3.12.2 Appliances producing hot water or steam

Confirm that electric appliances producing hot water or steam have been protected against overheating.

9.3.12.3 Cables and conductors for low voltage

Confirm that flexible and non-flexible cables, and flexible cords (and conductors used as an overhead line), operating at low voltage comply with the Relevant British or Harmonised Standard.

9.3.12.4 Electric surface heating systems

Confirm that the equipment, system design, installation and testing of all electric surface heating systems meet the requirements of BS 6351.

9.3.12.5 Emergency switching – verification tests

For any exterior installation, confirm that the switch is placed outside the building, adjacent to the equipment (where this is not possible, a notice showing the position of the switch shall be placed adjacent to the equipment and a notice fixed near the switch shall indicate its use).

9.3.12.6 Forced air heating systems

Confirm (by inspection and test) that electric heating elements of forced air heating systems (other than those of central-storage heaters):

- are incapable of being activated until the prescribed air flow has been established;
- deactivate when the air flow is reduced or stopped;
- do not have two, independent, temperature-limiting devices;
- have frames and enclosures constructed out of non-ignitable material.

9.3.12.7 Heating cables

Check that:

- heating conductors and cables that pass through (or are in close proximity to) a fire hazard:
  - are enclosed in material with an ignitability characteristic 'P' as specified in BS 476;
  - are protected from any mechanical damage;

- heating cables that have been laid (directly) in soil, concrete, cement screed, or other material used for road and building construction are:
  - capable of withstanding mechanical damage;
  - constructed of material that will be resistant to damage from dampness and/or corrosion;
- heating cables that have been laid (directly) in soil, a road, or the structure of a building are installed so that it:
  - is completely embedded in the substance it is intended to heat;
  - is not damaged by movement (by it or the substance in which it is embedded);
  - complies in all respects with the maker's instructions and recommendations;
- the maximum loading of floor-warming cable under operating conditions is no greater than the temperatures shown in Table 9.9.

9.3.12.8 Heating and ventilation systems

In locations where heating and ventilation systems containing heating elements are installed and where there is a risk of fire due to the nature of processed or stored materials, check to ensure that:

- the dust or fibre content and the temperature of the air does not present a fire hazard;
- temperature limiting devices have a manual reset;
- heating appliances are fixed;
- heating appliances mounted close to combustible materials are protected by barriers to prevent the ignition of such materials;
- heat storage appliances are incapable of igniting combustible dust and/or fibres;
- enclosures of equipment such as heaters and resistors do not attain higher surface temperatures than:
  - 90°C under normal conditions; and
  - 115°C under fault conditions.

9.3.12.9 Identification of conductors by letters and/or numbers

Test to confirm that all individual conductors and groups of conductors:

- have been identified by a label containing either letters or numbers that are clearly legible;
- have numerals that contrast, strongly, with the colour of the insulation;
- have numerals 6 and 9 underlined.

9.3.12.10 Plug and socket outlets

Inspect and test to ensure that any plug and socket outlet used in single-line a.c. or two-wire d.c. circuits that does **not** comply with BS 1363, BS 546, BS 196 or BS EN 60309-2 has either been designed specially for that purpose or:

- the plug and socket outlet used for an electric clock has been specially designed for that purpose and the plug has a fuse not exceeding 3 amperes which complies with BS 646 or BS 1362;
- the plug and socket outlet used for an electric shaver, is either part of the shaver supply unit that complies with BS 3535 (as amended) or, in a room (other than a bathroom); that complies with BS 4573.

9.3.12.11 Water heaters

Confirm that water heaters (or boilers) having immersed and uninsulated heating elements are permanently connected to the electricity supply via a double-pole linked switch, which is either:

- separate from and within easy reach of the heater/ boiler; or
- part of the boiler/heater (provided that the wiring from the heater or boiler is directly connected to the switch without use of a plug and socket outlet).

Functional testing

The following are among the most important functional tests that should be completed:

- Verify the effectiveness of residual current devices providing protection against indirect contact (or supplementary protection against direct contact) by a test simulating a typical fault condition.
- Functionally test assemblies (such as switchgear and controlgear assemblies, drives, controls and interlocks) to show that they are properly mounted, adjusted and installed in accordance with the Regulations.

9.3.13 Electrical connections

Of main concern are:

- connections between conductors and between a conductor and equipment;
- main earthing terminals or bars;
- final and distribution circuits.

Test and verify that:

- connections between conductors and between a conductor and equipment provide durable electrical continuity and adequate mechanical strength;

- the earthing conductor of main earthing terminals (or bars) is capable of being disconnected to enable the resistance of the earthing arrangements to be measured;

 For convenience (and if required) this may be combined with the main earthing terminal or bar.

- all joints:
  - are capable of disconnection only by means of a tool;
  - are mechanically strong;
  - ensure the maintenance of electrical continuity;
- the wiring of final and distribution circuits to equipment with a protective conductor current exceeding 10 mA have a protective connection complying with one or more of the following:
  - a single protective conductor with a cross-section greater than $10\,mm^2$;
  - a single (mechanically protected) copper protective conductor with a cross-section greater than $4\,mm^2$;
  - two individual protective conductors;
  - a BS 4444 earth monitoring system that will automatically disconnect the supply to the equipment in the event of a continuity fault;
- connection (i.e. of the equipment) to the supply by means of a double wound transformer having its secondary winding connected to the protective conductor of the incoming supply and the exposed conductive parts of the above.

9.3.14 Tests for compliance with the Building Regulations

As shown in Table 9.10, there are four types of installation that have to be inspected and tested for compliance with the Building Regulations.

Table 9.10 Types of installation

Type of inspection	When is it used?	What should it contain?	Remarks
Minor Electrical Installation Works Certificate	For new work associated with an alteration or addition to an existing installation	Relevant provisions of Part 6 of BS 7671	
Full Electrical Installation	For the design, construction, inspection and testing of an installation	A schedule of inspections and test results as required by Part 6 of BS 7671A certificate, including guidance for recipients (standard form from Appendix 6 of BS 7671)	For safety reasons, the electrical installation will need to be inspected at appropriate intervals by a competent person

(continued)

Table 9.10 (*continued*)

Type of inspection	When is it used?	What should it contain?	Remarks
Electrical Installation Certificate (short form)	For use when one person is responsible for the design, construction, inspection and testing of an installation	A schedule of inspections and a schedule of test results as required by Part 6 of BS 7671	For safety reasons, the electrical installation will need to be inspected at appropriate intervals by a competent person
Periodic Inspection Report	For the inspection of an existing electrical installation	A schedule of inspections and a schedule of test results as required by Part 6 of BS 7671	For safety reasons, the electrical installation will need to be inspected at appropriate intervals by a competent person

Figure 9.8 indicates how to choose what type of inspection is required.

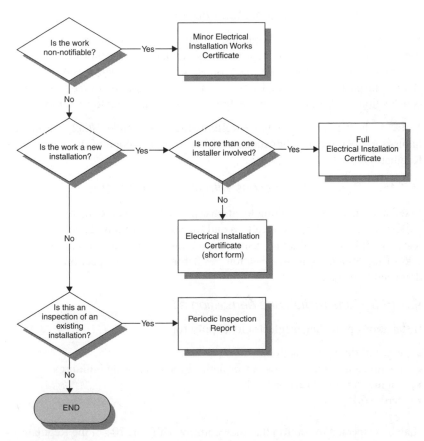

Figure 9.8 Choosing the correct Inspection Certificate

Part P applies only to fixed electrical installations that are intended to operate at low voltage or extra-low voltage which are not controlled by the Electricity Supply Regulations 1988 as amended, or the Electricity at Work Regulations 1989 as amended.

9.3.15 Additional tests required for special installations and locations

Part 7 of the Regulations contains additional requirements in respect of installations where the risk of electric shock is increased by a reduction in body resistance or by contact with earth potential (e.g. locations containing a bath or shower, swimming pools, hot air sauna, construction installations, agricultural and horticultural premises, caravans, motor caravans and highway power supplies).

Additional tests and inspections for these special installations and locations have been listed at the end of the following sections in the form of check sheets.

9.3.15.1 Agricultural and horticultural premises

All fixed agricultural and horticultural installations (outdoors and indoors) and locations where livestock is kept (such as stables, chicken houses, piggeries, feed-processing locations, lofts and storage areas for hay, straw and fertilisers) shall be inspected to confirm that they comply with Part 705 of the Regulations (see Section 7.3.1 for a list of inspections and tests that need to be completed).

Note: If these premises include dwellings that are intended solely for human habitation, then the dwellings are excluded from the scope of these particular Regulations.

9.3.15.2 Conducting locations with restricted movement

Fixed equipment in conducting locations (particularly where the movement of persons is restricted by the location) and the supplies to mobile equipment for use in such locations shall be inspected to confirm that they comply with Part 706 of the Regulations (see Section 7.3.2 for a list of inspections and tests that need to be completed).

9.3.15.3 Construction and demolition sites

Installations providing an electricity supply for:

- new building construction;
- repairs, alterations, extensions or demolition of existing buildings;
- engineering construction; and
- earthworks

shall be inspected to confirm that they comply with Part 704 of the Regulations (see Section 7.3.3 for list of inspections and tests that need to be completed).

These requirements do **not** apply to:

- construction site offices, cloakrooms, meeting rooms, canteens, restaurants, dormitories and toilets;
- installations covered by BS 6907.

9.3.15.4 Electrical installations in caravan/camping parks and similar locations

Electrical installations in caravan/camping parks and similar locations providing facilities for supplying leisure accommodation vehicles (including caravans) or tents, shall be inspected to confirm that they comply with Part 708 of the Regulations (see Section 7.3.4 for a list of inspections and tests that need to be completed).

9.3.15.5 Electrical installations in caravans and motor caravans

All electrical installations in caravans and motor caravans shall be inspected to confirm that they comply with Part 721 of the Regulations (see Section 7.3.5 for a list of inspections and tests that need to be completed).

It should be noted that the requirements of this section do not apply to:

- electrical circuits and equipment covered by the Road Vehicles Lighting Regulations 1989;
- installations covered by BS EN 1648-1 and BS EN 1648-2;
- internal electrical installations of mobile homes, fixed recreational vehicles, transportable sheds and the like, temporary premises or structures.

9.3.15.6 Exhibitions, shows and stands

Temporary electrical installations in exhibitions, shows and stands (including mobile and portable displays and equipment) shall be inspected to confirm that they comply with Part 711 of the Regulations (see Section 7.3.6 for a list of inspections and tests that need to be completed).

9.3.15.7 Floor and ceiling heating systems

Electric floor and ceiling heating systems which are erected as either thermal storage heating systems or direct heating systems, shall be inspected to confirm that they comply with Part 753 of the Regulations (see Section 7.3.7 for a list of inspections and tests that need to be completed).

9.3.15.8 Locations containing a bath or shower

All locations containing a bath or shower shall be inspected to confirm that they comply with Part 701 of the Regulations (see Section 7.3.8 for a list of inspections and tests that need to be completed).

Locations containing baths or showers for medical treatment, or for disabled persons, may have special requirements.

9.3.15.9 Marinas and similar locations

Circuits intended to supply pleasure craft or houseboats in marinas and similar locations, shall be inspected to confirm that they comply with Part 709 of the Regulations (see Section 7.3.9 for a list of inspections and tests that need to be completed).

9.3.15.10 Mobile and transportable units

A vehicle and/or mobile (self-propelled or towed) or transportable structure (such as a container or cabin) in which all or part of an electrical installation is contained and which is provided with a temporary supply by means of, for example, a plug and socket outlet, shall be inspected to confirm compliance with Part 717 of the Regulations (see Section 7.3.10 for a list of inspections and tests that need to be completed).

9.3.15.11 Rooms and cabins containing saunas

Installations supplying electricity for locations in which hot air sauna heating equipment is fitted (in accordance with BS EN 60335-2-53) shall be inspected to confirm that they comply with Part 703 of the Regulations (see Section 7.3.11 for a list of inspections and tests that need to be completed).

9.3.15.12 Solar, photovoltaic (PV) power supply systems

Electrical installations of PV power supply systems (including subsystems with a.c. modules) shall be inspected to confirm that they comply with Part 712 of the Regulations (see Section 7.3.12 for a list of inspections and tests that need to be completed).

9.3.15.13 Swimming pools and other basins

Requirements applicable to basins of swimming pools, paddling pools and other basins plus their surrounding zones shall be inspected to confirm that they comply with Part 702 of the Regulations (see Section 7.3.13 for a list of inspections and tests that need to be completed).

9.3.15.14 Temporary electrical installations for structures, amusement devices and booths at fairgrounds, amusement parks and circuses

Electrical installation required for the safe design, installation and operation of temporarily erected mobile or transportable electrical machines and structures which incorporate electrical equipment, shall be inspected to confirm that they comply with Part 740 of the Regulations (see Section 7.3.14 for a list of inspections and tests that need to be completed).

9.4 Identification and notices

For safety purposes, the Regulations require a number of notices and labels to be used for electrical installations and these will need to be checked during initial and periodic inspections.

9.4.1 General

Verify that:

- labels (or other means of identification) indicate the purpose of each item of switchgear and control gear;
- wiring is marked and/or arranged so that it can be quickly identified for inspection, testing, repair or alteration of the installation;
- unambiguous marking has been provided at the interface between conductors identified in accordance with these Regulations and conductors identified to previous versions of the Regulations;

Note: Appendix 7 of the Regulations provides guidance on how this can be achieved.

- orange-coloured conduits have been used to distinguish an electrical conduit from other services or other pipelines.

9.4.2 Conductors

Verify that:

- conductor cable cores are identified by colour and/or lettering and/or numbering;
- binding and sleeves used for identifying protective conductors comply with BS 3858;
- neutral or mid-point conductors are coloured blue;
- protective conductor cable cores are identifiable at all terminations (and preferably throughout their length);
- protective conductors are a bi-colour combination of green and yellow and that neither colour covers more than 70% of the surface being coloured;

Verify that this combination of colours has **not** been used for any other purpose.

- single-core cables used as protective conductors are coloured green-and-yellow throughout their length;
- bare conductors or busbars used as protective conductors are identified by equal green and yellow stripes that are 15 mm to 100 mm wide;

If adhesive tape is used, then it has been bicoloured.

- PEN conductors (when insulated) are either:
  - green-and-yellow throughout their length, with blue markings at the terminations; or
  - blue throughout their length, with green-and-yellow markings at the terminations;
- bare conductors are painted or identified by a coloured tape, sleeve or disc as per Table 9.11.

ALL other conductors (including those used to identify conductors and switchboard busbars) **shall** be coloured as shown in accordance with Table 9.11.

Table 9.11 Identification of conductors

Function	Alphanumeric	Colour
Protective conductors		Green-and-yellow
Functional earthing conductor		Cream
a.c. power circuit (Note 1)		
Phase of single-line circuit	L	Brown
Neutral of single- or three-line circuit	N	Blue
Phase 1 of three-line a.c. circuit	L1	Brown
Phase 2 of three-line a.c. circuit	L2	Black
Phase 3 of three-line a.c. circuit	L3	Grey
Two-wire unearthed d.c. power circuit		
Positive of two-wire circuit	L+	Brown
Negative of two-wire circuit	L−	Grey
Two-wire earthed d.c. power circuit		
Positive (of negative earthed) circuit	L+	Brown
Negative (of negative earthed) circuit (Note 2)	M	Blue
Positive (of positive earthed) circuit (Note 2)	M	Blue
Negative (of positive earthed) circuit	L−	Grey
Three-wire d.c. power circuit		
Outer positive of two-wire circuit derived from three-wire system	L+	Brown
Outer negative of two-wire circuit derived from three-wire system	L−	Grey
Positive of three-wire circuit	L+	Brown
Mid-wire of three-wire circuit (Notes 2 & 3)	M	Blue
Negative of three-wire circuit	L−	Grey
Control circuits, ELV and other applications		
Phase conductor	L	Brown, Black, Red, Orange, Yellow, Violet, Grey, White, Pink or Turquoise
Neutral or mid-wire (Note 4)	N or M	Blue

The colour green shall **not** be used on its own.

 Note:

(1)   Power circuits include lighting circuits.

(2)   M identifies either the mid-wire of a three-wire d.c. circuit, or the earthed conductor of a two-wire earthed d.c. circuit.

(3)   Only the middle wire of three-wire circuits may be earthed.

(4)   An earthed PELV conductor is blue.

Further information, concerning cable identification colours for extra-low-voltage and d.c. power circuits, is available from the IEE website at www.iee.org/cablecolours.

9.4.3 Identification of conductors by letters and/or numbers

Where letters and/or numbers are used to identify conductors, check to confirm that:

- individual conductors and/or groups of conductors are identified by either letters or numbers that are clearly legible;
- all numerals contrast, strongly, with the colour of the insulation;
- numerals 6 and 9 are underlined;
- protective devices are arranged and identified so that the circuit protected is easily recognisable;
- protective conductors coloured green-and-yellow are not numbered other than for the purpose of circuit identification;
- the alphanumeric numbering system is in accordance with Table 9.11.

9.4.4 Omission of identification by colour or marking

Colour or marking is **not** required for:

- concentric conductors of cables;
- metal sheath or armour of cables (when used as a protective conductor);
- bare conductors (where permanent identification is impracticable);
- extraneous conductive parts used as a protective conductor;
- exposed conductive parts used as a protective conductor.

9.4.5 Diagrams

Verify that all available diagrams, charts, tables or schedules that have been used, indicate:

- the type and composition of each circuit (points of utilisation served, number and size of conductors, type of wiring);
- the method used for protection against indirect contact;
- the data (where appropriate) and characteristics of each protective device used for automatic disconnection;

- the identification (and location) of all protection, isolation and switching devices;
- the circuits or equipment that are susceptible to a particular test.

 Verify that distribution board schedules have been provided within (or adjacent to) each distribution board.

 All symbols used shall comply with BS EN 60617 (see Appendix A to this book).

9.4.6 Warning notices

Check to confirm that the following warning notices are appropriate to the situation and correctly positioned, and contain the right information.

9.4.6.1 Earth-free local equipotential bonding

Where earth-free local equipotential bonding has been used confirm that an appropriate notice has been fixed in a prominent position adjacent to every point of access to the location concerned.

9.4.6.2 Earthing and bonding connections

Confirm that a permanent label with the words shown in Figure 9.9 has been permanently fixed at or near:

- the connection point of every earthing conductor to an earth electrode;
- the connection point of every bonding conductor to an extraneous conductive part;
- the main earth terminal (when separated from the main switchgear).

> **Safety Electrical Connection – Do Not Remove**

Figure 9.9 Earthing and bonding notice

Where protection is by earth-free local equipotential bonding or electrical separation, confirm that the warning notice reads as per Figure 9.10.

> Equipotential protective bonding conductors associated with the electrical installation in this location
>
> and
>
> Equipment having exposed conductive parts
>
> **MUST NOT BE CONNECTED TO EARTH**

Figure 9.10 Earthing, bonding and electrical separation notice

9.4.6.3 Electrical installations in caravans, motor caravans and caravan parks

Confirm that:

- a notice has been fixed on (or near) the electrical inlet recess of the caravan or motor caravan installation containing details concerning the:
  - nominal (design) voltage and frequency;
  - current rating;
- a notice, worded as shown in Figure 9.11, has been permanently fixed near the main isolating switch inside the caravan or motor caravan.

1.1 INSTRUCTIONS FOR ELECTRICITY SUPPLY

TO CONNECT

1. Before connecting the caravan installation to the mains supply, check that:
    (a) the supply available at the caravan pitch supply point is suitable for the caravan electrical installation and appliances, and
    (b) the caravan main switch is in the OFF position.

2. Open the cover to the appliance inlet provided at the caravan supply point and insert the connector of the supply flexible cable.

3. Raise the cover of the electricity outlet provided on the pitch supply point and insert the plug of the supply cable.

THE CARAVAN SUPPLY FLEXIBLE CABLE MUST BE FULLY UNCOILED TO AVOID DAMAGE BY OVERHEATING.

4. Switch on at the caravan main switch.

5. Check the operation of residual current devices, if any, fitted in the caravan by depressing the test buttons.

IN CASE OF DOUBT OR, IF AFTER CARRYING OUT THE ABOVE PROCEDURE THE SUPPLY DOES NOT BECOME AVAILABLE, OR IF THE SUPPLY FAILS, CONSULT THE CARAVAN PARK OPERATOR OR THE OPERATOR'S AGENT OR A QUALIFIED ELECTRICIAN.

TO DISCONNECT

6. Switch off at the caravan main isolating switch, unplug both ends of the cable.

1.2 PERIODIC INSPECTION

Preferably not less than once every three years and more frequently if the vehicle is used more than normal average mileage for such vehicles, the caravan electrical installation and supply cable should be inspected and tested and a report on their condition obtained as prescribed in BS 7671 Requirements for Electrical Installations published by the Institution of Electrical Engineers.

Figure 9.11 Electrical safety notice (reproduced with permission of IET)

9.4.6.4 Emergency switching

For any exterior installation for emergency switching, confirm that the switch is placed outside the building, adjacent to the equipment (where this is not possible, a notice showing the position of the switch shall be placed adjacent to the equipment and a notice fixed near the switch shall indicate its use).

9.4.6.5 Final circuit distribution boards

Confirm that distribution boards for socket outlet final circuits have a notice that clearly indicates circuits which have a high protective conductor current and that this information is positioned so as to be plainly visible to a person employed in modifying or extending the circuit.

9.4.6.6 Inspection and testing

Confirm that:

• a notice has been fixed in a prominent position at or near the origin of every installation upon completion of the work (e.g. initial verification, alterations and additions to an installation and periodic inspection and testing);
• the notice has been inscribed in indelible characters and reads as shown in Figure 9.12.

IMPORTANT

This installation should be periodically inspected and tested and a report on its condition obtained, as prescribed in BS 7671:2008 Requirements for Electrical Installations.

Date of last inspection......................

Recommended date of next inspection...................

Figure 9.12 Inspection and testing notice (reproduced with permission of IET)

If an installation includes a residual current device then it has a notice (fixed in a prominent position) that reads as shown in Figure 9.13.

This installation, or part of it, is protected by a device that automatically switches off the supply if an earth fault develops.

Test quarterly by pressing the button marked 'T' or 'Test'. The device should switch off the supply and should then be switched on to restore the supply.

If the device does not switch off the supply when the button is pressed, seek expert advice.

Figure 9.13 Inspection and testing notice (reproduced with permission of IET)

- Confirm that following initial verification, an Electrical Installation Certificate (together with a schedule of inspections and a schedule of test results), was given to the person ordering the work and that:
  - o the schedule of test results identified every circuit and its related protective device(s) and recorded the results of the appropriate tests and measurements;
  - o the Certificate took account of the respective responsibilities for the safety of that installation and the relevant schedules;
  - o defects or omissions revealed during inspection and testing of the installation work covered by the Certificate were rectified before the Certificate was issued.

Confirm that:

- an Electrical Installation Certificate (containing details of the installation together with a record of the inspections made and the test results) or a Minor Electrical Installation Works Certificate (for all minor electrical installations that do not include the provision of a new circuit) was provided for all alterations or additions to electrical circuits; and that
- any defects found in an existing installation were recorded on an Electrical Installation Certificate or a Minor Electrical Installation Works Certificate.

9.4.6.7 Isolation

Verify that a notice is fixed in each position where there are live parts which are not capable of being isolated by a single device.

 Note: The location of each isolator shall be indicated unless there is no possibility of confusion.

If an installation is supplied from more than one source confirm that:

- a main switch is provided for each source of supply;
- a notice has been placed warning operators that more than one switch needs to be operated.

Unless an interlocking arrangement is provided, confirm that a notice has been provided warning people that they will need to use the appropriate isolating devices if an item of equipment or enclosure that contains live parts cannot be isolated by a single device.

9.4.6.8 Voltage

Verify that:

- items of equipment (or enclosures) within which a nominal voltage exceeding 230 V exists (and where the presence of such a voltage would not normally be expected) are arranged so that before access is gained to a live part, a warning of the maximum voltage present is clearly visible;

- where terminals (or other fixed live parts between which a nominal voltage exceeding 230 V exists) are housed in separate enclosures (or items of equipment which, although separated, can be reached simultaneously by a person) a notice has been secured in a position such that anyone, before gaining access to such live parts, is warned of the maximum voltage which exists between those parts;
- means of access to all live parts of switchgear and other fixed live parts where different nominal voltages exist are marked to indicate the voltages present.

9.4.6.9 Warning notice – non-standard colours

If an installation that was wired to a previous version of the Regulations is partially altered or rewired according to the current Regulations (i.e. in accordance with Table 9.11: Identification of conductors) confirm that a warning notice (see Figure 9.14) has been placed at (or near) the appropriate distribution board.

> **WARNING**
> This installation has wiring colours to two versions of BS 7671.
> Great care should be taken before undertaking extension, alteration or repair that all conductors are correctly identified.

Figure 9.14 Non-standard colours (reproduced with permission of IET)

9.5 What types of certificate and report are there?

There are three types of Certificates associated with electrical installations. These are:

The Electrical Installation Certificate	For of the design, construction, inspection and testing of electrical installation work
The Minor Works Certificate	For the inspection and testing of an electrical installation
The Periodic Inspection Report	For the regular inspection and testing of an electrical installation

All Certificates need to be made out and signed (or otherwise authenticated) by a competent person or persons. If the design, construction and inspection and testing is the responsibility of one person, then a certificate (similar to that shown in Figure 9.15) may be used as a replacement for the multiple signatures section of the model form.

FOR DESIGN, CONSTRUCTION, INSPECTION & TESTING

I being the person responsible for the Design, Construction, Inspection & Testing of the electrical installation (as indicated by my signature below), particulars of which are described above, having exercised reasonable skill and care when carrying out the Design, Construction, Inspection & Testing, hereby CERTIFY that the said work for which I have been responsible is to the best of my knowledge and belief in accordance with BS 7671:2008 except for the departures, if any, detailed as follows:

. . . .

. . . .

. . . .

Name:. . . . .   Signature:. . . . .   Date:. . . . .

Figure 9.15 Single signature declaration form

Schedules of inspections and test results (see Figures 9.16 and 9.17) shall be issued along with the associated Electrical Installation Certificate and/or Periodic Inspection Report.

9.5.1 Electrical Installation Certificate

The Electrical Installation Certificate (containing details of the installation together with a record of the inspections made and the test results) is only used for the initial certification of a new installation or for the alteration or addition to an existing installation where new circuits have been introduced.

The Certificate is **only** valid if accompanied by the Schedule of Inspections and the Schedule(s) of Test Results that clearly show when the first periodic inspection must be completed.

 Note: The original Certificate shall be given to the person ordering the work and a duplicate retained by the contractor.

The following notice shall be attached to the Certificate (see Figures 9.18 and 9.19).

Electrical Installation Certificates:

- may be produced in any durable medium, including written and electronic media;
- shall be compiled and signed by (a) competent person(s).

SCHEDULE OF INSPECTIONS

Methods of protection against electric shock	Prevention of mutual detrimental influence

(a) Protection against both direct and indirect contact:

☐ (i)   SELV

☐ (ii) Limitation of discharge of energy

(b) Protection against direct contact:

☐ (i)   Insulation of live parts

☐ (ii)   Barriers or enclosures

☐ (iii) Obstacles

☐ (iv)   Placing out of reach

☐ (v)   PELV

☐ (vi)   Presence of RCD for supplementary protection

(c) Protection against indirect contact:

☐ (i)   EEBADS including:

☐ Presence of earthing conductor

☐ Presence of circuit protective conductors

☐ Presence of main equipotential bonding conductors

☐ Presence of supplementary equipotential bonding conductors

☐ Presence of earthing arrangements for combined protective and functional purpose

☐ Presence of adequate arrangements for alternative source(s), where applicable

☐ Presence of residual current device(s)

☐ (ii)   Use of Class II equipment or equivalent insulation

☐ (iii) Non-conducting location: Absence of protective conductors

☐ (iv)   Earth-free equipotential bonding: Presence of earth-free equipotential bonding conductors

☐ (v)   Electrical separation

Prevention of mutual detrimental influence

☐ (a) Proximity of non-electrical services and other influences

☐ (b) Segregation of band I and bandI circuits or band II insulation used

☐ (c) Segregation of safety circuits

Identification

☐ (a) Presence of diagrams, instructions, circuit charts and similar information

☐ (b) Presence of danger notices and other warning notices

☐ (c) Labelling of protective devices, switches and terminals

☐ (d) Identification of conductors

Cables and conductors

☐ (a) Routing of cables in prescribed zones or within mechanical protection

☐ (b) Connection of conductors

☐ (c) Erection methods

☐ (d) Selection of conductors for current-carrying capacity and voltage drop

☐ (e) Presence of fire barriers, suitable seals and protection against thermal effects

General

☐ (a) Presence and correct location of appropriate devices for isolation and switching

☐ (b) Adequacy of access to switchgear and other equipment

☐ (c) Particular protective measures for special installations and locations

☐ (d) Connection of single-pole devices for protection or switching in phase conductors only

☐ (e) Correct connection of accessories and equipment

☐ (f) Presence of undervoltage protective devices

☐ (g) Choices and setting of protective and monitoring devices for protection against indirect contact and/or overcurrent

☐ (h) Selection of equipment and protective measures appropriate to external influences

☐ (i) Selection of appropriate functional switching devices

Inspected by ....   Date ....

Notes:
✓   to indicate an inspection has been carried out and the result is satisfactory
✗   to indicate an inspection has been carried out and the result was unsatisfactory
N/A   to indicate the inspection is not applicable
LIM   to indicate that, exceptionally, a limitation agreed with the person ordering the work prevented the inspection or test being carried out.

Figure 9.16 Schedule of Inspections

9.5.2 Minor Electrical Installation Works Certificate

A Minor Electrical Installation Works Certificate is used for additions and alterations to an installation such as an extra socket outlet or lighting point for an existing circuit, the relocation of a light switch etc. This Certificate may also be used for the replacement of equipment such as accessories or luminaires, but **not** for the replacement of distribution boards (or similar items) or the provision of a new circuit.

SCHEDULE OF TEST RESULTS

Contractor: ....
Test Date: ....
Signature: ....
Method of protection against indirect contact: ....
Equipment vulnerable to testing: ....

Address/Location of distribution board: ....
....

Type of Supply: TN-S/TN-C-S/TT
Z_e at origin: ....ohms
PFC: ....kA

Instruments
loop impedance: ....
continuity: ....
insulation: ....
RCD tester: ....

Description of Work: ....

Circuit Description	Overcurrent Device Short-circuit capacity: ......kA			Wiring Conductors		Continuity			Insulation Resistance		Polarity	Earth Loop Imped-ance	Functional Testing		Remarks
	Type	Rating I_n		Live	CPC	$(R_1 + R_2)^*$	R_2^*	$R_i n^*$	Live/ Live	Live/ Earth		Z_s	RCD time	Other	
				mm^2	mm^2	Ω	Ω	Ω	$M\Omega$	$M\Omega$		Ω	ms		
1	2	3		4	5	6	7	8	9	10	11	12	13	14	15

Test Results

Deviations from Wiring Regulations and special notes:

* Complete column 6 or 7.

Figure 9.17 Schedule of Test Results

GUIDANCE FOR RECIPIENTS

This safety Certificate has been issued to confirm that the electrical installation work to which it relates has been designed, constructed and inspected and tested in accordance with British Standard 7671:2008 (the IEEE Wiring Regulations).

You should have received an original Certificate and the contractor should have retained a duplicate Certificate.

The 'original' Certificate should be retained in a safe place and be shown to any person inspecting or undertaking further work on the electrical installation in the future. If you later vacate the property, this Certificate will demonstrate to the new owner that the electrical installation complied with the requirements of British Standard 7671:2008 at the time the Certificate was issued. The Construction (Design and Management) Regulations require that for a project covered by those Regulations, a copy of this Certificate, together with schedules, is included in the project health and safety documentation.

For safety reasons, the electrical installation will need to be inspected at appropriate intervals by a competent person. The maximum time interval recommended before the next inspection is stated on Page 1 under 'Next Inspection'.

This Certificate is intended to be issued only for a new electrical installation or for new work associated with an alteration or addition to an existing installation. It should not have been issued for the inspection of an existing electrical installation. A 'Periodic Inspection Report' should be issued for such a periodic inspection.

Figure 9.18 Guidance notice to accompany the Electrical Installation Certificate

The following notice shall be attached to the certificate (see Figures 9.20 and 9.21).

Minor Electrical Installation Works Certificates:

- may be produced in any durable medium, including written and electronic media;
- shall be compiled and signed by a competent person(s).

9.5.3 Periodic inspection

A Periodic Inspection Report is used for reporting on the condition of an existing installation and shall include schedules of both the inspection and the test results.

The following notice shall be attached to the certificate on completion of the inspection.

Periodic Inspection Reports:

- may be produced in any durable medium, including written and electronic media;
- shall be compiled and signed by (a) competent person(s).

Form 2   Form No /2

ELECTRICAL INSTALLATION CERTIFICATE (notes 1 and 2)

(REQUIREMENTS FOR ELECTRICAL INSTALLATIONS - BS 7671 [IEE WIRING REGULATIONS])

DETAILS OF THE CLIENT (note 1) ....

....

....

INSTALLATION ADDRESS

....

....

....Postcode...

DESCRIPTION AND EXTENT OF THE INSTALLATION Tick boxes as appropriate (note 1) Description of installation: .... Extent of installation covered by this Certificate:	New installation ☐
	Addition to an existing installation ☐
	Alteration to an existing installation ☐

FOR DESIGN

I/We being the person(s) responsible for the design of the electrical installation (as indicated by my/our signatures below), particulars of which are described above, having exercised reasonable skill and care when carrying out the design, hereby CERTIFY that the design work for which I/we have been responsible is to the best of my/our knowledge and belief in accordance with BS 7671: ...., amended to........... (date) except for the departures, if any, detailed as follows:

> Details of departures from BS 7671 (Regulations 120-01-03, 120-02):

The extent of liability of the signatory or the signatories is limited to the work described above as the subject of this Certificate. For the DESIGN of the installation.   **(Where there is mutual responsibility for the design)

Signature: .... Date ....   Name (BLOCK LETTERS): .... Designer No. 1

Signature: .... Date ....   Name (BLOCK LETTERS): .... Designer No. 2**

FOR CONSTRUCTION

I/We being the person(s) responsible for the construction of the electrical installation (as indicated by my/our signatures below), particulars of which are described above, having exercised reasonable skill and care when carrying out the construction, hereby CERTIFY that the construction work for which I/we have been responsible is to the best of my/our knowledge and belief in accordance with BS 7671: ....amended to ....(date) except for the departures, If any, detailed as follows:

> Details of departures from BS 7671 (Regulations 120-01-03, 120-02):

The extent of liability of the signatory is limited to the work described above as the subject of this Certificate. For CONSTRUCTION of the installation:

Signature: .... Date ....

Name (BLOCK LETTERS): .... Constructor

FOR INSPECTION & TESTING

I/We being the person(s) responsible for the inspection & testing of the electrical installation (as indicated by my/our signatures below), particulars of which are described above, having exercised reasonable skill and care when carrying out the inspection & testing, hereby CERTIFY that the work for which I/we have been responsible is to the best of my/our knowledge and belief in accordance with BS 7671:........., amended to .... (date) except for the departures, if any, detailed as follows:

> Details of departures from BS 7671 (Regulations 120-01-03, 120-02):

The extent of liability of the signatory is limited to the work described above as the subject of this Certificate. For INSPECTION & TEST of the installation: **(Where there is mutual responsibility for the design)

Signature: .... Date ....

Name (BLOCK LETTERS): .... Inspector

NEXT INSPECTION (notes 4 and 7)

I/We the designer(s) recommend that this installation is further inspected and tested after an interval of not more than............ years/months

Figure 9.19 Electrical Installation Certificate

PARTICULARS OF THE SIGNATORIES TO THE ELECTRICAL INSTALLATION CERTIFICATE (note 3)

Designer (No. 1)
Name: .... Company: ....
Address: ....
.... Postcode: .... Tel No.: ....

Designer (No. 2)
(if applicable)   Name: .... Company: ....
Address: ....
.... Postcode: .... Tel No.: ....

Constructor
Name: .... Company: ....
Address: ....
.... Postcode: .... Tel No.: ....

Inspector
Name: .... Company: ....
Address: ....
.... Postcode: .... Tel No.: ....

SUPPLY CHARACTERISTICS AND EARTHING ARRANGEMENTS Tick boxes and enter details, as appropriate

Earthing arrangements	Number and Type of Live Conductors		Nature of Supply Parameters	Supply Protective Device Characteristics
TN-C ☐	a.c. ☐	d.c. ☐	Nominal voltage, U/U_o[(1)] V	
TN-S ☐	1-line, 2-wire ☐	2-pole ☐	Nominal frequency, f[(1)] .... Hz	Type: ....
TN-C-S ☐	1-line, 3-wire ☐	3-pole ☐	Prospective fault current, I_{pf}[(2)] (note 6)....... kA	....
TT ☐	2-line, 3-wire ☐	other ☐	External loop impedance, Z_e[(2)] ....Ω	
IT ☐	3-line, 3-wire ☐		(Note: (1) by enquiry, (2) by enquiry or by measurement)	Nominal current rating
	3-line, 4-wire ☐			...............................A
Alternative source ☐ of supply (to be detailed on attached schedules)				

PARTICULARS OF INSTALLATION REFERRED TO IN THE CERTIFICATE Tick boxes and enter details, as appropriate

Means of Earthing	Maximum Demand
Distributor's facility ☐	Maximum demand (load) ....Amps per phase

Details of Installation Earth Electrode (where applicable)

Installation ☐   Type   Location   Electrode resistance to earth
earth electrode   (e.g. rod(s), tape etc.)
....Ω

Main Protective Conductors

Earthing conductor: material .... csa ....mm^2 connection verified ☐

Main equipotential
bonding conductors: material .... csa ....mm^2 connection verified ☐

To incoming water and/or gas service ☐   To other elements ....

Main Switch or Circuit-breaker

BS. Type .... No. of poles .... Current rating ....A   Voltage rating V

Location ....   Fuse rating or setting ....A

Rated residual operating current $I_{\Delta n}$ = ... mA, and operating time of ....ms (at $I_{\Delta n}$)
(applicable only where an RCD is suitable and is used as a main circuit-breaker)

COMMENTS ON EXISTING INSTALLATION: (In the case of an alteration or addition see Section 743)
....
....
....
....

SCHEDULES (note 2)
The attached Schedules are part of this document and this Certificate is valid only when they are attached to it.
.... Schedules of Inspections and .... Schedules of Test Results are attached. (Enter quantities of schedules attached)

Figure 9.19 (*continued*)

MINOR ELECTRICAL INSTALLATION WORKS CERTIFICATE GUIDANCE FOR RECIPIENTS

This Certificate has been issued to confirm that the electrical installation work to which it relates has been designed, constructed and inspected and tested in accordance with British Standard 7671:2008 (IEE Wiring Regulations).

You should have received an 'original' Certificate and the contractor should have retained a duplicate. If you were the person ordering the work, but not the owner of the installation, you should pass this Certificate, or a copy of it, to the owner. A separate Certificate should have been received for each existing circuit on which minor works have been carried out. This Certificate is not appropriate if you requested the contractor to undertake more extensive installation work, for which you should have received an Electrical Installation Certificate.

The Certificate should be retained in a safe place and be shown to any person inspecting or undertaking further work on the electrical installation in the future. If you later vacate the property, this Certificate will demonstrate to the new owner that the minor electrical installation work carried out complied with the requirements of British Standard 7671:2008 at the time the Certificate was issued.

Figure 9.20 Guidance notice to accompany the Minor Electrical Installation Certificate

The inspection and test schedule must include a verification that a Periodic Inspection Report (Together with a Schedule of Inspections and a Schedule of Test Results) was given to the person ordering the inspection and that this Report included details of:

- any damage, deterioration, defects, dangerous conditions and non-compliance with the requirements of the Regulations which may give rise to danger;
- any limitations of the inspection and testing.

9.6 Test requirements specific for compliance with the Building Regulations

9.6.1 Mandatory requirements

9.6.1.1 Part P – electrical safety

Confirm that:

- reasonable provision has been made in the design, installation, inspection and testing of electrical installations to protect persons from fire or injury;
- sufficient information has been provided so that persons wishing to operate, maintain or alter an electrical installation can do so with reasonable safety.

MINOR ELECTRICAL INSTALLATION WORKS CERTIFICATE
(REQUIREMENTS FOR ELECTRICAL INSTALLATIONS - BS 7671 [IEE WIRING REGULATIONS])
To be used only for minor electrical work which does not include the provision of a new circuit

---

PART 1: Description of minor works

1. Description of the minor works: ....

2. Location/Address: ....

3. Date minor works completed: ....

4. Details of departures, if any, from BS 7671

....

....

....

PART 2: Installation details

1. System earthing arrangement: TN-C-S ☐   TN-S ☐   TT ☐

2. Method of protection against indirect contact: ....

3. Protective device for the modified circuit: Type BS ....   Rating ....A

4. Comments on existing installation, including adequacy of earthing and bonding arrangements:
(see Regulation 130-07) ....

....

....

....

PART 3: Essential Tests

1. Earth continuity: satisfactory ☐

2. Insulation resistance:

                 Phase/neutral ....$M\Omega$

                 Phase/earth ....$M\Omega$

                 Neutral/earth ....$M\Omega$

3. Earth fault loop impedance: ....Ω

4. Polarity: satisfactory   ☐

5. RCD operation (if applicable): Rated residual operating current $I_{\Delta n}$ ....mA and operating time of ....ms (at $I_{\Delta n}$)

PART 4: Declaration

1. I/We CERTIFY that the said works do not impair the safety of the existing installation, that the said works have been designed, constructed, inspected and tested in accordance with BS 7671: .... (IEE Wiring Regulations), amended to .... and that the said works, to the best of my/our knowledge and belief, at the time of my/our inspection, complied with BS 7671 except as detailed in Part 1.

2. Name: ....   3. Signature: ....

    For and on behalf of: ....   Position: ....

    Address: ....

    ....   Date: ....

    ....

    ....Postcode: ....

Figure 9.21 Minor Electrical Installation Works Certificate

9.6.1.2 Part M – access and facilities for disabled people

Confirm that (in addition to the requirements of the Disability Discrimination Act 1995) precautions have been taken to ensure that:

- new non-domestic buildings and/or dwellings (e.g. houses and flats used for student living accommodation etc.);

PERIODIC INSPECTION REPORT – GUIDANCE FOR RECIPIENTS

This Periodic Inspection Report form is intended for reporting on the condition of an existing electrical installation.

You should have received an original Report and the contractor should have retained a duplicate. If you were the person ordering this Report, but not the owner of the installation, you should pass this Report, or a copy of it, immediately to the owner.

The original Report is to be retained in a safe place and be shown to any person inspecting or undertaking work on the electrical installation in the future. If you later vacate the property, this Report will provide the new owner with details of the condition of the electrical installation at the time the Report was issued.

The 'Extent and Limitations' box should fully identify the extent of the installation covered by this report and any limitations on the inspection and tests. The contractor should have agreed these aspects with you and with any other interested parties (Licensing Authority, Insurance Company, Building Society etc.) before the inspection was carried out.

The Report will usually contain a list of recommended actions necessary to bring the installation up to the current standard. For items classified as 'requires urgent attention', the safety of those using the installation may be at risk, and it is recommended that a competent person undertakes the necessary remedial work without delay.

For safety reasons, the electrical installation will need to be re-inspected at appropriate intervals by a competent person. The maximum time interval recommended before the next inspection is stated in the Report under 'Next Inspection'.

Figure 9.22 Guidance notice to accompany the Periodic Inspection Report

- extensions to existing non-domestic buildings; and
- non-domestic buildings that have been subject to a material change of use (e.g. so that they become a hotel, boarding house, institution, public building or shop)

are capable of allowing people, regardless of their disability, age or gender to:

- gain access to buildings;
- gain access within buildings;
- be able to use the facilities of the buildings (both as visitors and as people who live or work in them);
- use sanitary conveniences in the principal storey of any new dwelling.

9.6.1.3 Part L – conservation of fuel and power

Confirm that energy efficiency measures have been provided which:

- ensure that lighting systems utilise energy-efficient lamps with:
  - manual switching controls; or
  - automatic switching (in the case of external lighting fixed to the building); or
  - both manual and automatic switching controls

PERIODIC INSPECTION REPORT FOR AN ELECTRICAL INSTALLATION
(REQUIREMENTS FOR ELECTRICAL INSTALLATIONS - BS 7671 [IEE WIRING REGULATIONS])

DETAILS OF THE CLIENT
Client: ....
Address: ....
Purpose for which this Report is required: ....
DETAILS OF THE INSTALLATION Tick boxes as appropriate
Occupier: ....
Installation: ....
Address: ....
Description of Premises: Domestic ☐   Commercial ☐ Industrial ☐   Other ☐
Estimated age of the Electrical   ....years Installation:
Evidence of Alterations or Additions: Yes ☐   No ☐   Not apparent ☐
If 'Yes', estimate age: .... years
Date of last inspection: ....   Records available ?   Yes ☐   No ☐
EXTENT AND LIMITATIONS OF THE INSPECTION
Extent of electrical installation covered by this report: ....
Limitations: ....
This inspection has been carried out in accordance with BS 7671: 2001 (IEE Wiring Regulations), amended to ...... Cables concealed within trunking and conduits, or cables and conduits concealed under floors, in roof spaces and generally within the fabric of the building or underground have not been inspected.
NEXT INSPECTION
I/We recommend that this installation is further inspected and tested after an interval of not more than .... months/years, provided that any observations 'requiring urgent attention' are attended to without delay.
DECLARATION
INSPECTED AND TESTED BY
Name: ....   Signature: ....
For and on behalf of: ....   Position: ....
Address: ....   Date: ....

Figure 9.23 Periodic Inspection Report

SUPPLY CHARACTERISTICS AND EARTHING ARRANGEMENTS Tick boxes and enter details, as appropriate

Earthing arrangements	Number and Type of Live Conductors	Nature of Supply Parameters	Supply Protective Device Characteristics
TN-C ☐	a.c. ☐ d.c. ☐	Nominal voltage, U/U_0[(1)] V	Type: ....
TN-S ☐	1-line, 2-wire ☐ 2-pole ☐	Nominal frequency, f[(1)] ....Hz	
TN-C-S ☐	2-line, 3-wire ☐ 3-pole ☐	Prospective fault current, I_{pf}[(2)] ....kA	Nominal current rating ....A
TT ☐	3-line, 3-wire ☐ other ☐	External loop impedance, Z_e[(2)] ....Ω	
IT ☐	3-line, 4-wire ☐	(Note: (1) by enquiry, (2) by enquiry or by measurement)	

PARTICULARS OF INSTALLATION REFERRED TO IN THE REPORT Tick boxes and enter details, as appropriate

Means of Earthing
Distributor's facility ☐

Details of Installation Earth electrode (where applicable)

	Type	Location	Electrode resistance
Installation earth electrode ☐	(e.g. rod(s), tape etc)	....	to earth .... Ω

Main Protective Conductors

Earthing conductor: material .... csa ....

Main equipotential bonding
conductors   material .... csa ....

To incoming water service ☐   To incoming gas service ☐   To incoming oil service ☐   To structural steel ☐
To lightning protection ☐   To other incoming service(s) ☐   (state details ....)

Main Switch or Circuit-breaker

BS, type and number of poles ....   Current rating ....A   Voltage rating V

Location ....   Fuse rating or setting ....A

Rated residual operating current $I_{\Delta n}$ = ... mA, and operating time of .... ms (at $I_{\Delta n}$) (applicable only where an RCD is suitable and is used as a main circuit-breaker)

OBSERVATIONS AND RECOMMENDATIONS Tick boxes as appropriate   Recommendations detailed below

Referring to the attached Schedule(s) of Inspection and Test Results, and subject to the limitations specified at the Extent and Limitations of the Inspection section
☐ No remedial work is required   ☐ The following observations are made:

....

....

....

....

....

....

One of the following number, as appropriate, is to be allocated to each of the observations made above to indicate to the person(s) responsible for the installation the action recommended.

| 1 | requires urgent attention | 2 | requires improvement | 3 | requires further investigation |

| 4 | does not comply with BS 7671: 2001 amended to .... This does not imply that the electrical installation inspected is unsafe.

SUMMARY OF THE INSPECTION
Date(s) of the inspection: ....
General condition of the installation: ....
....
....
....
Overall assessment: Satisfactory/Unsatisfactory

SCHEDULE(S)
The attachment Schedules are part of this document and this Report is valid only when they are attached to it.
.... Schedules of Inspections and .... Schedules of Test Results are attached.
(Enter quantities of schedules attached).

Figure 9.23 (*continued*)

so that the lighting systems can be operated effectively with regard to the conservation of fuel and power.

Confirm that building occupiers have been supplied with sufficient information (including results of performance tests carried out during the works) to show how the heating and hot water services can be operated and maintained.

9.6.2 Inspection and test

Verify that all electrical installations have been inspected and tested during and at the end of installation and before they are taken into service and that they:

- are reasonably safe and that they comply with BS 7671:2008;
- meet the relevant equipment and installation standards.

Confirm that all components that are part of an electrical installation have been inspected (during installation as well as on completion) to verify that the components have:

- been selected and installed in accordance with BS 7671:2008;
- been made in compliance with appropriate British Standards or harmonised European Standards;
- been evaluated against external influences (such as the presence of moisture);
- been checked to see that they have not been visibly damaged (or are defective) so as to be unsafe;
- been tested to check satisfactory performance with respect to continuity of conductors, insulation resistance, separation of circuits, polarity, earthing and bonding arrangements, earth fault loop impedance and functionality of all protective devices including residual current devices;
- been inspected for conformance with BS 7671:2008;
- had their test results recorded;
- had their test results compared with the relevant performance criteria to confirm compliance.

 Note: Inspections and testing of DIY work should **also** meet the above requirements.

9.6.3 Extensions, material alterations and material changes of use

Where any electrical installation work is classified as an extension, a material alteration or a material change of use, confirm that:

- the existing fixed electrical installation in the building is capable of supporting the amount of additions and alterations that will be required;
- the earthing and bonding systems are satisfactory and meet the requirements;
- the mains supply equipment is suitable and can carry the additional loads envisaged;
- any additions and alterations to the circuits which feed them comply with the requirements of the Regulations;
- the protective measures meet the requirements;
- the rating and the condition of existing equipment (belonging to both the consumer and the electricity distributor) is sufficient.

9.6.4 Design

Confirm that electrical installations have been designed and installed so that they:

- are suitably enclosed (and appropriately separated) to provide mechanical and thermal protection;
- do not present an electric shock or fire hazard to people;
- meet the requirements of the Building Regulations;
- provide adequate protection for persons against the risks of electric shock, burn or fire injuries;
- provide adequate protection against mechanical and thermal damage.

9.6.5 Electricity distributor's responsibilities

Prior to starting works, confirm that the electricity distributor has:

- accepted responsibility for ensuring that the supply is mechanically protected and can be safely maintained;
- evaluated and agreed the proposal for a new (or significantly altered) installation;
- installed the cut-out and meter in a safe location.

Confirm that the distributor has:

- provided an earthing facility for new connections;
- maintained the supply within defined tolerance limits;
- provided certain technical and safety information to the consumer to enable them to design their installation(s);
- ensured that their equipment on consumers' premises:
  - ○ is suitable for its purpose;
  - ○ is safe in its particular environment;
  - ○ clearly shows the polarity of the conductors.

9.6.6 Consumer units

Ensure that accessible consumer units have been fitted with a child-proof cover or installed in a lockable cupboard.

9.6.7 Earthing

Inspect and confirm that:

- electrical installations have been properly earthed;
- lighting circuits include a circuit protective conductor;
- socket outlets which have a rating of 32 A or less and which may be used to supply portable equipment for use outdoors, are protected by an RCD;

- the distributor has provided an earthing facility for all new connections;
- new or replacement, non-metallic light fittings, switches or other components do not require earthing (e.g. non-metallic varieties) unless new circuit protective (earthing) conductors have been provided;
- socket outlets that will accept unearthed (2-pin) plugs do not use supply equipment that needs to be earthed.

9.6.8 Electrical installations

Verify (by inspection and test) that during installation, at the end of installation and before they are taken into service that all electrical installations:

- have been designed and installed (suitably enclosed and appropriately separated) to provide mechanical and thermal protection;
- provide adequate protection for persons against the risks of electric shock, burn or fire injuries;
- meet the requirements of the Building Regulations.

Confirm that all electrical installation work:

- has been carried out professionally;
- complies with the Electricity at Work Regulations 1989 as amended;
- has been carried out by persons who are competent to prevent danger and injury while doing it, or who are appropriately supervised.

9.6.9 Wiring and wiring systems

Confirm that:

- cables concealed in floors and walls have (if required):
  - an earthed metal covering; or
  - are enclosed in steel conduit; or
  - have some form of additional mechanical protection;
- cables to an outside building (e.g. garage or shed) if run underground, have been routed and positioned so as to give protection against electric shock and fire as a result of mechanical damage to a cable;
- heat-resisting flexible cables (if required) have been supplied for the final connections to certain equipment (see maker's instructions).

9.6.9.1 Equipotential bonding conductors

Confirm that:

- main equipotential bonding conductors for water service pipes, gas installation pipes, oil supply pipes plus and certain other 'earthy' metalwork have been provided;

- where there is an increased risk of electric shock (e.g. such as bathrooms and shower rooms), supplementary equipotential bonding conductors have been installed;
- the minimum size of supplementary equipotential bonding conductors (without mechanical protection) is $4\,mm^2$.

9.6.10 Socket outlets

Confirm by inspection that:

- older types of socket outlet designed non-fused plugs are not connected to a ring circuit;
- RCD protection has been provided for all socket outlets which have a rating of 32 A or less and which may be used to supply portable equipment for use outdoors;
- switched socket outlets indicate whether they are 'ON';
- socket outlets that will accept unearthed (2-pin) plugs are not used to supply equipment that needs to be earthed;
- the following requirements (Table 9.12) for wall sockets have been met:

Table 9.12 Building Regulations requirements for wall sockets

Type of wall	Requirement
Timber framed	Power points: • that have been set in the linings have a similar thickness of cladding behind the socket box • have not been placed back-to-back across the wall.
Solid masonry	Deep sockets and chases have **not** been used in separating walls The position of sockets has been staggered on opposite sides of the separating wall
Cavity masonry	The position of sockets has been staggered on opposite sides of the separating wall Deep sockets and chases have **not** been used in a separating wall Deep sockets and chases in a separating wall have **not** been placed back-to-back
Framed walls with absorbent material	Sockets have: • been positioned on opposite sides of a separating wall • not been connected back-to-back • been staggered a minimum of 150 mm edge-to-edge

Confirm (by inspection and testing) that all socket outlets used for lighting:

- are wall-mounted;
- are easily reachable;
- have been installed between 450 mm and 1200 mm from the finished floor level;

- are located no nearer than 350 mm from room corners;
- indicate whether they are 'ON'.

9.6.11 Switches

Ensure that all controls and switches:

- are be easy to operate, visible and free from obstruction;
- have been located between 750 mm and 1200 mm above the floor;
- do not require the simultaneous use of both hands (unless necessary for safety reasons) to operate;

and that:

- mains and circuit isolator switches clearly indicate whether they 'ON' or 'OFF';
- individual switches on panels and on multiple socket outlets have been well separated;
- front plates should contrast visually with their backgrounds;
- controls that need close vision (e.g. thermostats) have been located between 1200 mm and 1400 mm above the floor;
- the operation of switches, outlets and controls does not require the simultaneous use of both hands (unless necessary for safety reasons);
- where possible, light switches with large push pads shave been used in preference to pull cords;
- the colours red and green have not been used in combination as indicators of 'ON' and 'OFF' for switches and controls;
- **all** switches used for lighting:
  - are easily reachable;
  - have been installed between 450 mm and 1200 mm from the finished floor level.

Confirm that light switches:

- have large push pads (in preference to pull cords);
- align horizontally with door handles;
- are within the 900 to 1100 mm from the entrance door opening;
- are located between 750 mm and 1200 mm above the floor;
- are not coloured red and green (i.e. as a combination) as indicators for 'ON' and 'OFF'.

9.6.12 Telephone points and TV sockets

Confirm that all telephone points and TV sockets have been located between 400 mm and 1000 mm above the floor (or 400 mm and 1200 mm above the floor for permanently wired appliances).

9.6.13 Equipment and components

9.6.13.1 Emergency alarms

Test and inspect to ensure that:

- emergency assistance alarm systems have:
  - visual and audible indicators to confirm that an emergency call has been received;
  - a reset control reachable from a wheelchair, WC, or from a shower/ changing seat;
  - a signal that is distinguishable visually and audibly from the fire alarm;
- emergency alarm pull cords should be:
  - coloured red;
  - located as close to a wall as possible;
  - have two red 50 mm diameter bangles;
- front plates contrast visually with their backgrounds;
- the colours red and green shave not been used (in combination) to indicate 'ON' and 'OFF' for switches and controls.

9.6.13.2 Fire alarms

Verify (by test and inspection) that fire detection and fire-warning systems have been properly designed, installed and maintained and that:

- all buildings have arrangements for detecting fire;
- all buildings have been fitted with a suitable (electrically operated) fire warning system (in compliance with BS 5839) or have means of raising an alarm in case of fire (e.g. rotary gongs, handbells or shouting '**FIRE**');
- fire alarms emit an audio and visual signal to warn occupants with hearing or visual impairments;
- the fire warning signal is distinct from other signals which may be in general use;
- in premises that are used by the general public (e.g. large shops and places of assembly) a staff alarm system (complying with BS 5839) has been used.

9.6.13.3 Heat emitters

Check that heat emitters:

- are either screened or have their exposed surfaces kept at a temperature below 43°C;
- that are located in toilets and bathrooms, do not restrict:
  - the minimum clear wheelchair manoeuvring space;
  - the space beside a WC used to transfer from the wheelchair to the WC.

9.6.13.4 Portable equipment for use outdoors

Verify that RCDs have been provided for all socket outlets which have a rating of 32 A or less and which may be used to supply portable equipment for use outdoors.

9.6.13.5 Power operated entrance doors

- Confirm that all power operated doors have been provided with:
  - safety features to prevent injury to people who are stuck or trapped (such as a pressure sensitive door edge which operates the power switch);
  - a readily identifiable (and accessible) stop switch;
  - a manual or automatic opening device in the event of a power failure where and when necessary for health or safety.

Confirm that:

- all doors to accessible entrances have been provided with a power operated door opening and closing system if a force greater than 20N is required to open or shut a door;
- once open, all doors to accessible entrances are wide enough to allow unrestricted passage for a variety of users, including wheelchair users, people carrying luggage, people with assistance dogs, and parents with pushchairs and small children;
- power operated entrance doors:
  - have a sliding, swinging or folding action controlled manually (by a push pad, card swipe, coded entry, or remote control) or automatically controlled by a motion sensor or proximity sensor such as contact mat;
  - open towards people approaching the doors;
  - provide visual and audible warnings that they are operating (or about to operate);
  - incorporate automatic sensors to ensure that they open early enough (and stay open long enough) to permit safe entry and exit;
  - incorporate a safety stop that is activated if the doors begin to close when a person is passing through;
  - revert to manual control (or fail safe) in the open position in the event of a power failure;
  - when open, do not project into any adjacent access route;
- its manual controls are:
  - located between 750 mm and 1000 mm above floor level;
  - operable with a closed fist;
  - set back 1400 mm from the leading edge of the door when fully open;
  - clearly distinguishable against the background;
  - contrasted visually with the background.

9.6.14 Thermostats

Check that all controls that need close vision (e.g. thermostats) are located between 1200 mm and 1400 mm above the floor.

9.6.14.1 Smoke alarms – dwellings

Confirm by test and inspection that smoke alarms have been positioned:

- in the circulation space within 7.5 m of the door to every habitable room;
- in the circulation spaces between sleeping spaces and places where fires are most likely to start (e.g. kitchens and living rooms);
- on every storey of a house (including bungalows).

Confirm by test and inspection that:

- kitchen areas that are not separated from the stairway or circulation space by a door, have been equipped with an additional heat detector in the kitchen, that is interlinked to the other alarms;
- if more than one smoke alarm has been installed in a dwelling then they have been linked so that if a unit detects smoke it will operate the alarm signal of all the smoke detectors.

Verify by inspection that smoke alarms:

- have ideally been mounted 25 mm and 600 mm below the ceiling (25–150 mm in the case of heat detectors) and at least 300 mm from walls and light fittings;
- have not been fixed over a stair shaft or any other opening between floors;
- have not been fitted:
  o in places that get very hot (such as a boiler room);
  o in a very cold area (such as an unheated porch);
  o in bathrooms, showers, cooking areas or garages, or any other place where steam, condensation or fumes could give false alarms;
  o next to or directly above heaters or air conditioning outlets;
  o on surfaces which are normally much warmer or colder than the rest of the space.

Test, inspect and confirm that the power supply for a smoke alarm system:

- has been derived from the dwelling's mains electricity supply via a single independent circuit at the dwelling's main distribution board (consumer unit);
- includes a stand-by power supply that will operate during mains failure;
- is not (preferably) protected by an RCD.

9.6.15 Lighting

9.6.15.1 External lighting fixed to the building

Confirm (by test and inspection) that all external lighting (including lighting in porches, but not lighting in garages and carports):

- automatically extinguishes when there is enough daylight, and when not required at night;
- has sockets that can only be used with lamps having an efficacy greater than 40 lumens per circuit-watt (such as fluorescent or compact fluorescent lamp types, and **not** GLS tungsten lamps with bayonet cap or Edison screw bases).

9.6.15.2 Fittings, switches and other components

Confirm that new or replacement, non-metallic light fittings, switches or other components that require earthing (e.g. non-metallic varieties) have been provided with new circuit protective (earthing) conductors.

9.6.15.3 Fixed lighting

Ensure that in locations where lighting can be expected to have most use, fixed lighting (e.g. fluorescent tubes and compact fluorescent lamps – but **not** GLS tungsten lamps with bayonet cap or Edison screw bases) with a luminous efficacy greater than 40 lumens per circuit-watt have been made available.

9.6.15.4 Lighting circuits

Verify that all lighting circuits include a circuit protective conductor.

9.6.16 Lecture/conference facilities

In lecture halls and conference facilities, confirm that artificial lighting has been designed to:

- give good colour rendering of all surfaces;
- be compatible with other electronic and radio frequency installations.

9.6.17 Cellars or basements

Ensure that LPG storage vessels and LPG fired appliances that are fitted with automatic ignition devices or pilot lights have not been installed in cellars or basements.

9.7 What about test equipment?

Figure 9.24 A selection of test equipment

Although BS 7671 lays great emphasis on the requirements for 'Inspection and Testing' in Chapter 6 of the Regulations, the only reference (as far as the author can tell!) to actual test equipment concerns insulation monitoring devices which states that:

- an insulation monitoring device shall be so designed or installed that it can only be possible to modify the setting with the use of a key or a tool;
- in an IT system, an insulation monitoring device shall be provided so as to indicate the occurrence of a first fault from a live part to an exposed conductive part, or to earth;
- installations forming part of an IT system can make use of an insulation monitoring device.

Of course, the actual choice of test equipment that the electrician chooses to use will normally be based on personal preference and experience. Nevertheless, it is essential that any piece of test equipment (including software) that is used when installing or inspecting electrical installations for compliance with the Regulations can be relied on to produce accurate results.

ISO 9001:2000 (i.e. the internationally recognised Standard for Quality Management) specifies the requirements for the control of test equipment (although it actually refers to them as 'measuring and monitoring devices') as follows (see Figure 9.25).

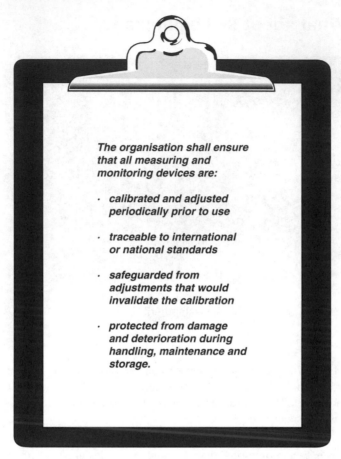

The organisation shall ensure that all measuring and monitoring devices are:

· *calibrated and adjusted periodically prior to use*

· *traceable to international or national standards*

· *safeguarded from adjustments that would invalidate the calibration*

· *protected from damage and deterioration during handling, maintenance and storage.*

Figure 9.25 Mandatory requirements from the Building Regulations

Proof	The controls that an organisation has in place to ensure that equipment (including software) used for conformance to specified requirements is properly maintained
Likely documentation	Equipment records of maintenance and calibration; Work instructions

Although the majority of electricians probably work on an individual basis and the requirement to operate as an accredited and registered ISO 9001:2000 company doesn't really apply, following the recommendations of this Standard can only help to improve the quality of any organisation – no matter what its size.

In general, therefore:

- all measuring and test equipment that is used by an electrician needs to be well maintained, in good condition and capable of safe and effective operation within a specified tolerance of accuracy;
- all measuring and test equipment should be regularly inspected and/or calibrated to ensure that it is capable of accurate operation (and where necessary by comparison with external sources traceable back to national standards);
- any electrostatic protection equipment that is utilised when handling sensitive components is regularly checked to ensure that it remains fully functional;
- the control of measuring and test equipment (whether owned by the electrician, on loan, hired or provided by the customer) should always include a check that the equipment:
  - is exactly what is required;
  - has been initially calibrated before use;
  - operates within the required tolerances;
  - is regularly recalibrated; and that;
  - facilities exist (either within the organisation or via a third party) to adjust, repair or recalibrate as necessary.

if the measuring and test equipment is used to verify process outputs against a specified requirement, then the equipment needs to be maintained and calibrated against national and international standards and the results of any calibrations carried out **must** be retained and the validity of previous results re-assessed if they are subsequently found to be out of calibration.

9.7.1 Control of inspection, measuring and test equipment

Measuring and test equipment should always be stored correctly and satisfactorily protected between use (to ensure their bias and precision) and should be verified and/or recalibrated at appropriate intervals.

9.7.2 Computers

Special attention should be paid to computers if they are used in controlling processes and particularly to the maintenance and accreditation of any related software.

9.7.3 Software

Software used for measuring, monitoring and/or testing specified requirements should be validated prior to use.

9.7.4 Calibration

Without exception, all measuring instruments can be subject to damage, deterioration or just general wear and tear when they are in regular use. The electrician should, therefore, take account of this fact and ensure that all of his test equipment is regularly calibrated against a known working standard.

The accuracy of the instrument will depend very much on what items it is going to be used to test and the frequency of use of the test instrument, and the electrician will have to decide on the maximum tolerance of accuracy for each item of test equipment.

Of course, calibrating against a 'working standard' is pretty pointless if that particular standard cannot be relied upon and so the workshop standard must also be calibrated, on a regular basis, at either a recognised calibration centre or at the UK Physical Laboratory against one of the national standards.

The electrician will, therefore, have to make allowances for:

- the calibration and adjustment of all measuring and test equipment that can affect product quality of their inspection and/or test;
- the documentation and maintenance of calibration procedures and records;
- the regular inspection of all measuring or test equipment to ensure that they are capable of the accuracy and precision that is required;
- the environmental conditions being suitable for the calibrations, inspections, measurements and tests to be completed.

If the instrument is found to be outside of its tolerance of accuracy, any items previously tested with the instrument must be regarded as suspect. In these circumstances, it would be wise to review the test results obtained from the individual instrument. This could be achieved by compensating for the extent of inaccuracy to decide if the acceptability of the item would be reversed.

9.7.5 Calibration methods

There are various possibilities, such as:

- sending all working equipment to an external calibration laboratory;
- sending one of each item (i.e. a 'workshop standard') to a calibration laboratory, and then sub-calibrating each working item against the workshop standard;
- testing by attributes – i.e. take a known 'faulty' product, and a known 'good' product and then test each one to ensure that the test equipment can identify the faulty and the good product correctly.

9.7.6 Calibration frequency

The calibration frequency depends on how much the instrument is used, its ability to retain its accuracy and how critical the items being tested are. Infrequently used instruments are often only calibrated prior to their use whilst frequently used items would normally be checked and re-calibrated at regular intervals depending, again, on product criticality, cost, availability etc. Normally 12 months is considered as about the maximum calibration interval.

9.7.7 Calibration ideals

- Each instrument should be uniquely identified, allowing it to be traced.
- The calibration results should be clearly indicated on the instrument.
- The calibration results should be retained for reference.
- The instrument should be labelled to show the next 'calibration due' date to easily avoid its use outside of the period of confidence.
- Any means of adjusting the calibration should be sealed, allowing easy identification if it has been tampered with (e.g. a label across the joint of the casing).

Note: Examples of test equipment normally used by electricians is shown at Annex 9A.

9.8 Requirements from the Regulations – testing

Every electrical connection and joint shall be accessible for testing and maintenance, except for the following:	WR-526.3 WR-543.3.3

- a joint designed to be buried in the ground
- a compound filled or encapsulated joint
- a connection between a cold tail and the heating element as in ceiling heating, floor heating or a trace heating system
- a joint made by welding, soldering, brazing or appropriate compression tool
- a joint forming part of the equipment complying with the appropriate product standard.

When undertaking testing in a potentially explosive atmosphere, appropriate safety precautions in accordance with BS EN 60079-17 and BS EN 61241-17 are necessary.	WR-612.1
The tests of Part 6 of the Regulations, where relevant, shall be carried out in that order before the installation is energised. Where the installation incorporates an earth electrode, the test of Regulation 612.7 shall also be carried out before the installation is energised.	WR-612.1
If any test indicates a failure to comply, that test and any preceding test, the results of which may have been influenced by the fault indicated, shall be repeated after the fault has been rectified.	WR-612.1

9.8.1 Protective systems

9.8.1.1 SELV

> The separation of the live parts from those of other circuits shall be confirmed by a measurement of the insulation resistance. The resistance values obtained shall be in accordance with Table 9.13.　　WR-612.4.1

Table 9.13 Minimum values of insulation resistance

Circuit nominal voltage (V)	Test voltage d.c. (V)	Minimum insulation resistance
SELV and PELV	250	>0.5
Up to and including 500 V with the exception of the	500	>1.0
Above 500 V	1000	>1.0

9.8.1.2 PELV

> The separation of the live parts from other circuits shall be confirmed by a measurement of the insulation resistance. The resistance values obtained shall be in accordance with Table 9.13.　　WR-612.4.2

9.8.1.3 FELV

> Functional extra-low-voltage circuits shall meet all the test requirements for low-voltage circuits.　　WR-612.4.4
>
> Basic protection by a barrier or an enclosure shall be provided during erection.　　WR-612.4.4
>
> Where basic protection is intended to be afforded by a barrier or an enclosure provided during erection, it shall be verified by test that each barrier or enclosure affords a degree of protection not less than IP2X or IPXXB, or IP4X or IPXXD as appropriate, where that Regulation so requires.　　WR-612.4.5

9.8.1.4 Protection by automatic disconnection of the supply

> Where RCDs are also used for protection against fire,   WR-612.8
> the conditions for protection by automatic disconnection
> of the supply shall be verified.

The verification of the effectiveness of the measures for fault protection by automatic disconnection of supply depends on the type of system used and is as follows.

TN system

Compliance shall be verified by:

- measurement of the earth fault loop impedance;
- verification of the characteristics and/or the effectiveness of the associated protective device.

TT system

Compliance shall be verified by:

- measurement of the resistance of the earth electrode for exposed conductive parts of the installation;
- confirmation of the characteristics and/or effectiveness of the associated protective device.

IT system

Compliance shall be verified by calculation or measurement of the current (I_d) in case of a first fault at the line conductor or at the neutral.

Where conditions that are similar to the conditions of a TT system occur, in the event of a second fault in another circuit verification shall be made according to a TT system.

Where conditions that are similar to the conditions of a TN system occur, in the event of a second fault in another circuit verification shall be made according to a TN system.

9.8.1.5 Protection by electrical separation

> The separation of the live parts from those of other   WR-612.4.3
> circuits and from earth, shall be confirmed by a
> measurement of the insulation resistance. The
> resistance values obtained shall be in accordance with
> Table 9.13.

9.8.1.6 Additional protection

The verification of the effectiveness of the measures applied for additional protection is fulfilled by visual inspection and test.	WR-612.10
Where RCDs are required for additional protection, the effectiveness of automatic disconnection of supply by RCDs shall be verified using suitable test equipment according to BS EN 61557-6 to confirm that the relevant requirements are met.	WR-612.10

9.8.1.7 Earth fault loop impedance

Where protective measures are used which require knowledge of earth fault loop impedance, the relevant impedances shall be measured, or determined by an alternative method.	WR-612.9

 Note: Further information on measurement of earth fault loop impedance can be found in Appendix 14 of BS 7671:2008.

9.8.1.8 Functional testing

Where fault protection and/or additional protection is to be provided by an RCD, the effectiveness of any test facility incorporated in the device shall be verified.	WR-612.13.1
Assemblies, such as switchgear and controlgear assemblies, drives, controls and interlocks, shall be subjected to a functional test to show that they are properly mounted, adjusted and installed in accordance with the relevant requirements of these Regulations.	WR-612.13.2

9.8.2 Other tests

9.8.2.1 Continuity of protective conductors, including main and supplementary equipotential bonding

A continuity test shall be made. It is recommended that the test be carried out with a supply having a no-load voltage between 4 V and 24 V, d.c. or a.c., and a short-circuit current of not less than 200 mA.	WR-612.2.1

9.8.2.2 Continuity of ring final circuit conductors

A test shall be made to verify the continuity of each   WR-612.2.2
conductor, including the protective conductor, of every
ring final circuit.

9.8.2.3 Earth electrode resistance

Where the earthing system incorporates an earth   WR-612.7
electrode as part of the installation, the electrode
resistance to earth shall be measured.

9.8.2.4 Insulation resistance

The insulation resistance shall be measured between   WR-612.3.1
live conductors and between live conductors and
the protective conductor connected to the earthing
arrangement.

The insulation resistance measured with the test   WR-612.3.2
voltages indicated in Table 9.13 shall be considered
satisfactory if the main switchboard and each
distribution circuit tested separately, with all its final
circuits connected but with current-using equipment
disconnected, has an insulation resistance not less than
the appropriate value given in Table 9.13.

 More stringent requirements are applicable for the wiring of fire alarm systems in buildings; see BS 5839-1.

Where surge protective devices (SPD) or other   WR-612.3.2
equipment are likely to influence the verification
test, or be damaged, such equipment shall be
disconnected before carrying out the insulation
resistance test.

In locations exposed to fire hazard, a measurement of   WR-612.3.2
the insulation resistance between the live conductors
should be applied.

Insulation resistance values are usually much higher than those of Table 9.13.

Where the circuit includes electronic devices which are likely to influence the results or be damaged, only a measurement between the live conductors connected together and the earthing arrangement shall be made.	WR-612.3.3

9.8.2.5 Insulation resistance/impedance of floors and walls

In a non-conducting location at least three measurements shall be made in the same location, one of these measurements being approximately 1 m from any accessible extraneous conductive part in the location. The other two measurements shall be made at greater distances. The measurement of resistance/impedance of insulating floors and walls is carried out with the system voltage to earth at nominal frequency. The above series of measurements shall be repeated for each relevant surface of the location.	WR-612.5.1

Note: Further information on measurement of the insulation resistance/impedance of floors and walls can be found in Appendix 13 of BS 7671:2008.

Any insulation or insulating arrangement of extraneous conductive parts: • when tested at 500 V d.c. shall be not less than 1 megohm; and • shall be able to withstand a test voltage of at least 2 kV a.c. rms; and • shall not pass a leakage current exceeding 1 mA in normal conditions of use.	WR-612.5.2

9.8.2.6 Polarity

A test of polarity shall be made and it shall be verified that: • every fuse and single-pole control and protective device is connected in the line conductor only; and	WR-612.6

- except for E14 and E27 lampholders to BS EN 60238, in circuits having an earthed neutral conductor, centre contact bayonet and Edison screw lampholders have the outer or screwed contacts connected to the neutral conductor; and
- wiring has been correctly connected to socket outlets and similar accessories.

9.8.2.7 Line sequence

For multiphase circuits, it shall be verified that the line sequence is maintained.   WR-612.12

9.8.2.8 Prospective fault current

The prospective short-circuit current and prospective earth fault current shall be measured, calculated or determined by another method, at the origin and at other relevant points in the installation.   WR-612.11

9.8.2.9 Verification of voltage drop

When required, compliance to the Regulations may be confirmed by using the following options: WR-612.14

- The voltage drop may be evaluated by measuring the circuit impedance.
- the voltage drop may be evaluated by using calculations, for example, by diagrams or graphs showing maximum cable length vs. load current for different conductor cross-sectional areas with different percentage voltage drops for specific nominal voltages, conductor temperatures and wiring systems.

 Note: Verification of voltage drop is not normally required during initial verification.

9.8.3 Certification and reporting

Electrical Installation Certificates, Periodic Inspection Reports and Minor Electrical Installation Works Certificates shall be compiled and signed or otherwise authenticated by a competent person or persons.

Electrical Installation Certificates, Periodic Inspection Reports and Minor Electrical Installation Works Certificates may be produced in any durable medium, including written and electronic media.	WR-631.5
Regardless of the media used for original certificates, reports or their copies, their authenticity and integrity shall be verified by a reliable process or method.	WR-631.5
The process or method shall also verify that any copy is a true copy of the original.	WR-631.5
An Electrical Installation Certificate or a Minor Electrical Installation Works Certificate (as appropriate) shall apply to all work completed including of the additions or alterations.	WR-633.1
The contractor or other person responsible for the new work, or a person authorised to act on their behalf, shall record on the Electrical Installation Certificate or the Minor Electrical Installation Works Certificate, any defects found, so far as is reasonably practicable, in the existing installation.	WR-633.2

9.8.3.1 Initial verification

During erection, on completion of an installation, addition or alteration to an installation (and before being put into service) appropriate inspection and testing shall be carried out by competent persons to verify that the requirements of BS 7671:2008 have been met.	WR-134.2.1

Certification shall be completed in accordance with Section 631.

The designer of the installation is responsible for recommending the interval to the first periodic inspection and test as detailed in Part 6 of BS 7671:2008.	WR-134.2.2

9.8.3.2 Electrical Installation Certificate

Upon completion of the verification of a new installation or changes to an existing installation, an Electrical Installation Certificate (based on the model shown in Figure 9.19) shall be provided, together with a schedule of inspections and a schedule of test results, to the person ordering the work.	WR-631.1
The schedule of test results shall identify every circuit, including its related protective device(s), and shall record the results of the appropriate tests and measurements.	WR-632.2
The person or persons responsible for the design, construction, inspection and testing of the installation shall provide the person ordering the work a Certificate which takes account of their respective responsibilities for the safety of that installation, together with the schedules described in Regulation 632.1 of BS 7671:2008.	WR-632.3

 Defects or omissions revealed during inspection and testing of the installation work covered by the Certificate **shall** be made good before the Certificate is issued.

9.8.3.3 Periodic Inspection Report

Where required, periodic inspection and testing of every electrical installation shall be carried out (in accordance with Regulations 621.2 to 621.5 of BS 7671:2008) in order to determine, so far as is reasonably practicable, whether the installation is in a satisfactory condition for continued service.	WR-621.1
The documentation arising from the initial certification and any previous periodic inspection and testing shall be taken into account. Where no previous documentation is available, investigation of the electrical installation shall be undertaken prior to carrying out the periodic inspection and testing.	WR-621.1
Periodic inspection comprising a detailed examination of the installation shall be carried out without dismantling, or with partial dismantling as required, supplemented by appropriate tests to	WR-621.2

show that the requirements for disconnection times for protective devices are complied with, to provide for:

- safety of persons and livestock against the effects of electric shock and burns
- protection against damage to property by fire and heat arising from an installation defect
- confirmation that the installation is not damaged or has deteriorated so as to impair safety
- the identification of installation defects and departures from the requirements of these Regulations that may give rise to danger.

Precautions shall be taken to ensure that the periodic inspection and testing shall not cause danger to persons or livestock and shall not cause damage to property and equipment even if the circuit is defective.

WR-621.3

Measuring instruments and monitoring equipment and methods shall be chosen in accordance with relevant parts of BS EN 61557.

WR-621.3

The extent and results of the periodic inspection and testing of an installation, or any part of an installation, shall be recorded.

WR-621.4

Periodic inspection and testing shall be undertaken by a competent person.

WR-621.5

Upon completion of the periodic inspection and testing of an existing installation, a Periodic Inspection Report, based on the model shown in Figure 9.23, shall be provided.

WR-631.2

Such documentation shall include details of the extent of the installation and limitations of the inspection and testing.

WR-631.2

A copy of the Periodic Inspection Report, together with a schedule of inspections and a schedule of test results, shall be given by the person carrying out the inspection, or a person authorised to act on their behalf, to the person ordering the inspection.

WR-634.1

Any damage, deterioration, defects, dangerous conditions and non-compliance with the requirements of the Regulations, which may give rise to danger, together with any significant limitations of the inspection and testing, including their reasons, shall be recorded.

WR-634.2

Note:

1.   The frequency of periodic inspection and testing of an installation will depend on the type of installation and equipment, its use and operation, the frequency and quality of maintenance and the external influences to which it is subjected.
2.   The results and recommendations of the previous report, if any, shall be taken into account.
3.   In the case of an installation under an effective management system for preventive maintenance in normal use, periodic inspection and testing may be replaced by an adequate regime of continuous monitoring and maintenance of the installation and all its constituent equipment by skilled persons, competent in such work.
4.   Appropriate records shall be kept.

9.8.3.4 Minor Electrical Installation Works Certificate

> Where minor electrical installation work does not   WR-631.3
> include the provision of a new circuit, a Minor Electrical
> Installation Works Certificate, based on the model given
> in Figure 9.21, may be provided for each circuit altered
> or extended as an alternative to an Electrical Installation
> Certificate.

Appendix 9.1: Examples of test equipment used to test electrical installations

The following are examples of instruments that are required to test electrical installations for compliance with the requirements of BS 7671.

 Quite a lot of test equipment manufacturers now produce dual or multifunctional instruments and so it is quite common to find an instrument that is capable of measuring a number of different types of tests – for example, continuity and insulation resistance, loop impedance and prospective fault current. It is, therefore, wise to carry out a little research before purchasing!

A9.1.1 Continuity tester

All protective and bonding conductors must be tested to ensure that they are electrically safe and correctly connected. Low-resistance ohmmeters and simple multimeters are normally used for continuity testing. Ideally they should have a no load voltage of between 4 V and 24 V, be capable of producing an a.c. or d.c. short-circuit voltage of not less than 200 mA and have a resolution of at least 0.01 milliohms.

TM INS1600 Digital insulation/ continuity tester by TLC www. tlc-direct. co.uk	 **Figure 9.26** Continuity testing	A compact, easy-to-read battery operated insulation and continuity tester covering a wide measuring range up to 2000 M ohm/1000 V, a.c. voltage (up to 600 V) and continuity beeper Capable of testing all requirements of the 17th edition and guidance notes	• Easy-to-read 0.65 LCD display • Data hold switch • Power ON lock facility for hands-free operation • Rotary switch for easy range selection • Overload protection • Dimensions: 165 × 100 × 57 mm • a.c. voltage range: up to 600 V • Batteries: 6 × AA • Sampling rate: 2.5 times/s • Weight: 500 g

A9.1.2 Insulation resistance tester

A low resistance between line and neutral conductors, or from live conductors to earth, will cause a leakage current which will cause weakening of the insulation, as well as involving a waste of energy which would increase the running costs of the installation. To overcome this problem, the resistance between poles or to earth must never be less than $0.5\,\text{m}\Omega$ for the usual supply voltages.

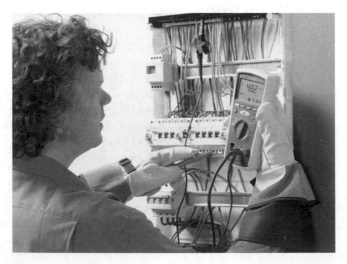

Figure 9.27 Insulation tester

A9.1.3 Loop impedance tester

Loop testing is a quick, convenient, and highly specific method of testing an electrical circuit for its ability to engage protective devices (circuit-breakers and fuses etc.) by simulating a fault from live to earth or from live to neutral (short-circuit). The tester first measures the unloaded voltage, and then connects a known resistance between the conductors, thereby simulating a fault. The voltage drop is measured across the known resistor, in series with the loop, and the proportion of the supply voltage that appears across the resistor will be dependent on the impedance of the loop.

KMP4120DL by Robin Electronics www.robin electronics.com  **Figure 9.28** Loop impedence tester	An earth loop impedance tester with a 0.01 ohm resolution, capable of performing loop tests without tripping most passive RCDs	• 3 pre-selectable loop impedance ranges of 20, 200 & 2000 ohms • 3 prospective short-circuit (PSC) ranges (200 A, 2000 A & 20 kA) • ability to test the majority of passive RCDs without tripping • 'lock down' test button allowing 'hands free' operation

A9.1.4 RCD tester

The standard method for protecting electrical installations is to make sure that an earth fault results in a fault current that is high enough to operate the protective device quickly so that fatal shock is prevented. However, there are cases where the impedance of the earth-fault loop, or the impedance of the fault itself, are too high to enable enough fault current to flow. In such a case, either:

- the current will continue to flow to earth, perhaps generating enough heat to start a fire; or
- metalwork which can be touched may be at a high potential relative to earth, resulting in severe shock danger.

Either or both of these possibilities can be removed by the installation of a residual current device (RCD).

Note: RCDs are also, sometimes, referred to as:

RCCD	Residual current operated circuit breaker
SRCD	Socket outlet incorporating an RCD
PRCD	Portable RCD, usually an RCD incorporated into a plug
RCBO	An RCCD which includes overcurrent protection
SRCBO	A socket outlet incorporating an RCBO

An RCD tester allows a selection of out of balance currents to flow through the RCD and cause its operation.

 The RCD tester should not be operated for longer than 2 s.

| TM-RC-70 RCD Tester by TLC www.tlc-direct.co.uk |  **Figure 9.29** RCD tester | A compact and simple instrument for monitoring 10, 30, 100, 300 and 500 mA RCD devices as well as selective 100, 300 and 500 mA RCDs | • Display: LCD
• RCD current: 10, 30, 100, 300 mA
• Power supply: 230V ± 10%
• Measurement range: 10–300 mA
• Test sequence: 10–300 mA
• Dimensions: 235 × 103 × 70 mm
• Weight: approx. 700 g |

A9.1.5 Prospective fault current tester

A prospective fault current (PFC) tester is used to measure the prospective line neutral fault current.

| PROFiTEST 0100S- by GMC Instrumentation Ltd techinfo@ gmciuk.com |  **Figure 9.30** Prospective fault current tester | Equipment designed to test prospective fault current as well as other insulation and impedance tests | • Overcurrent protection devices
• Loop and line impedance
• Earth resistance, earth leakage resistance
• Insulation resistance
• Line sequence indicator |

It is usual to find PFC testers are part of a combined PFC/Loop Impedance tester.

A9.1.6 Test lamp or voltage indicator

This type of tester (often referred to as a *tetrascope* or *neon screwdriver*) is frequently used by electricians and might look like the examples shown in Figure 9.31.

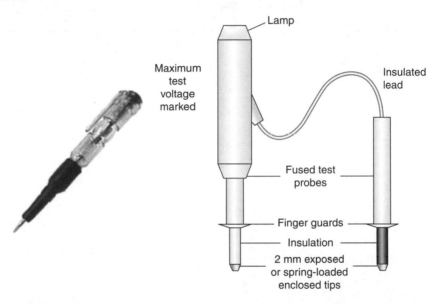

Figure 9.31 Typical test lamp and voltage indicators (courtesy StingRay)

These compact screwdriver multi-testers are normally water- and impact-resistant, with a.c. voltage test, contact test 70–250 V a.c., non-contact 100–1000 V a.c., polarity test 1.5 V d.c.–36 V d.c., continuity check 0–5 ohm and auto power on/off.

A9.1.7 Earth electrode resistance

The earth electrode (when used) is the means of making contact with the general mass of earth and should be regularly tested to ensure that good contact is made. In all cases, the aim is to ensure that the electrode resistance is not so high that the voltage from earthed metalwork to earth exceeds 50 V.

 Note: Acceptable electrodes are rods, pipes, mats, tapes, wires, plates and structural steelwork buried or driven into the ground. The pipes of other services such as gas and water must **not** be used as earth electrodes (although they must be bonded to earth).

Fluke 1620 Series
GEO Earth Ground
Testers by Fluke (UK)
Ltd www.fluke.co.uk

Figure 9.32 Earth
electrode resistance
tester

10

Installation, maintenance and repair

According to studies recently completed by CEN/CENELEC, the installation and maintenance engineer is the primary cause of reliability degradations during the in-service stage of most electrical installations. The problems associated with poorly trained, poorly supported, and/or poorly motivated personnel with respect to reliability and dependability, therefore, require careful assessment and quantification.

The most important factor, however, that affects the overall reliability of a modern product (system or installation) is the increased number of individual components that are required in that product. Since most system failures are actually caused by the failure of a single component (equipment or subassembly), the reliability of each individual component must be considerably better than the overall system reliability.

Because of this requirement, quality standards for the installation, maintenance, repair and inspection of in-service products have had to be laid down in engineering standards, handbooks and local operating manuals (written for specific items and equipment). These publications are used by maintenance engineers and should always include the most recent amendments. It is essential that assurance personnel also use the same procedures for their inspections as were used for the installation.

 Note: Schedules for installing wiring systems (together with guidance for selecting the appropriate size of cable, current ratings and so on) are shown in Appendix 4, Tables 4A1 and 4A2 of the Regulations.

Although this final chapter is a comparatively small one, it is nevertheless extremely important as it provides some guidance on the requirements for installation, maintenance and repair. It also lists the Regulations' requirements for maintenance etc. with respect to electrical installations – but as per the previous chapters which contain similar lists, please remember that this is **only** the author's impression of the most important aspects of the Wiring Regulations and electricians should **always** consult BS 7671 to satisfy compliance.

Finally, in Appendix 10.1 in this chapter there is a complete set of checklists for the quality control of electrical equipment and electrical installations.

The Regulations devote a complete part to inspection and testing (i.e. Part 6) and emphasise the need for continual improvement by stating:

> Every installation (or alteration to an existing installation) shall, during erection and on completion before being put into service, be inspected and tested to verify, so far as is reasonably practicable, that the requirements of the Regulations have been met.
>
> The verification shall be made by a competent person and on completion of the verification, a certificate shall be prepared.

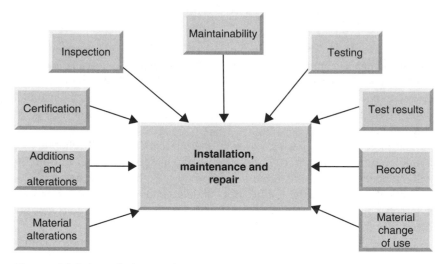

Figure 10.1 Installation, maintenance and repair

10.1 General

As previously shown, the requirements of BS 7671:2008 apply to the design, erection and verification of electrical installations such as those of:

- agricultural and horticultural premises;
- caravans, caravan parks and similar sites;
- commercial premises;
- construction sites, exhibitions, shows, fairgrounds and other installations for temporary purposes including professional stage and broadcast applications;
- domestic buildings;
- external lighting and similar installations;
- industrial premises;

- highway equipment and street furniture;
- low voltage generating sets;
- marinas;
- mobile or transportable units;
- photovoltaic systems;
- prefabricated buildings;
- public premises;
- residential premises.

Note: 'Premises' covers the land and all facilities including buildings belonging to it.

The Regulations do **not** apply to any of the following installations:

- aircraft equipment;
- electrical equipment of machines covered by BS EN 60204;
- equipment for mobile and fixed offshore installations;
- equipment for motor vehicles;
- equipment on board ships covered by BS 8450;
- lightning protection systems for buildings and structures covered by BS EN 62305;
- systems for the distribution of electricity to the public;
- railway traction equipment, rolling stock and signalling equipment;
- those aspects of lift installations covered by relevant parts of BS 5655 and BS EN 81-1;
- those aspects of mines and quarries specifically covered by Statutory Regulations;
- radio interference suppression equipment (except so far as it affects safety of the electrical installation).

10.2 Installation

Many requirements and recommendations for the installation of electrical equipment are to be found in BS 7671:2008. Most of these have already been mentioned in other parts of this book and the intention of this particular section is to bring to your attention some of the most important ones. It should **not**, however, be viewed as a complete list!

Note: Also see Chapter 7 for specific requirements concerning special installations and locations and Chapter 9 for inspections and tests.

10.2.1 Design

Two of the most important requirements for the installation of electrical equipment are that:

- installed equipment and their connections must be accessible for operational, inspection and maintenance purposes; and

- equipment must be arranged to allow easy access for periodic inspection, testing and maintenance.

Every installation shall be divided into circuits, as necessary, to:	WR-314.1

- avoid hazards and minimise inconvenience in the event of a fault
- ensure safe inspection, testing and maintenance
- prevent the indirect energising of a circuit that is intended to be isolated
- reduce the possibility of unwanted tripping of RCDs due to excessive protective conductor currents produced by equipment
- reduce the effects of electromagnetic interferences (EMI)
- take account of any danger that may arise from the failure of a single circuit (e.g. such as a lighting circuit).

Separate circuits shall be provided for parts of the installation which need to be separately controlled.	WR-314.2
The electrical installation shall be designed to provide for:	WR-132.1

- the protection of persons, livestock and property
- the proper functioning of the electrical installation for the intended use
- the maximum demand of an installation.

The electrical installation shall be arranged in such a way that no mutual detrimental influence will occur between electrical installations and non-electrical installations.	WR-132.11
Equipment shall be selected and installed to provide for the safety and proper functioning for the intended use of the installation.	WR-530.3
Equipment installed shall be appropriate to the external influences foreseen.	WR-530.3
Where the use of a new material or invention leads to departures from the Regulations, the resulting degree of safety of the installation shall be not less than that obtained by compliance with the Regulations.	WR-120.4

10.2.2 Installation of equipment

Electrical equipment not installed on or in a combustible wall shall be enclosed with a suitable thickness of non-flammable material.	WR-422.4.4
Electrical equipment shall be arranged so that:	WR-132.12

- there is sufficient space for the initial installation and later replacement of individual items of electrical equipment
- the equipment is accessible for operation, inspection, testing, fault detection, maintenance and repair.

Electrical equipment shall be installed so that design temperatures are not exceeded.	WR-134.1.5
Electrical joints and connections shall be properly constructed with regard to conductance, insulation, mechanical strength and protection.	WR-134.1.4
If an item of equipment (such as a capacitor) is installed behind a barrier or in an enclosure (and that equipment could retain a dangerous electrical charge after it has been switched off) a warning label shall be provided.	WR-416.2.5
The characteristics of the electrical equipment shall not be impaired by the process of erection.	WR-134.1.2
Switchgear or controlgear shall be installed outside the location unless:	WR-422.3.3

- it is suitable for the location; or
- it is installed in an enclosure providing a degree of protection of at least IP4X or, in the presence of dust, IP5X.

10.2.3 Power supplies

Electrical sources for safety services shall be installed as fixed equipment, in such a manner that they cannot be adversely affected by failure of the normal source.	WR-560.6.1
In a TN, TT or IT system an RCD with a rated residual operating current of not more than 30 mA shall be installed to protect every circuit.	WR-551.4.4.2

Stationary batteries shall be installed so that they are accessible **only** to skilled or instructed persons.	WR-551.8.1
Where danger or damage is expected to arise due to an interruption of supply, suitable provisions shall be made in the installation or installed equipment.	WR-131.7
Where in case of danger there is the necessity for immediate interruption of supply, an interrupting device shall be installed in such a way that it can be easily recognised and effectively and rapidly operated.	WR-132.9

10.2.4 Earthing

The type of earthing system to be used for the installation shall be determined taking into consideration the characteristics of the source of energy and (in particular) any earthing facilities.	WR-312.3.1

 Note: The types of earthing system are: TN-C, TN-S, TN-C-S, TT and IT (see Chapter 3 for further details concerning these systems).

10.2.5 Cables

A cable buried in the ground (that is not installed in a conduit or duct) shall incorporate an earthed armour or metal sheath or both, suitable for use as a protective conductor.	WR-522.8.10
A cable installed under a floor or above a ceiling shall be run in such a position that it is not liable to be damaged by contact with the floor or the ceiling or their fixings.	WR-522.6.5
A cable should preferably not be installed in a location where it is liable to be covered by thermal insulation.	WR-523.7
Cables complying with the requirements of BS EN 60332-1-2 may be installed without special precautions.	WR-527.1.3
Where multicore cables are installed in parallel each cable shall contain one conductor of each line.	WR-521.8.1

10.2.6 Conductors

A bare live conductor **shall** be installed on insulators.

A device protecting a conductor against overload may be installed along the run of that conductor provided that part of the run (i.e. between the point where a change occurs and the position of the protective device), has neither branch circuits nor outlets for connection of current-using equipment and is protected against fault current; or

WR-433.2.2

- its length does not exceed 3 m
- it is installed so as to reduce the risk of fault to a minimum; and
- it is installed so as to reduce to a minimum the risk of fire or danger to persons

A device protecting a conductor may be installed on the supply side of the point where a change occurs provided that it possesses an operating characteristic such that it protects the wiring situated on the load side against fault current.

WR-434.2.2

The number and type of live conductors (e.g. single-line two-wire a.c., three-line four-wire a.c.) shall be determined, both for the source of energy and for each circuit to be used within the installation.

WR-312.2.1

10.2.7 Wiring systems

The choice of the type of wiring system and the method of installation shall include consideration of the following:

WR-132.7

- the nature of the location
- the nature of the structure supporting the wiring
- accessibility of wiring to persons and livestock
- voltage
- the electromechanical stresses likely to occur due to short-circuit and earth fault currents
- electromagnetic interference

- other external influences (e.g. mechanical, thermal and those associated with fire) to which the wiring is likely to be exposed during the erection of the electrical installation or in service.

The installation of wiring systems will meet the requirements if: WR-412.2.4.1

- the rated voltage of the cable(s) is not less than the nominal voltage of the system and at least 300/500 V; and
- adequate mechanical protection of the basic insulation is provided by (one or more of) the following:
  - the non-metallic sheath of the cable
  - non-metallic trunking or ducting (complying with BS EN 50085)
  - non-metallic conduit (complying with BS EN 61386).

A wiring system which passes through the location but is not intended to supply electrical equipment in the location shall: WR-422.3.5

- have no connection or joint within the location, unless the connection or joint is installed in an enclosure; and
- is protected against overcurrent; and
- does **not** use bare live conductors.

A wiring system shall be installed so that the general building structural performance and fire safety are not reduced.   WR-527.1.2

A wiring system shall not be installed in the vicinity of services which produce heat, smoke or fumes likely to be detrimental to the wiring, **unless** it is protected from harmful effects by shielding arranged so as not to affect the dissipation of heat from the wiring.   WR-528.3.1

10.2.8 Protective devices

A device for protection against overload shall be installed at the point where a reduction occurs in   WR-433.2.1

the value of the current-carrying capacity of the conductors of the installation.

A device protecting a conductor against overload may be installed along the run of that conductor provided that part of the run (i.e. between the point where a change occurs and the position of the protective device), has neither branch circuits nor outlets for connection of current-using equipment and is protected against fault current; or

WR-433.2.2

- its length does not exceed 3 m
- it is installed so as to reduce the risk of fault to a minimum; and
- it is installed so as to reduce to a minimum the risk of fire or danger to persons.

A device protecting a conductor may be installed on the supply side of the point where a change occurs provided that it possesses an operating characteristic such that it protects the wiring situated on the load side against fault current.

WR-434.2.2

A device providing protection against fault current shall be installed at the point where a reduction in the cross-sectional area or other change causes a reduction in the current-carrying capacity of the conductors of the installation (except installations situated in locations presenting a fire risk or risk of explosion and where the requirements for special installations and locations specify different conditions).

WR-434.2

A device (such as a circuit-breaker with a short-circuit release, or a fuse) providing protection against fault current shall only be installed where overload protection is achieved by other means.

WR-432.3

Except where a protective device is installed to interrupt the supply in the event of the first earth fault, an RCM or an insulation fault location system shall be provided to indicate the occurrence of a first fault from a live part to an exposed conductive part or to earth.

WR-411.6.3.2

The installation of equipment (e.g. fixing, connection of conductors) shall not affect the protection afforded by an enclosure.

WR-412.2.3.1

10.2.9 Electromechanical stresses

Every conductor or cable shall have adequate strength and be so installed as to withstand the electromechanical forces that may be caused by any current, including fault current, it may have to carry in service.	WR-521.5.1

10.2.10 Electromagnetic comparability

Electromagnetic interference shall be taken into account.

All fixed installations shall be in accordance with the current and relevant EMC regulations.	WR-332.1
Consideration shall be given by the planner and designer of the electrical installation to measures reducing the effect of induced voltage disturbances and electromagnetic interferences (EMI).	WR-332.2
The installation design shall take into consideration the anticipated electromagnetic emissions generated by the installation or the installed equipment.	WR-131.6.4

10.2.11 Thermal effects

Electrical installation shall be so arranged that: • the risk of ignition of flammable materials due to high temperature or electric arc is minimised • during normal operation of the electrical equipment, there shall be minimal risk of burns to persons or livestock.	WR-131.3.1

10.2.12 Initial verification

During erection, or on completion of an installation, or making addition or alteration to an installation (and before being put into service) appropriate inspection and testing shall be carried out by competent persons to verify that the requirements of BS 7671:2008 have been met.	WR-134.2.1

10.3 Maintenance and repair

The aim of periodic inspection and testing is to:

- confirm that the installation is in a satisfactory condition for continued service;
- confirm that the safety of the installation has not deteriorated or has been damaged;
- ensure the continued safety of persons and livestock against the effects of electric shock and burns;
- identify installation defects and non-compliance with the requirements of the Regulations which may give rise to danger;
- protect property being damaged by fire and heat caused by a defective installation.

The Regulations are clear in the requirements that maintenance inspections shall consist of a careful scrutiny of the installation (dismantled or otherwise) using the appropriate tests described in Chapter 6 of the Regulations.

The frequency of maintenance inspections will depend on:

- the type of installation, its use and operation;
- the frequency and quality of maintenance and;
- the external influences to which it is subjected.

Precautions shall be taken to ensure that maintenance inspections do not cause:

- danger to persons or livestock;
- damage to property and equipment (even if the circuit is defective).

The maintenance inspection shall be made to verify that the installed electrical equipment:

- complies with the requirements of the Regulations and the appropriate National Standards and European Harmonised Directives;
- is correctly selected and erected in accordance with the Regulations;
- is not visibly damaged or defective so as to impair safety.

The inspection shall include the following items, where relevant:

- access to switchgear and equipment;
- cable routing in safe zones;
- choice and setting of protective and monitoring devices;
- connection of conductors;
- connection of single-pole devices for protection or switching in phase conductors only;
- correct connection of accessories and equipment;
- erection methods;

- identification of conductors;
- labelling of protective devices, switches and terminals;
- presence of danger notices and other warning signs;
- presence of diagrams, instructions and similar information;
- presence of fire barriers, suitable seals;
- presence of isolation and switching devices (and their correct location);
- prevention of mutual (i.e. detrimental) influence;
- presence of undervoltage protective devices;
- protection against electric shock (direct and indirect contact);
- protection against external influences;
- protection against direct and indirect contact;
- protection against mechanical damage (and compliance with Section 522);
- protection against overcurrent;
- protection against thermal effects selection of conductors for current-carrying capacity and voltage drop.

 Note: Generally speaking, the following CEN/CENELEC recommendations are relevant for all installed electrical and/or electronic equipment.

> *For ease of maintenance, all equipment provided should have:*
>
> - *easily accessible test points to facilitate fault location*
> - *modules that have been constructed so as to facilitate the connection of test equipment (e.g. logic analysers, emulators and test ROMs)*
> - *fault location provision to allow functional areas within each module or equipment to be isolated.*

Under workshop conditions it should be possible to gain access to all circuitry while it is operating, with a minimum of effort required, to partially dismantle the module concerned with the assumption that there will be a minimum of risk to the components, or to the testing maintenance staff.

Special connecting leads, printed wiring extension boards and any other special items required for maintenance purposes, together with the mating half of all necessary connectors, will probably have to be obtained from the manufacturer.

> An assessment shall be made of the frequency and   WR-341.1
> quality of maintenance the installation can reasonably
> be expected to receive during its intended life and to
> ensure that:
>
> - any periodic inspection and testing, maintenance and
>   repairs likely to be necessary during the intended life
>   can be readily and safely carried out; and

- the effectiveness of the protective measures for safety during the intended life shall not diminish; and
- the reliability of equipment for proper functioning of the installation is appropriate to the intended life.

10.3.1 Frequency of inspection and testing

The frequency of periodic inspection and maintenance of an installation will depend on the type of installation and equipment, its use and operation, the frequency and quality of maintenance and the external influences to which it is subjected.	WR-622.1
The results and recommendations of the previous report, if any, shall be taken into account.	WR-622.1
In the case of an installation under an effective management system for preventive maintenance in normal use, periodic inspection and testing may be replaced by an adequate regime of continuous monitoring and maintenance of the installation and all its constituent equipment by skilled persons, competent in such work.	WR-622.2

10.3.2 Accessibility of electrical equipment

Provision shall be made for safe and adequate access to all parts of a wiring system which may require maintenance.	WR-529.3
Electrical equipment shall be arranged so that:	WR-132.12
<ul><li>there is sufficient space for the initial installation and later replacement of individual items of electrical equipment</li><li>the equipment is accessible for operation, inspection, testing, fault detection, maintenance and repair.</li></ul>	
Every item of equipment shall be arranged so as to facilitate its operation, inspection and maintenance and access to each connection.	WR-513.1 and WR-543.3.3

Every installation shall be divided into circuits, as necessary, to: WR-314.1

- avoid hazards and minimise inconvenience in the event of a fault
- ensure safe inspection, testing and maintenance
- prevent the indirect energising of a circuit that is intended to be isolated
- reduce the possibility of unwanted tripping of RCDs due to excessive protective conductor currents produced by equipment
- reduce the effects of electromagnetic interferences (EMI)
- take account of any danger that may arise from the failure of a single circuit (e.g. such as a lighting circuit).

The selection and erection of equipment shall facilitate safe maintenance and shall not adversely affect provisions made by the manufacturer of the PV equipment to enable maintenance or service work to be carried out safely.   WR-712.513.1

10.3.3 Safety protection

Disconnecting devices shall be provided so as to allow electrical installations, circuits or individual items of equipment to be switched off or isolated for the purposes of operation, inspection, fault detection, testing, maintenance and repair.   WR-132.10

Where it is necessary to remove a protective measure in order to carry out maintenance, provision shall be made so that the protective measure can be reinstated without reduction of the degree of protection originally intended.   WR-529.2

The protective measures of placing out of reach and obstacles shall **not** be used except where: WR-559.10.1

- the maintenance of equipment is restricted to skilled persons who are specially trained
- items of street furniture are within 1.5 m of a low voltage overhead line.

10.3.4 Switching off for mechanical maintenance

The capability of switching off for mechanical maintenance shall be provided where mechanical maintenance could involve a risk of physical injury.

WR-537.3.1.1

Suitable means shall be provided to prevent electrically powered equipment from becoming unintentionally reactivated during mechanical maintenance.

WR-537.3.1.2

10.3.5 Devices for switching off for mechanical maintenance

A device such as a:

WR-537.3.2.1

- multipole switch
- circuit-breaker
- control and protective switching device (CPS)
- control switch operating a contactor
- plug and socket outlet

may be inserted in the main supply circuit for switching off for mechanical maintenance.

A device for switching off for mechanical maintenance (or a control switch for such a device) shall:

WR-537.3.2.2

- require manual operation
- be designed and/or installed so as to prevent inadvertent or unintentional switching on
- be so placed and durably marked so as to be readily identifiable and convenient for the intended use.

WR-537.3.2.4

- ensure that the open position of the contacts of the device shall be visible or be clearly and reliably indicated by the use of the symbols '0' and '1' to indicate the open and closed positions respectively.

WR-537.3.2.2

A plug and socket outlet or similar device of rating not exceeding 16 A may be used as a device for switching off for mechanical maintenance.

WR-537.3.2.6

Where a switch is used as a device for switching off for mechanical maintenance, it shall be capable of cutting off the full load current of the relevant
part of the installation.

10.3.6 Connections

Every connection shall be accessible for inspection,   WR-526.3
testing and maintenance, except for the following:

- a joint designed to be buried in the ground
- a compound filled or encapsulated joint
- a connection between a cold tail and the heating
  element as in ceiling heating, floor heating or a
  trace heating system
- a joint made by welding, soldering, brazing or
  appropriate compression tool
- a joint forming part of the equipment complying
  with the appropriate product standard.

10.3.7 Insulating enclosure

An insulating enclosure shall **not** contain any screws or other fixing means
which might need to be removed (e.g. during installation and maintenance)
and which 'could' be replaced by metallic screws or some other type of fixing
that could affect the enclosure's insulation.

10.3.8 Low power supply sources

The power output of a low power supply system is   WR-560.6.10
limited to 500 W for 3-hour duration or 1500 W for
1-hour duration.
However:

- the batteries may be of the gastight or valve-
  regulated maintenance-free type; and
- the minimum design life of the batteries shall be
  5 years.

10.3.9 Multiphase sequence

For multiphase circuits, it shall be verified that the   WR-612.12
phase sequence is maintained.

10.3.10 Environmental aspects

The wiring system shall be selected and erected so: WR-522.3.1

- that no damage is caused by condensation or ingress of water during installation, use and maintenance
- as to minimise the damage arising from mechanical   WR-522.6.1
  stress, e.g. by impact, abrasion, penetration,
  tension or compression during installation, use or
  maintenance
- as to avoid during installation, use or maintenance,   WR-522.8.1
  damage to the sheath or insulation of cables and
  their terminations.

10.4 Repair

Equipment is normally expected to have been designed to have a useful life of not less than 20 years. 'Useful life' normally means 'the period for which the equipment will continue to operate with the specified level of reliability'.

No components, modules or equipment should, therefore, be used (so far as can be ascertained at the time of manufacture), for which spares cannot be fully guaranteed to be available throughout the life of the equipment.

An assessment shall be made of the frequency and   WR-341.1
quality of maintenance the installation can reasonably
be expected to receive during its intended life and to
ensure that:

- any periodic inspection and testing, maintenance and
  repairs likely to be necessary during the intended life
  can be readily and safely carried out; and
- the effectiveness of the protective measures for
  safety during the intended life shall not diminish; and
- as the reliability of equipment for proper functioning
  of the installation is appropriate to the intended life.

10.4.1 Disconnecting devices

Disconnecting devices shall be provided so as to   WR-132.10
allow electrical installations, circuits or individual
items of equipment to be switched off or isolated for
the purposes of maintenance and repair.

10.4.2 Accessibility of electrical equipment

Electrical equipment shall be arranged so that: WR-132.12

- there is sufficient space for the later replacement
  of individual items of electrical equipment
- the equipment is accessible for operation,
  inspection, testing, fault detection, maintenance
  and repair.

10.4.3 Certification and reporting

As previously explained in Chapter 9 of this book, the following Certificates
shall be completed where appropriate.

Electrical Installation Certificate	Details of the installation together with a record of the inspections made and the test results shall be provided for new installations and changes to existing installations.	WR-631.1
Minor Electrical Installation Works Certificate	For all minor electrical installation work that does not include the provision of a new circuit.	WR-631.3
Periodic Inspection Report	Details of the installation together with a record of the inspections made and the test results shall be provided for all installations subject to periodic inspection and testing.	741-01-02

Although alterations and repairs may be completed during an installation,
normal repairs are completed following a Periodic Inspection, and reports of
these repairs shall be compiled and signed or otherwise authenticated by a
competent person or persons.

Following the periodic inspection, a Periodic Inspection   WR-634.1
Report, together with a Schedule of Inspections and a
Schedule of Test Results, shall be given by the person
carrying out the inspection, or a person authorised to act
on their behalf, to the person ordering the inspection.

Any damage, deterioration, defects, dangerous conditions and non-compliance with the requirements of the Regulations, which may give rise to danger, together with any significant limitations of the inspection and testing, including their reasons, shall be recorded.	WR-634.2
Periodic Inspection Reports may be produced in any durable medium, including written and electronic media.	WR-631.5
Regardless of the media used for original Certificates, reports or their copies, their authenticity and integrity shall be verified by a reliable process or method.	WR-631.5
The process or method shall also verify that any copy is a true copy of the original.	WR-631.5

 Records are an important part of the management of any electrical installation as they provide objective evidence of activities performed and/or results achieved; the following requirements come from the Regulations.

10.5 Additions and alterations to an installation

The Regulations require every electrical installation to be inspected and tested during erection and on completion, before being put into service, have it to verified that the requirements of BS 7671 have been met. By definition, this requirement also applies to alterations and/or additions to an existing installation, as well as entirely new installations.

The following is a summary of the relevant requirements for additions and alterations to an installation.

10.5.1 Additions and alterations to an installation

No addition or alteration, temporary or permanent, shall be made to an existing installation: • unless it has been ascertained that the rating and the condition of any existing equipment (including that of the distributor) will be adequate for the altered circumstances • unless the earthing and bonding arrangements, used as a protective measure for the safety of the addition or alteration, are adequate.	WR-131.8

> If wiring additions or alterations are made to an installation such that some of the wiring complies with the current Regulations but there is also wiring to previous versions of these Regulations, a warning notice shall be affixed at or near the appropriate distribution board with the following wording (see Figure 10.2). WR-514.14.1

CAUTION

This installation has wiring colours to two versions of BS 7671. Great care should be taken before undertaking extension, alteration or repair that all conductors are correctly identified.

Figure 10.2 Warning notice: non-standard colours

> An Electrical Installation Certificate or a Minor Electrical Installation Works Certificate (as appropriate) shall apply to all work competed including the additions or alterations. WR-633.1
>
> The contractor or other person responsible for the new work, or a person authorised to act on their behalf, shall record on the Electrical Installation Certificate or the Minor Electrical Installation Works Certificate, any defects found, so far as is reasonably practicable, in the existing installation. WR-633.2

10.6 Material changes of use

Where there is a material change of use of a building, any work carried out shall ensure that the building complies with the applicable requirements of the following paragraphs of Schedule 1 of the Building Act 1984:
(a) In all cases:

- means of warning and escape (B1);
- internal fire spread – linings (B2);
- internal fire spread – structure (B3);
- external fire spread (B4);
- access and facilities for the fire service (B5);
- resistance to moisture (C1)(2);
- dwelling houses and flats formed by material change of use (E4);
- ventilation (F1);

- sanitary conveniences and washing facilities (G1);
- bathrooms (G2);
- foul water drainage (H1);
- solid waste storage (H6);
- combustion appliances (J1, J2 & J3);
- conservation of fuel and power – dwellings (L1);
- conservation of fuel and power – buildings other than dwellings (L2);
- electrical safety (P1, P2).

(b) In other cases:

Table 10.1 Building Act requirements

Material change of use	Requirement	Approved document
The building is used as a dwelling, where previously it was not	Resistance to moisture	C2; E1, E2, E3
The public building consists of a new school	Acoustic conditions in schools	E4
The building contains a flat, where previously it did not	Resistance to the passage of sound	E1, E2 & E3
The building is used as a hotel or a boarding house, where previously it was not	Structure	A1, A2 & A3; E1, E2, E3
The building is used as an institution, where previously it was not	Structure	A1, A2 & A3
The building is used as a public building, where previously it was not		A1, A2 & A3
The building is not a building described in Classes I to VI in Schedule 2, where previously it was not	Structure	A1, A2 & A3
The building, which contains at least one room for residential purposes, contains a greater or lesser number of dwellings than it did previously	Structure	A1, A2 & A3; E1, E2, E3
The building, which contains at least one dwelling, contains a greater or lesser number of dwellings than it did previously	Resistance to the passage of sound	E1, E2 & E3

 Note: In some circumstances (particularly when a historic building is undergoing a material change of use and where the special characteristics of the building need to be recognised) it may **not** be practical to improve sound insulation to the standards set out in Part E1 or resistance to contaminants and water as set out in Part C. In these cases, the aim should be to improve the insulation and resistance where it is practically possible – always provided that the work does not prejudice the character of the historic building, or increase the risk of long-term deterioration to the building fabric and/or fittings.

Appendix 10.1: Example stage audit checks

Design stage

Item	Related item	Remark
1 Requirements	1.1 Information	Has the customer fully described his requirement?
		Has the customer any mandatory requirements?
		Are the customer's requirements fully understood by all members of the design team?
		Is there a need to have further discussions with the customer?
		Are other suppliers or subcontractors involved?
		If yes, who is the prime contractor?
	1.2 Standards	What international Standards need to be observed?
		Are they available?
		What national Standards need to be observed?
		Are they available?
1		What other information and procedures are required?
		Are they available?
	1.3 Procedures	Are there any customer-supplied drawings, sketches or plans?
		Have they been registered?
2 Quality Procedures	2.1 Procedures manual	Is one available? If so: Does, it contain detailed procedures and instructions for the control of all drawings within the drawing office?
	2.2 Planning Implementation and Production	Is the project split into a number of Work Packages? If so:
		• are the various Work Packages listed?
		• have Work Package Leaders been nominated?
		• is their task clear?
		• is their task achievable?
		Is a time plan available?
		Is it up to date?
		Is it regularly maintained?

(*continued*)

Design stage (*continued*)

Item	Related item	Remark
		Is it relevant to the task?
3 Drawings	3.1 Identification	Are all drawings identified by a unique number?
		Is the numbering system strictly controlled?
	3.2 Cataloguing	Is a catalogue of drawings maintained?
		If so, this catalogue regularly reviewed and up to date?
	3.3 Amendments and Modifications	Is there a procedure for authorising the issue of amendments, changes to drawings?
		Is there a method for withdrawing and disposing of obsolete drawings?
4 Components	4.1 Availability	Are complete lists of all the relevant components available?
	4.2 Adequacy	Are the selected components currently available and adequate for the task?
		If not, how long will they take to procure?
		Is this acceptable?
	4.3 Acceptability	If alternative components have to be used are they acceptable to the task?
5 Records	5.1 Failure reports	Has the Design Office access to all records, failure reports and other relevant data?
	5.2 Reliability data	Is reliability data correctly stored, maintained and analysed?
	5.3 Graphs, diagrams, plans	In addition to drawings, is there a system for the control of all graphs, tables, plans etc.?
		Are CAD facilities available?
		(If so, go to 6.1)
6 Reviews and Audits	6.1 Computers	If a processor is being used: • are all the design office personnel trained in its use? • are regular back-ups taken? • is there an anti-virus system in place?
	6.2 Manufacturing division	Is a close relationship being maintained between the design office and the manufacturing division?
	6.3	Is notice being taken of the manufacturing division's exact requirements, their problems and their choices of components etc.?

(*continued*)

Installation stage

Item	Related item	Remark
1 Degree of quality	1.1 Quality control procedures	Are quality control procedures available?
		Are they relevant to the task?
		Are they understood by all members of the manufacturing team?
		Are they regularly reviewed and up to date?
		Are they subject to control procedures?
	1.2 Quality control checks	What quality checks are being observed?
		Are they relevant?
		Are there laid down procedures for carrying out these checks?
		Are they available?
		Are they regularly updated?
2 Reliability of product design	2.1 Statistical data	Is there a system for predicting the reliability of the product's design?
		Is sufficient statistical data available to be able to estimate the actual reliability of the design, before a product is manufactured?
		Is the appropriate engineering data available?
	2.2 Components and parts	Are the reliability ratings of recommended parts and components available?
		Are probability methods used to examine the reliability of a proposed design?
		If so, have these checks revealed design deficiencies such as:
		• assembly errors? • operator learning, motivational, or fatigue factors? • latent defects? • Improper part selection?

Note: If necessary, use additional sheets to list actions taken

Acceptance stage

Item	Related item	Remark
1 Product performance		Does the product perform to the required function?
		If not what has been done about it?
2 Quality level	2.1 Workmanship	Does the workmanship of the product fully meet the level of quality required or stipulated by the user?
	2.2 Tests	Is the product subjected to environmental tests?
		If so, which ones?
		Is the product field tested as a complete system?
		If so, what were the results?
3 Reliability	3.1 Probability function	Are individual components and modules environmentally tested?
		If so, how?
	3.2 Failure rate	Is the product's reliability measured in terms of probability function?
		If so, what were the results?
		Is the product's reliability measured in terms of failure rate?
		If so, what were the results?
	3.3 Mean time between failures	Is the product's reliability measured in terms of mean time between failure?
		If so, what were the results?

In-service stage

Item	Related item	Remark
1 System reliability	1.1 Product basic design	Are statistical methods being used to prove the product's basic design?
		If so, are they adequate?
		Are the results recorded and available?
		What other methods are used to prove the product's basic design?
		Are these methods appropriate?
2 Equipment reliability	2.1 Personnel	Are there sufficient trained personnel to carry out the task?
		Are they sufficiently motivated?
		If not, what is the problem?

(*continued*)

In-service stage (*continued*)

Item	Related item	Remark
		Have individual job descriptions been developed?
		Are they readily available?
	2.1.1 Operators	Are all operators capable of completing their duties?
	2.1.2 Training	Do all personnel receive appropriate training?
		Is a continuous on-the-job training (OJT) programme available to all personnel?
		If not, why not?
	2.2 Product dependability	What proof is there that the product is dependable?
		How is product dependability proved?
		Is this sufficient for the customer?
	2.3 Component reliability	Has the reliability of individual components been considered?
		Does the reliability of individual components exceed the overall system reliability?
	2.4 Faulty operating procedures	Are operating procedures available?
		Are they appropriate to the task?
		Are they regularly reviewed?
	2.5 Operational abuses	Are there any obvious operational abuses?
		If so, what are they?
		How can they be overcome?
	2.5.1 Extended duty cycle	Do the staff have to work shifts?
		If so, are they allowed regular breaks from their work?
		Is there a senior shift worker?
		If so, are his duties and responsibilities clearly defined?
		Are computers used?
		If so, are screen filters available?
		Do the operators have keyboard wrist rests?
	2.5.2 Training	Do the operational staff receive regular on-the-job training?
		Is there any need for additional in-house or external training?
3 Design capability	3.1 Faulty operating procedures	Are there any obvious faulty operating procedures?
		Can the existing procedures be improved upon?

Appendix A

Symbols used in electrical installations

SYMBOLS

Socket-outlet	Microphone	Operating device (coil)
Switched socket-outlet	Loudspeaker	Make contact - normally open
Switch	Antenna	Break contact - normally closed
Two-way switch, single-pole	Machine * Function M = Motor G = Generator	Manually operated switch
Intermediate switch	Generator	Three-phase winding - delta
Pull switch, single-pole	Indicating instrument * function V = Voltmeter A = Ammeter	Three-phase winding - Star
Lighting outlet position		Changer, Converter
Fluorescent luminaire	Integrating instrument or Energy meter * function Wh = Watt-hour VArh = Volt ampere reactive hour	Rectifier
Wall mounted luminaire		Invertor
Emergency lighting luminaire (or special circuit)	Load *details	Primary cell - longer line positive, shorter line negative
Self-contained emergency lighting luminaire	Motor starter *indicates type	Battery
Push button	Class II appliance	Transformer - general symbol
Clock	Class III appliance	
Bell	Safety isolating transformer	
Buzzer	Isolating transformer	
Horn	Fuse link, rated current in amperes	
Telephone handset		

10^9 giga G
10^6 mega M
10^3 kilo k
10^{-3} milli m
10^{-6} micro μ
10^{-9} nano n

Appendix B

List of electrical and electromechanical symbols

Symbol	Description
$\beta°$	tube oscillating angle
°C	degrees Celsius
Ω	ohm
μg	microgram (10^{-6} g)
$\mu g/m^3$	micrograms per cubic metre
μm	micrometer (10^{-6} m)
μs	microsecond (10^{-6} s)
a	amplitude
A	ampere
A/m	amperes per metre
am	attometre
atm	standard atmosphere
C	coulomb
cd	candela
cd/m^2	candelas per square metre
dB	decibels
dB(A)	decibel amps
dBm	decibel metres
dm^3	cubic decimetre
dm^3/mm	cubic decimetres/millimetre – flow
Em	exametre
eV	electron volt
f	frequency
F	farad
fm	femtometre
ft	foot
g	gram
G	gauss
G	shock
g^2/Hz	accelerated spectral density
GHz	gigahertz (10^9 Hz)
Gm	gigametre (10^9 m)
g/m^3	grams per cubic metre
g_n	peak acceleration
G_s	setting value of a characteristic quantity
h	hour
H	henry
ha	hectare
hp	horsepower

(*continued*)

Symbol	Description
hr(s)	hour(s) – alternative to h
Hz	hertz
I	current
I^2R	power
in	inch
J	joule
k	constant of the relay
K	kelvin
kA	kiloamps
kA/μs	kiloamps per microsecond
kg	kilogram
kg/m^3	kilograms per cubic metre
kgf	kilogram force
kHz	kilohertz
kPa	kilopascal – pressure
ks	kilosecond
kV	kilovolts
kW	kilowatt
kW/m^2	kilowatts per square metre – irradiance
l	litre
lb	pound
lb/in	pounds per square inch
m	metre
m/s	metres per second
m/s^2	metres per second per second – amplitude
m^2	square metres
m^3	cubic metres
mbar	millibar – pressure
MHz	megahertz (10^6 Hz)
min	minute
mm	millimeter (10^{-6} m)
Mm	megametre (10^6 m)
mm/h	millimetres per hour
mm/m^2	millimetres/square metre – exposure
mol	mole
ms	millisecond
mV	millivolts
MVA	megavolt amps
N	newton
N/m^2	newtons per square metre
NaCl	sodium chloride
nF	nanofarad (10^{-9} F)
nm	nanometre (10^{-9} m)
pH	alkalinity/acidity value
pm	picometre
Pm	petametre
R	intensity of dropfield in mm/h
R	resistance
rad/s	radians per second
s	second
S	siemens
t	tonne
T	time
T	tesla
Tm	terametre

(*continued*)

Symbol	Description
u	amplitude of voltage surge
U_n	nominal voltage
V	volt
V/μs	volts per microsecond
V/km	volts per kilometre
Vm	volts per metre
W	watt
Wb	weber
W/m^2	watts per square metre – irradiance
yd	yard
ym	yocotmetre
Ym	yottametre
zm	zeptometre
Zm	zettametre

Appendix C
SI units for existing technology

As Gregor M. Grant explained in his article published in the April/May 1997 issue of *ElectroTechnology*, the *Système International d'Unités* (SI) was a child of the 1960s, a creation of the 11th General Conference on Weights and Measures, *Conférence Générale des Poids et Mesures* (CGPM). This assembly endorsed the Italian physicist Professor Giovanni Giorgi's MKS (i.e. metre-kilogram-second) system of 1901 and decided to base the SI system on it. Seven basic units were adopted, as shown in Table C.1, each of which was harmonised to a standard value.

Of the seven units, only the kilogram (kg) is represented by a physical object, namely a cylinder of platinum-iridium kept at the International Bureau of Weights and Measures at Sèvres, near Paris, with a duplicate at the US Bureau of Standards.

The metre (m), on the other hand, is 'the length of the path travelled by light in a vacuum during a time interval of 1/299,792,458 of a second'.

The second (s) has been defined as 'the duration of 9,192,631,770 periods of radiation corresponding to the energy-level change between the two hyperfine levels of the ground state of caesium-133 atom'.

The ampere (A) is 'that constant current which, if maintained in two straight parallel conductors of infinite length, of negligible circular cross section and placed 1 m apart in vacuum, would produce between these conductors a force equal to 2×10^{-7} newtons per metre length'.

The unit of temperature is the kelvin (K), which is a thermodynamic measurement as opposed to one based on the properties of real material. Its origin is at absolute zero and there is a fixed point where the pressure and temperature of water, water vapour and ice are in equilibrium, which is defined as 273.16 K.

The mole (mol) is 'that quantity of substance of a system which contains as many elementary entities as there are atoms in 0.012 kg of carbon-12'.

For definition purposes the entities *must* be specified (e.g. atoms, electrons, ions or any other particles or groups of such particles).

Finally there is the candela (cd), the unit of light intensity. This is defined as 'the luminous intensity, in the perpendicular direction, of a surface of $1/600,000\,m^2$ of a black body at the temperature of freezing platinum under a pressure of $101,325\,N/m^2$'.

Two years before the creation of SI units, another international agreement had made the prefixes mega and micro official and introduced some new ones, such as the nano, whose name derives from the Greek 'nanos' meaning dwarf (see Table C.2). Its symbol is n, and its mathematical representation is 10^{-9}, indicating the number of *digits* to the right of the decimal point, in this case 0.000 000 001.

Even these minute quantities, however, soon became inadequate and, by 1962, it was decided that a thousandth of a picometre be designated a femtometre and one thousandth of this new measurement be termed an attometre. Later on, the zeptometre and yoctometre were introduced.

C.1 Basic SI units

Many SI units are named after people but when these units are written in full, they do not necessarily require initial capital letters, e.g. amperes, coulombs, newtons, siemens.

All the above examples are expressed in the plural, but note that siemens does not drop the final 's' in the singular as this was derived from a person's name (i.e. Siemens) thus we have one newton, but one siemens.

Table C.1 Basic SI units

SI nomenclature	Abbreviation	Quantity
metre	m	length
kilogram	kg	mass
second	s	time
ampere	A	electrical current
kelvin	K	temperature
mole	mol	amount of substance
candela	cd	luminous intensity

C.2 Small number SI prefixes

Within the SI units there is a distinction between a quantity and a unit. Length is a quantity, but metres (abbreviated to m) is a unit.

Table C.2 Small number SI units

Measurement	Symbol	Equivalent to
millimetre	mm	0.001 m or 10^{-3} m
micrometre	μm	0.000 001 m or 10^{-6} m
nanometre	nm	0.000 000 001 m or 10^{-9} m
picometre	pm	0.000 000 000 001 m or 10^{-12} m
femtometre	fm	0.000 000 000 000 001 m or 10^{-15} m
attometre	am	0.000 000 000 000 000 001 m or 10^{-18} m
zeptometre	zm	0.000 000 000 000 000 000 001 m or 10^{-21} m
yoctometre	ym	0.000 000 000 000 000 000 000 001 m or 10^{-24} m

C.3 Large number SI prefixes

Table C.3 Large number SI prefixes

Measurement	Symbol	Equivalent to
megametre	Mm	1 000 000 m or 10^{6} m
gigametre	Gm	1 000 000 000 m or 10^{9} m
terametre	Tm	1 000 000 000 000 m or 10^{12} m
petametre	Pm	1 000 000 000 000 000 m or 10^{15} m
exametre	Em	1 000 000 000 000 000 000 m or 10^{18} m
zettametre	Zm	1 000 000 000 000 000 000 000 m or 10^{21} m
yottametre	Ym	1 000 000 000 000 000 000 000 000 m or 10^{24} m

C.4 Deprecated prefixes

Some non-SI fractions and multiples are occasionally used (see below), but they are not encouraged.

Table C.4 Deprecated prefixes

Fractions	Prefix	Abbreviation	Multiple	Prefix	Abbreviation
10^{-1}	deci	d	10	deca	da
10^{-2}	centi	c	10^{2}	hecto	h

C.5 Derived units

Some units, derived from the basic SI units, have been given special names, many of which originate from a person's name (i.e. Siemens).

Table C.5 Derived units

Quantity	Name of unit	Abbreviation (symbol)	Expression in terms of other SI units
Energy	joule	J	Nm
Force	newton	N	—
Power	watt	W	J/s
Electric charge	coulomb	C	As
Potential difference (voltage)	volt	V	W/A
Electrical resistance (or reactance or impedance)	ohm	Ω	V/A
Electrical capacitance	farad	F	C/V
Magnetic flux	weber	Wb	Vs
Inductance (note that the plural of henry is henrys)	henry	H	Wb/A
Magnetic flux density	tesla	T	Wb/m^2
Admittance (electrical conductance)	siemens	S	$A/V (=\Omega^{-1})$
Frequency	hertz	Hz	cycles per second (or events per second)

C.6 Units without special names

Other derived units, without special names, are listed below.

Table C.6 Units without special names

Quantity	Unit	Abbreviation
Area	square metre	m^2
Volume	cubic metre	m^3
Density	kilograms per cubic metre	kg/m^3
Velocity	metres per second	m/s
Angular velocity (angular frequency)	radians per second	rad/s
Acceleration	metres per second per second	m/s^2
Pressure	newtons per square metre	N/m^2
Electric field strength	volts per metre	V/m
Magnetic field strength	amperes per metre	A/m
Luminance	candelas per square metre	cd/m^2

C.7 Tolerated units

Some non-SI units are tolerated in conjunction with SI units.

Table C.7 Tolerated units

Quantity	Unit	Abbreviation (symbol)	Definition
Area	hectare	ha	$10^4\,m^2$
Volume	litre	l	$10^{-3}\,m^3$
Pressure	standard atmosphere	atm	101,325 Pa
Mass	tonne	t	$10^3\,kg$ (Mg)
Energy	electronvolt	eV	1.6021×10^{19} J
Magnetic induction	gauss	G	10^{-4} T

C.8 Obsolete units

For historical interest (as well as for completeness), the following table gives a list of obsolete units.

Table C.8 Obsolete units

Quantity	Unit	Abbreviation (symbol)	Definition
Length	inch	in	0.0254 m
	foot	ft	0.3048 m
	yard	yd	0.9144 m
	mile	mi	1.60394 km
Mass	pound	lb	0.4539237 kg
Force	dyne	dyn	10^{-5} N
	poundal	pdl	0.138255 N
	pound force	lbf	4.44822 N
	kilogram force	kgf	9.80665 N
Pressure	atmosphere	atm	101.325 kN/m^2
	torr	torr	133.322 N/m^2
	pounds per square inch	lb/in^2	6894.76 N/m^2
Energy	erg	erg	10^{-7} J
Power	horsepower	hp	745.700 W

Appendix D

Acronyms and abbreviations

a.c.	Alternating current
ACS	Assemblies For Construction Sites
ADP	Automatic Data Processing
BRE	Building Research Establishment Ltd
BS	British Standard
BSI	British Standards Institution
CAD	Computer-Aided Design
CE	Conformity Europe
CECC	CENELEC Electronic Components Committee
CEN	Comité Européen de Normalisation
CENELEC	Comité Européen de Normalisation Electrotechnique
CNE	Combined Neutral and Earth
CORGI	Council for Registered Gas Installers
CPC	Circuit Protective Conductor
CPS	Control and Protective Switching Device
d.c.	Direct Current
DCL	Device for Connecting a Luminaire
DDA	Disability Discrimination Act
DIY	Do It Youself
DTI	Department of Trade & Industry
EBADS	Equipotential Bonding and Automatic Disconnection of Supplies
ECA	Electrical Contractors Association
EEBAD	Earthed Equipotential Bonding and Automatic Disconnection
EEC	European Economic Commission
ELECSA	Fenestration Self-Assessment Scheme
ELV	Exra Low Voltage
EMC	Electromagnetic Compatibility
EMI	Electromagnetic Interference
EN	European Normalisation
ESQCR	Electricity Safety, Quality and Continuity Regulations 2002
EC	European Community
EU	European Union
FE	Functional Earth
FELV	Functional Extra Low Voltage
GLS	As in tungsten lights

(*continued*)

HBES	Home and Building Electronic Systems
HD	Harmonised Directive
HELA	Health and Safety Executive/Local Authorities
HEMP	High Altitude Electromagnetic Pulse
HSE	Health & Safety Executive
HV	High Voltage
I/O	Input/Output
ICM	Insulation Current Monitoring device
IEC	International Electrotechnical Commission
IEE	Institution of Electrical Engineers
IET	Institution of Engineering and Technology
IIE	Institution on of Incorporated Engineers
ILU	Integrated Logistic Unit
IMD	Insulation Monitoring Device
IPC	Implant Point of Coupling
ISM	Industrial, Scientific and Medical
ISO	International Standards Organisation
IT	Information Technology
ITCZ	International Conveyance Zone
ITE	Information Technology Equipment
LUM	Luminaire Supporting Coupler
LUR	Logical User Requirement
LV	Low Voltage
MDD	Medical Devices Directive
MKS	Metre-Kilogram-Second
MMI	Man-Machine Interface
MTBF	Mean Time Between Failures
N	Neutral
NAPIT	National Association of Professional Inspectors and Testers
NICEIC	National Inspection Council for Elerctrical Installation Counselling
NSO	National Standards Organisation
OFTEC	Oil Firing Technical Association
OPSI	Office of Public Sector Information
PCB	Printed Circuit Board
PE	Protective Earth
PELV	Protective Extra Low Voltage
PEN	Combined Protective and Neutral conductors
PME	Protective Multiple Earthing
prEN	European draft standards
PV	Photovoltaic
PVC	PolyVinyl Chloride
QA	Quality Assurance
QC	Quality Control
QMS	Quality Management System
QP	Quality Procedure
rms	Root Mean Square
RAH	Relative Air Humidity
RAM	Reliability, Availability and Maintainability
RCBO	Residual Current Operated Circuit Breaker without integral overcurrent protection
RCCB	Residual Current Operated Circuit Breaker with integral overcurrent protection

(*continued*)

RCD	Residual Current Device
RCM	Residual Current Monitor
RF	Radio Frequency
RH	Relative Humidity
S/N	Signal to Noise Ratio
SELV	Safety Extra Low Voltage
SI	Statutory Instrument
SI	Système International d'Unités
SPD	Surge Protective Device
SSEG	Small-Scale Embedded Generators
T&E	Time and Expense
TDS	Time Delay Switches
TLV	Threshold Limit Values
TQM	Total Quality Management
TVL-C	Threshold Limit Values – Ceiling List
UPS	Uninterruptible Power System
VDU	Visual Display Unit
WAUILF	Workplace Applied Uniform Indicated Low Frequency (application)
WI	Work Instruction
YFR	Yearly Forecast Rationale

Appendix E

British Standards currently used with the Wiring Regulations

E.1 Listed by standard

BS or EN number	Title
BS 67:1987 (1999)	Specification for ceiling roses.
BS 88	Cartridge fuses for voltages up to and including 1000V a.c. and 1500V d.c.
BS 88-2:2007	Low-voltage fuses. Supplementary requirements for fuses for use by authorised persons (fuses mainly for industrial application). Examples of standardised systems of fuses A to I.
BS 88-2.2:1988	Specification for fuses for use by authorised persons (mainly for industrial application). Additional requirements for fuses with fuse-links for bolted connections.
BS 88-6:1988	Specification of supplementary requirements for fuses of compact dimensions for use in 240/415V a.c. industrial and commercial electrical installations.
BS 196:1961	Specification for protected-type non-reversible plugs, socket-outlets cable-couplers and appliance-couplers with earthing contacts for single phase a.c. circuits up to 250 volts.
BS 476	Fire tests on building materials and structures.
BS 476-4:1970	Non-combustible test for materials
BS 476-12:1991	Method of test for ignitability of products by direct flame impingement.
BS 546:1950 (1988)	Specification. Two-pole and earthing-pin plugs, socket-outlets and socket-outlet adaptors.
BS 559:1998 (2005)	Specification for design, construction and installation of signs.
BS 646:1958 (1991)	Specification. Cartridge fuse-links (rated up to 5 amperes) for a.c. and d.c. service.
BS 951:1999	Electrical earthing. Clamps for earthing and bonding. Specification.
BS 1361:1971 (1986)	Specification for cartridge fuses for a.c. circuits in domestic and similar premises.

(*continued*)

E.1 Listed by standard *(continued)*

BS or EN number	Title
BS 1362:1973 (1992)	Specification for general purpose fuse links for domestic, and similar purposes (primarily for use in plugs).
BS 1363	13A plugs, socket-outlets, connection units and adaptors.
BS 1363-1:1995	Specification for rewirable and non-rewirable 13A fused plugs.
BS 1363-2:1995	Specification for 13A switched and unswitched socket-outlets.
BS 1363-3:1995	Specification for adaptors.
BS 1363-4:1995	Specification for 13A fused connection units switched and unswitched.
BS 3036:1958 (1992)	Specification. Semi-enclosed electric fuses (ratings up to 100 amperes and 230 volts to earth).
BS 3676	Switches for household and similar fixed electrical installations. Specification for general requirements.
BS 3858:1992 (2004)	Specification for binding and identification sleeves for use on electric cables and wires.
BS 4177:1992	Specification for cooker control units.
BS 4444:1989 (1995)	Guide to electrical earth monitoring and protective conductor proving.
BS 4573:1970 (1979)	Specification for 2-pin reversible plugs and shaver socket-outlets.
BS 4662:2006	Boxes for flush mounting of electrical accessories. Requirements and test methods and dimensions.
BS 4727	Glossary of electrotechnical power, telecommunications, electronics, lighting and colour terms.
BS 5266	Emergency lighting.
BS 5467:1997	Electric cables. Thermosetting insulated, armoured cables for voltages of 600/1000 V and 1900/3300 V.
BS 5499	Graphical symbols and signs. Safety signs, including fire safety signs.
BS 5655	Lifts and service lifts.
BS 5655-1:1986	Safety rules for the construction and installation of electric lifts (applicable only to the modernisation of existing lift installations).
13S 5655-2:1988	Safety rules for the construction and installation of hydraulic lifts (applicable only to the modernisation of existing lift installations).
BS 5655-11:2005	Code of practice for the undertaking of modifications to existing electric lifts (applicable only to the modernisation of existing lift installations).
BS 5655-12:2005	Code of practice for the undertaking of modifications to existing hydraulic lifts (applicable only to the modernisation of existing lift installations).
BS 5733:1995	Specification for general requirements for electrical accessories.
BS 5803-5:1985	Thermal insulation for use in pitched roof spaces in dwellings. Specification for installation of man-made mineral fibre and cellulose fibre insulation.
BS 5839	Fire detection and fire alarm systems for buildings.

(continued)

E.1 Listed by standard (*continued*)

BS or EN number	Title
BS 5839-1:2002	Code of practice for system design, installation, commissioning and maintenance.
BS 6004:2000 (2006)	Electric cables. PVC insulated, non-armoured cables for voltages up to and including 450/750V, for electric power, lighting and internal wiring.
BS 6007:2006	Electric cables. Single core unsheathed heat resisting cables for voltages up to and including 450/750V, for internal wiring.
BS 6220:1983 (1999)	Electric cables. Single core PVC insulated flexible cables of rated voltage 600/1000V for switchgear and controlgear wiring.
BS 6231:2006	Electric cables. Single core PVC insulated flexible cables of rated voltage 600/1000V for switchgear and controlgear wiring.
BS 6346:1997 (2005)	Electric cables. PVC insulated, armoured cables for voltages of 600/1000V and 1900/3300V.
BS 6351	Electric surface heating.
BS 6351-1:1983 (2007)	Specification for electric surface heating devices.
BS 6351-2:1983 (2007)	Guide to the design of electric surface heating systems.
BS 6351-3:1983 (2007)	Code of practice for the installation, testing and maintenance of electric surface heating systems.
BS 6500:2000 (2005)	Electric cables. Flexible cords rated up to 300/500V, for use with appliances and equipment intended for domestic, office and similar environments.
BS 6701:2004	Telecommunications equipment and telecommunications cabling. Specification for installation, operation and maintenance.
BS 6724:1997 (2007)	Electric cables. Thermosetting insulated, armoured cables for voltages of 600/1000V and 1900/3300V, having low emission of smoke and corrosive gases when affected by fire.
BS 6907	Electrical installations for open-cast mines and quarries.
BS 6972:1988	Specification for general requirements for luminaire supporting couplers for domestic, light industrial and commercial use.
BS 6991:1990	Specification for 6/10A, two-pole weather-resistant couplers for household, commercial and light industrial equipment.
BS 7001:1988	Specification for interchangeability and safety of a standardised luminaire supporting coupler.
BS 7211:1998 (2005)	Electric cables. Thermosetting insulated, non-armoured cables for voltages up to and including 450/750V, for electric power, lighting and internal wiring, and having low emission of smoke and corrosive gases when affected by fire.
BS 7361:1991	Cathodic protection. Code of practice for land and marine applications. (Current but partially replaced by BS EN 15112:2006 and BS EN 13636:2004).
BS 7375:1996	Code of practice for distribution of electricity on construction and building sites.
BS 7430:1998	Code of practice for earthing.

(*continued*)

E.1 Listed by standard (*continued*)

BS or EN number	Title
BS 7454:1991 (2003)	Method for calculation of thermally permissible short-circuit currents, taking into account non-adiabatic heating effects.
BS 7629-1:1997 (2007)	Specification for 300/500V fire resistant electric cables having low emission of smoke and corrosive gases when affected by fire. Multicore cables.
BS 7697:1993 (2004)	Nominal voltages for low voltage public electricity supply systems.
BS 7698-12:1998	Reciprocating internal combustion engine driven alternating current generating sets. Emergency power supply to safety devices.
BS 7769	Electric cables. Calculation of the current rating.
BS 7769-1.1:1997	Has been superseded/withdrawn and replaced by BS IEC 60287-1-1:2006.
BS 7769-1.2:1994 (2005)	Current rating equations (100% load factor) and calculation of losses. Sheath eddy current loss factors for two circuits in flat formation.
BS 7769-2.2:1997 (2005)	Thermal resistance. A method for calculating reduction factors for groups of cables in free air, protected from solar radiation.
BS 7769-2-2.1:1997 (2006)	Thermal resistance. Section 2.1: Calculation of thermal resistance.
BS 7769-3.1:1997 (2005)	Sections on operating conditions. Reference operating conditions and selection of cable type.
BS 7846:2000 (2005)	Electric cables. 600/1000V armoured fire-resistant cables having thermosetting insulation and low emission of smoke and corrosive gases when affected by fire.
BS 7889:1997	Electric cables. Thermosetting insulated, unarmoured cables for a voltage of 600/1000V.
BS 7909	Code of practice for design and installation of temporary distribution systems delivering a.c. electrical supplies for lighting, technical services and other entertainment related purposes.
BS 7919:2001 (2006)	Electric cables. Flexible cables rated up to 450/750V, for use with appliances and equipment intended for industrial and similar environments.
BS 8436:2004	Electric cables. 300/500V screened electric cables having low emission of smoke and corrosive gases when affected by fire, for use in walls, partitions and building voids. Multicore cables.
BS 8450:2006	Code of practice for installation of electrical and electronic equipment in ships.
BS 61535:2006	Installation couplers intended for permanent connection in fixed installations.
BS AU 149a:1980 (1987)	Specification for electrical connections between towing vehicles and trailers with 6V or 12V electrical equipment: type 12N (normal).
BS AU 177a:1980 (1987)	Specification for electrical connections between towing vehicles and trailers with 6V or 12V electrical equipment: type 12 S (supplementary).
BS EN 81	Safety rules for the construction and installation of lifts.

(*continued*)

E.1 Listed by standard (*continued*)

BS or EN number	Title
BS EN 81-1:1998	Electric lifts (also known as BS 5655-1:1986 Lifts and service lifts ... etc.)
BS EN 1648	Leisure accommodation vehicles.
BS EN 1648-1:2004	12 V direct current extra low voltage electrical installations. Caravans.
BS EN 1648-2:2005	12 V direct current extra low voltage electrical installations. Motor caravans.
BS EN 6100-1	Glossary of building and civil engineering terms.
BS EN 50085	Cable trunking and cable ducting systems for electrical installations.
BS EN 50085-1:1999 (2005)	General requirements. BS EN 50085-1:1999 remains current.
BS EN 50085-2-1:2006	Cable trunking systems and cable ducting systems intended for mounting on walls and ceilings.
BS EN 50085-2-3:2001	Particular requirements for slotted cable trunking systems intended for installation in cabinets. Section 3: Slotted in cabinets.
BS EN 50086	Specification for conduit systems for cable management.
BS EN 50086-1:1994	General requirements. Replaced by BS EN 61386-1:2004 but remains current.
BS EN 50086-2-1:1996	Particular requirements. Rigid conduit systems. Replaced by BS EN 61386-21:2004 but remains current.
BS EN 50086-2-2:1996	Particular requirements. Pliable conduit systems. Replaced by BS EN 61386-22:2004 but remains current.
BS EN 50086-2-3:1996	Particular requirements. Flexible conduit systems. Replaced by BS EN 61386-23:2004 but remains current.
BS EN 50086-2-4:1994	Particular requirements. Conduit systems buried underground.
BS EN 50107	Signs and luminous-discharge-tube installations operating from a no-load rated output voltage exceeding 1 kV but not exceeding 10 kV.
BS EN 50107-1:2002	General requirements.
BS EN 50107-2:2005	Requirements for earth-leakage and open-circuit protective devices.
BS EN 50171:2001	Central power supply systems.
BS EN 50174	Information technology – cabling installation.
BS EN 50200:2006	Method of test for resistance to fire of unprotected small cables for use in emergency circuits.
BS EN 50266	Common test methods for cables under fire conditions. Test for vertical flame spread of vertically mounted bunched wires or cables.
BS EN 50266-1:2001 (2006)	Apparatus.
BS EN 50266-2-1:2001 (2006)	Procedures. Category A F/R.
BS EN 50266-2-2:2001 (2006)	Procedures. Category A.
BS EN 50266-2-3:2001 (2006)	Procedures. Category B.
BS EN 50266-2-4:2001 (2006)	Procedures. Category C.
BS EN 50266-2-5:2001 (2006)	Procedures. Small cables. Category D.
BS EN 50281	Electrical apparatus for use in the presence of combustible dust.

(*continued*)

E.1 Listed by standard (*continued*)

BS or EN number	Title
BS EN 50281-1-1:1999	Electrical apparatus protected by enclosures. Construction and testing. Replaced by BS EN 60241-0:2006 and BS EN 61241-1:2004 but remains current.
BS EN 50281-1-2:1999	Electrical apparatus protected by enclosures. Selection, installation and maintenance. Partially replaced by BS EN 61241-14:2004 and BS EN 61241-17: 2005.
BS EN 50281-2-1:1999	Test methods. Methods of determining minimum ignition temperatures.
BS EN 50362:2003	Method of test for resistance to fire of larger unprotected power and control cables for use in emergency circuits.
EN 50438	Requirements for the connection of micro-cogenerators in parallel with public low-voltage distribution networks.
BS EN 60079	Electrical apparatus for explosive gas atmospheres.
BS EN 60079-10:2003	Classification of hazardous areas.
BS EN 60079-14:2003	Electrical installations in hazardous areas (other than mines).
BS EN 60079-17:2003	Inspection and maintenance of electrical installations in hazardous areas (other than mines).
BS EN 60092-507:2000	Electrical installations in ships – Pleasure craft.
BS EN 60146-2:2000	Semiconductor converters. General requirements and line commutated converters. Self-commutated semiconductor converters including direct d.c. converters.
BS EN 60204	Safety of machinery. Electrical equipment of machines.
BS EN 60204-1:2006	General requirements.
BS EN 60228:2005	Conductors of insulated cables.
BS EN 60238:1999 (2004)	Edison screw lampholders. BS EN 60238:1999 remains current.
BS EN 60255-22-1:2005	Electrical relays. Electrical disturbance tests for measuring relays and protection equipment. 1 MHz burst immunity tests.
BS EN 60269	Low-voltage fuses.
BS EN 60269-1:2007	General requirements.
BS EN 60269-2:1995	Supplementary requirements for fuses for use by authorised persons (fuses mainly for industrial application). Replaced by BS 88-2:2007 and BS EN 60269-1:2007 but remains current.
BS EN 60269-3:1995	Supplementary requirements for fuses for use by unskilled persons (fuses mainly for household and similar applications). Replaced by BS 88-3:2007 and BS EN 60269-1:2007 but remains current.
BS EN 60309	Plugs, socket-outlets and couplers for industrial purposes.
BS EN 60309-1:1999	General requirements.
BS EN 60309-2:1999	Dimensional interchangeability requirements for pin and contact-tube accessories.
BS EN 60320-1:2001	Appliance couplers for household and similar general purposes. General requirements.

(*continued*)

E.1 Listed by standard (*continued*)

BS or EN number	Title
BS EN 60332-1-2:2004	Tests on electric and optical fibre cables under fire conditions. Test for vertical flame propagation for a single insulated wire or cable. Procedure for 1 kW pre-mixed flame.
BS EN 60335-1:2002	Household and similar electrical appliances. Safety. General requirements.
BS EN 60335-2-29:2004	Particular requirements for battery chargers.
BS EN 60335-2-41:2003	Particular requirements for pumps.
BS EN 60335-2-53:2003	Particular requirements for sauna heating appliances.
BS EN 60335-2-71:2003	Particular requirements for electrical heating appliances for breeding and rearing animals.
BS EN 60335-2-76:2005	Particular requirements for electric fence energisers.
BS EN 60335-2-96:2002	Particular requirements for flexible sheet heating elements for room heating.
BS EN 60439	Low-voltage switchgear and controlgear assemblies.
BS EN 60439-1:1999	Type-tested and partially type-tested assemblies.
BS EN 60439-2:2000	Particular requirements for busbar trunking systems (busways).
BS EN 60439-3:1991	Particular requirements for low-voltage switchgear and controlgear assemblies intended to be installed in places where unskilled persons have access to their use. Distribution boards.
BS EN 60439-4:2004	Particular requirements for assemblies for construction sites (ACS).
BS EN 60445:2000	Basic and safety principles for man-machine interface, marking and identification. Identification of equipment terminals and of terminations of certain designated conductors, including general rules for an alphanumeric system.
BS EN 60446:2000	Basic and safety principles for man-machine interface, marking and identification. Identification of conductors by colours or numerals.
BS EN 60529:1992 (2004)	Specification for degrees of protection provided by enclosures (IP code).
BS EN 60570:2003	Electrical supply track systems for luminaires. Replaces BS EN 60570:1997 and BS EN 60570-2-1:1995 which remain current.
BS EN 60598	Luminaires.
BS EN 60598-1:2004	Luminaires. General requirements and tests.
BS EN 60598-2-18:1994	Particular requirements. Luminaires for swimming pools and similar applications.
BS EN 60598-2-23:1997	Particular requirements. Extra-low voltage lighting systems for filament lamps.
BS EN 60598-2-24:1999	Particular requirements. Luminaires with limited surface temperatures.
BS EN 60664-1:2003	Insulation coordination for equipment within low-voltage systems. Principles, requirements and tests.
BS EN 60669	Switches for household and similar fixed electrical installations.
BS EN 60669-1:2000	General requirements.
BS EN 60669-2-1:2004	Particular requirements. Electronic switches.

(*continued*)

E.1 Listed by standard (*continued*)

BS or EN number	Title
BS EN 60669-2-2:2006	Particular requirements. Electromagnetic remote-control switches (RCS).
BS EN 60669-2-3:2006	Particular requirements. Time delay switches (TDS).
BS EN 60669-2-4:2005	Particular requirements. Isolating switches.
BS EN 60670	Boxes and enclosures for electrical accessories for household and similar fixed electrical installations.
BS EN 60670-1:2005	General requirements.
BS EN 60670-22:2006	Particular requirements for connecting boxes and enclosures.
BS EN 60684	Flexible insulating sleeving.
BS EN 60702-1:2002	Mineral insulated cables and their terminations with a rated voltage not exceeding 750 V. Cables.
BS EN 60721	Classification of environmental conditions.
BS EN 60721-3-3:1995 (2005)	Classification of groups of environmental parameters and their severities. Stationary use at weather protected locations.
BS EN 60721-3-4:1995 (2005)	Classification of groups of environmental parameters and their severities. Stationary use at non-weather protected locations.
BS EN 60898:1991	Specification for circuit-breakers for overcurrent protection for household and similar installations. Replaced by BS EN 60898-1:2003 but remains current.
BS EN 60898-1:2003	Circuit breakers for a.c. operation.
BS EN 60898-2:2001	Circuit-breakers for a.c. and d.c. operation. BS EN 60898-2:2001 remains current. (It was withdrawn in error and has been reinstated.)
BS EN 60904-3:1993	Photovoltaic devices. Measurement principles for terrestrial photovoltaic (PV) solar devices with reference spectral irradiance data.
BS EN 60947	Low-voltage switchgear and controlgear.
BS EN 60947-2:2006	Circuit-breakers.
BS EN 60947-3:1999	Switches, disconnectors, switch-disconnectors and fuse-combination units.
BS EN 60947-4-1:2001	Contactors and motor starters – Electromechanical contactor and motor starters.
BS EN 60947-5-1:2004	Control circuit devices and switching elements – Electromechanical control circuit devices.
BS EN 60947-6-1:2005	Multiple function equipment – Transfer switching equipment.
BS EN 60947-6-2:2003	Multiple function equipment – Control and protective switching devices (or equipment) (CPS).
BS EN 60947-7	Specification for low-voltage switchgear and controlgear.
BS EN 60947-7-1:2002	Ancillary equipment – Terminal blocks for copper conductors.
BS EN 60947-7-2:2002	Ancillary equipment – Protective conductor terminal blocks for copper conductors.
BS EN 60998	Connecting devices for low-voltage circuits for household and similar purposes.
BS EN 60998-2-1:2004	Particular requirements for connecting devices as separate entities with screw-type clamping units.
BS EN 60998-2-2:2004	Particular requirements for connecting devices as separate entities with screwless-type clamping units.

(*continued*)

E.1 Listed by standard (*continued*)

BS or EN number	Title
BS EN 61000	Electromagnetic compatibility (EMC).
BS EN 61008-1:1995 (2004)	Residual current operated circuit-breakers without integral overcurrent protection for household and similar uses (RCCBs). General rules. BS EN 61008-1:1995 remains current.
BS EN 61009-1:1995 (2004)	Electrical accessories. Residual current operated circuit-breakers with integral overcurrent protection for household and similar uses (RCBOs). General rules. BS EN 61009-1:1995 remains current.
BS EN 61034-2:2005	Measurement of smoke density of cables burning under defined conditions. Test procedure and requirements.
BS EN 61095:1993	Specification for electromechanical contactors for household and similar purposes.
BS EN 61140:2002	Protection against electric shock. Common aspects for installation and equipment.
BS EN 61184:1997	Bayonet lampholders.
BS EN 61215:2005	Crystalline silicon terrestrial photovoltaic (PV) modules. Design qualification and type approval.
BS EN 61241	Electrical apparatus for use in the presence of combustible dust.
BS EN 61241-17:2005	Inspection and maintenance of electrical installations in hazardous areas (other than mines).
BS EN 61347	Lamp controlgear.
BS EN 61347-1:2001	General and safety requirements.
BS EN 61347-2-2:2001	Particular requirements for d.c. or a.c. supplied electronic step-down converters for filament lamps.
BS EN 61386	Conduit systems for cable management.
BS EN 61386-1:2004	General requirements.
BS EN 61386-21:2004	Particular requirements. Rigid conduit systems.
BS EN 61386-22:2004	Particular requirements. Pliable conduit systems.
BS EN 613860-23:2004	Particular requirements. Flexible conduit systems.
BS EN 61534	Powertrack systems.
BS EN 61534-1:2003	General requirements.
BS EN 61534-21:2006	Particular requirements for powertrack systems intended for wall and ceiling mounting.
BS EN 61537:2002 (2007)	Cable tray systems and cable ladder systems for cable management. BS EN 61537:2002 remains current.
BS EN 61557	Electrical safety in low voltage distribution systems up to 1000V a.c. and 1500V d.c. Equipment for testing, measuring or monitoring of protective measures. General requirements.
BS EN 61557-2:2007	Insulation resistance.
BS EN 61557-6:1998	Residual current devices (RCD) in TT, TN and IT systems.
BS EN 61557-8:1997	Insulation monitoring devices for IT systems.
BS EN 61557-9:2000	Equipment for insulation fault location in IT systems.
BS EN 61558-1:1998 (2005)	Safety of power transformers, power supply units and similar. General requirements and tests. BS EN 61558-1:1998 remains current.
BS EN 61558-2-4:1998	Particular requirements for isolating transformers for general use.
BS EN 61558-2-5:1998	Particular requirements for shaver transformers and shaver supply units.

(*continued*)

E.1 Listed by standard (*continued*)

BS or EN number	Title
BS EN 61558-2-6:1998	Particular requirements for safety isolating transformers for general use.
BS EN 61558-2-23:2001	Particular requirements for transformers for construction sites.
BS EN 62020:1999	Electrical accessories. Residual current monitors for household and similar uses (RCMs).
BS EN 62040	Uninterruptible power systems (UPS).
BS EN 62208:2003	Empty enclosures for low-voltage switchgear and controlgear assemblies. General requirements.
BS EN 62262:2002	Degrees of protection provided by enclosures for electrical equipment against external mechanical impacts (IK code).
BS EN 62305	Protection against lightning.
BS EN 62305-1:2006	General requirements.
BS EN 62305-2:2006	Risk management.
BS EN 62305-3:2006	Physical damage to structures and life hazard.
BS EN 62305-4:2006	Electrical and electronic systems within structures.
BS EN ISO 11446:2004	Road vehicles. Connectors for the electrical connection of towing and towed vehicles. 13-pole connectors for vehicles with 12V nominal supply voltage.

E.2 Listed by subject

Title	BS or EN number
2-pin reversible plugs and shaver socket-outlets.	BS 4573:1970 (1979)
13A fused plugs (specification for rewirable and non-rewirable).	BS 1363-1:1995
13A plugs, socket-outlets, connection units and adaptors.	BS 1363
13A switched and unswitched socket-outlets.	BS 1363-2:1995
13A fused connection units switched and unswitched.	BS 1363-4:1995
Adaptors.	BS 1363-3:1995
Ancillary equipment – Protective conductor terminal blocks for copper conductors.	BS EN 60947-7-2:2002
Ancillary equipment – Terminal blocks for copper conductors.	BS EN 60947-7-1:2002
Appliance couplers for household and similar general purposes. General requirements	BS EN 60320-1:2001
Assemblies for construction sites (ACS).	BS EN 60439-4:2004
Basic and safety principles for man-machine interface, marking and identification. Identification of equipment terminals and of terminations of certain designated conductors, including general rules for an alphanumeric system.	BS EN 60445:2000
Basic and safety principles for man-machine interface, marking and identification. Identification of conductors by colours or numerals.	BS EN 60446:2000

(*continued*)

E.2 Listed by subject (continued)

Title	BS or EN number
Battery chargers.	BS EN 60335-2-29:2004
Bayonet lampholders.	BS EN 61184:1997
Binding and identification sleeves for use on electric cables and wires.	BS 3858:1992 (2004)
Boxes and enclosures for electrical accessories for household and similar fixed electrical installations.	BS EN 60670
Boxes for flush mounting of electrical accessories. Requirements and test methods and dimensions.	BS 4662:2006
Busbar trunking systems (busways).	BS EN 60439-2:2000
Cable tray systems and cable ladder systems for cable management.	BS EN 61537:2002 (2007)
Cable trunking and cable ducting systems for electrical installations.	BS EN 50085
Cable trunking systems and cable ducting systems intended for mounting on walls and ceilings.	BS EN 50085-2-1:2006
Caravans – 12V direct current extra low voltage electrical installations.	BS EN 1648-1:2004
Cathodic protection. Code of practice for land and marine applications. (Current but partially replaced by BS EN 15112:2006 and BS EN 13636:2004.)	BS 7361:1991
Ceiling roses.	BS 67:1987 (1999)
Central power supply systems.	BS EN 50171:2001
Circuit-breakers for a.c. operation.	BS EN 60898-1:2003
Circuit-breakers.	BS EN 60947-2:2006
Circuit-breakers for a.c. and d.c. operation. BS EN 60898-2:2001 remains current. (It was withdrawn in error and has been reinstated.)	BS EN 60898-2:2001
Circuit-breakers for overcurrent protection for household and similar installations. Replaced by BS EN 60898-1:2003 but remains current.	BS EN 60898:1991
Classification of environmental conditions.	BS EN 60721
Classification of groups of environmental parameters and their severities. Stationary use at weather protected locations.	BS EN 60721-3-3:1995 (2005)
Classification of groups of environmental parameters and their severities. Stationary use at non-weather protected locations.	BS EN 60721-3-4:1995 (2005)
Classification of hazardous areas.	BS EN 60079-10:2003
Conductors of insulated cables.	BS EN 60228:2005
Conduit systems buried underground.	BS EN 50086-2-4:1994
Conduit systems for cable management.	BS EN 61386
Conduit systems for cable management.	BS EN 50086
Connecting boxes and enclosures.	BS EN 60670-22:2006
Connecting devices as separate entities with screwless-type clamping units.	BS EN 60998-2-2:2004
Connecting devices as separate entities with screw-type clamping units.	BS EN 60998-2-1:2004
Connecting devices for low-voltage circuits for household and similar purposes.	BS EN 60998
Contactors and motor starters – Electromechanical contactor and motor starters.	BS EN 60947-4-1:2001
Control and protective switching devices (or equipment) (CPS).	BS EN 60947-6-2:2003

(continued)

E.2 Listed by subject (*continued*)

Title	BS or EN number
Control circuit devices and switching elements – Electromechanical control circuit devices.	BS EN 60947-5-1:2004
Cooker control units.	BS 4177:1992
Crystalline silicon terrestrial photovoltaic (PV) modules. Design qualification and type approval.	BS EN 61215:2005
Current rating equations (100% load factor) and calculation of losses. Sheath eddy current loss factors for two circuits in flat formation.	BS 7769-1.2:1994 (2005)
Degrees of protection provided by enclosures for electrical equipment against external mechanical impacts (IK code).	BS EN 62262:2002
Design and installation of temporary distribution systems delivering a.c. electrical supplies for lighting, technical services and other entertainment related purposes.	BS 7909
Design, construction and installation of signs.	BS 559:1998 (2005)
Dimensional interchangeability requirements for pin and contact-tube accessories.	BS EN 60309-2:1999
Distribution of electricity on construction and building sites.	BS 7375:1996
Earthing.	BS 7430:1998
Earthing-pin plugs, socket-outlets and socket-outlet adaptors.	BS 546:1950 (1988)
Edison screw lampholders.	BS EN 60238:1999 (2004)
Electric cables. 300/500V screened electric cables having low emission of smoke and corrosive gases when affected by fire, for use in walls, partitions and building voids. Multicore cables.	BS 8436:2004
Electric cables. 600/1000V armoured fire-resistant cables having thermosetting insulation and low emission of smoke and corrosive gases when affected by fire.	BS 7846:2000 (2005)
Electric cables. Calculation of the current rating.	BS 7769
Electric cables. Flexible cables rated up to 450/750V, for use with appliances and equipment intended for industrial and similar environments.	BS 7919:2001 (2006)
Electric cables. Flexible cords rated up to 300/500V, for use with appliances and equipment intended for domestic, office and similar environments.	BS 6500:2000 (2005)
Electric cables. PVC insulated, armoured cables for voltages of 600/1000V and 1900/3300V.	BS 6346:1997 (2005)
Electric cables. PVC insulated, non-armoured cables for voltages up to and including 450/750V, for electric power, lighting and internal wiring.	BS 6004:2000 (2006)
Electric cables. Single core PVC insulated flexible cables of rated voltage 600/1000V for switchgear and controlgear wiring.	BS 6220:1983 (1999)
Electric cables. Single core PVC insulated flexible cables of rated voltage 600/1000V for switchgear and controlgear wiring.	BS 6231:2006
Electric cables. Single core unsheathed heat resisting cables for voltages up to and including 450/750V, for internal wiring.	BS 6007:2006

(*continued*)

E.2 Listed by subject (*continued*)

Title	BS or EN number
Electric cables. Thermosetting insulated, armoured cables for voltages of 600/1000V and 1900/3300V.	BS 5467:1997
Electric cables. Thermosetting insulated, armoured cables for voltages of 600/1000V and 1900/3300V, having low emission of smoke and corrosive gases when affected by fire.	BS 6724:1997 (2007)
Electric cables. Thermosetting insulated, non-armoured cables for voltages up to and including 450/750V, for electric power, lighting and internal wiring, and having low emission of smoke and corrosive gases when affected by fire.	BS 7211:1998 (2005)
Electric cables. Thermosetting insulated, unarmoured cables for a voltage of 600/1000V.	BS 7889:1997
Electric fence energisers.	BS EN 60335-2-76:2005
Electric lifts (also known as BS 5655-1:1986 Lifts and service lifts etc.).	BS EN 81-1:1998
Electric surface heating devices.	BS 6351-1:1983 (2007)
Electric surface heating systems.	BS 6351-2:1983 (2007)
Electric surface heating.	BS 6351
Electrical accessories.	BS 5733:1995
Electrical accessories. Residual current monitors for household and similar uses (RCMs).	BS EN 62020:1999
Electrical accessories. Residual current operated circuit-breakers with integral overcurrent protection for household and similar uses (RCBOs). General rules. BS EN 610091:1995 remains current.	BS EN 61009-1:1995 (2004)
Electrical and electronic systems within structures.	BS EN 62305-4:2006
Electrical apparatus for explosive gas atmospheres.	BS EN 60079
Electrical apparatus for use in the presence of combustible dust.	BS EN 50281
Electrical apparatus for use in the presence of combustible dust.	BS EN 61241
Electrical apparatus protected by enclosures. Construction and testing. Replaced by BS EN 60241-0:2006 and BS EN 61241-1:2004 but remains current.	BS EN 50281-1-1:1999
Electrical apparatus protected by enclosures. Selection, installation and maintenance. Partially replaced by BS EN 61241-14:2004 and BS EN 61241-17: 2005.	BS EN 50281-1-2:1999
Electrical connections between towing vehicles and trailers with 6V or 12V electrical equipment: type 12N (normal).	BS AU 149a:1980 (1987)
Electrical connections between towing vehicles and trailers with 6V or 12V electrical equipment: type 12 S (supplementary).	BS AU 177a:1980 (1987)
Electrical earth monitoring and protective conductor proving.	BS 4444:1989 (1995)
Electrical earthing. Clamps for earthing and bonding. Specification.	BS 951:1999
Electrical heating appliances for breeding and rearing animals.	BS EN 60335-2-71:2003

(*continued*)

E.2 Listed by subject (*continued*)

Title	BS or EN number
Electrical installations for open-cast mines and quarries.	BS 6907
Electrical installations in hazardous areas (other than mines).	BS EN 60079-14:2003
Electrical installations in ships – Pleasure craft.	BS EN 60092-507:2000
Electrical relays. Electrical disturbance tests for measuring relays and protection equipment. 1 MHz burst immunity tests.	BS EN 60255-22-1:2005
Electrical safety in low voltage distribution systems up to 1000V a.c. and 1500V d.c. Equipment for testing, measuring or monitoring of protective measures. General requirements.	BS EN 61557
Electrical supply track systems for luminaires Replaces BS EN 60570:1997 and BS EN 60570-2-1:1995 which remain current.	BS EN 60570:2003
Electromagnetic compatibility (EMC).	BS EN 61000
Electromagnetic remote-control switches (RCS).	BS EN 60669-2-2:2006
Electromechanical contactors for household and similar purposes.	BS EN 61095:1993
Electronic switches.	BS EN 60669-2-1:2004
Emergency lighting.	BS 5266
Empty enclosures for low-voltage switchgear and controlgear assemblies. General requirements.	BS EN 62208:2003
Enclosures (IP code).	BS EN 60529:1992 (2004)
Equipment for insulation fault location in IT systems.	BS EN 61557-9:2000
Extra-low voltage lighting systems for filament lamps.	BS EN 60598-2-23:1997
Filament lamps (d.c. or a.c. supplied electronic step-down converters).	BS EN 61347-2-2:2001
Fire detection and fire alarm systems for buildings.	BS 5839
Fire resistant electric cables (300/500V) having low emission of smoke and corrosive gases when affected by fire. Multicore cables.	BS 7629-1:1997 (2007)
Fire tests on building materials and structures.	BS 476
Flexible conduit systems. Replaced by BS EN 61386-23:2004 but remains current.	BS EN 50086-2-3:1996
Flexible conduit systems.	BS EN 61386-23:2004
Flexible insulating sleeving.	BS EN 60684
Fuse links for domestic, and similar purposes (primarily for use in plugs).	BS 1362:1973 (1992)
Fuse-links (rated up to 5 amperes) for a.c. and d.c. service.	BS 646:1958 (1991)
Fuses – semi-enclosed electric fuses (ratings up to 100 amperes and 240 volts to earth).	BS 3036:1958 (1992)
Fuses for a.c. circuits in domestic and similar premises.	BS 1361:1971 (1986)
Fuses for use by authorized persons (fuses mainly for industrial application). Replaced by BS 88-2:2007 and BS EN 60269-1:2007 but remains current.	BS EN 60269-2:1995
Fuses for use by authorised persons (mainly for industrial application). Additional requirements for fuses with fuse-links for bolted connections.	BS 88-2.2:1988

(*continued*)

E.2 Listed by subject (*continued*)

Title	BS or EN number
Fuses for use by unskilled persons (fuses mainly for household and similar applications). Replaced by BS 88-3:2007 and BS EN 60269-1:2007 but remains current.	BS EN 60269-3:1995
Fuses for voltages up to and including 1000V a.c. and 1500V d.c.	BS 88
General and safety requirements.	BS EN 61347-1:2001
Glossary of building and civil engineering terms.	BS EN 6100-1
Glossary of electrotechnical power, telecommunications, electronics, lighting and colour terms.	BS 4727
Graphical symbols and signs. Safety signs, including fire safety signs.	BS 5499
Household and similar electrical appliances. Safety. General requirements.	BS EN 60335-1:2002
Information technology – Cabling installation.	BS EN 50174
Inspection and maintenance of electrical installations in hazardous areas (other than mines).	BS EN 60079-17:2003
Inspection and maintenance of electrical installations in hazardous areas (other than mines).	BS EN 61241-17:2005
Installation couplers intended for permanent connection in fixed installations.	BS 61535:2006
Installation of electrical and electronic equipment in ships.	BS 8450:2006
Installation, testing and maintenance of electric surface heating systems.	BS 6351-3:1983 (2007)
Insulation coordination for equipment within low-voltage systems. Principles, requirements and tests.	BS EN 60664-1:2003
Insulation monitoring devices for IT systems.	BS EN 61557-8:1997
Insulation resistance.	BS EN 61557-2:2007
Isolating switches.	BS EN 60669-2-4:2005
Isolating transformers for general use.	BS EN 61558-2-4:1998
Lamp controlgear.	BS EN 61347
Leisure accommodation vehicles.	BS EN 1648
Lifts and service lifts.	BS 5655
Low-voltage public electricity supply systems.	BS 7697:1993 (2004)
Low-voltage fuses.	BS EN 60269
Low-voltage fuses. Supplementary requirements for fuses for use by authorised persons (fuses mainly for industrial application). Examples of standardised systems of fuses A to I.	BS 88-2:2007
Low-voltage switchgear and controlgear.	BS EN 60947
Low-voltage switchgear and controlgear.	BS EN 60947-7
Low-voltage switchgear and controlgear assemblies intended to be installed in places where unskilled persons have access to their use. Distribution boards.	BS EN 60439-3:1991
Low-voltage switchgear and controlgear assemblies.	BS EN 60439
Luminaire supporting coupler (specification for interchangeability and safety of a standardised luminaire supporting coupler).	BS 7001:1988
Luminaire supporting couplers for domestic, light industrial and commercial use.	BS 6972:1988
Luminaires.	BS EN 60598

(*continued*)

E.2 Listed by subject (*continued*)

Title	BS or EN number
Luminaires for swimming pools and similar applications.	BS EN 60598-2-18:1994
Luminaires with limited surface temperatures.	BS EN 60598-2-24:1999
Luminaires. General requirements and tests.	BS EN 60598-1:2004
Measurement of smoke density of cables burning under defined conditions. Test procedure and requirements.	BS EN 61034-2:2005
Method for calculation of thermally permissible short-circuit currents, taking into account non-adiabatic heating effects.	BS 7454:1991 (2003)
Method of test for ignitability of products by direct flame impingement.	BS 476-12:1991
Method of test for resistance to fire of larger unprotected power and control cables for use in emergency circuits.	BS EN 50362:2003
Method of test for resistance to fire of unprotected small cables for use in emergency circuits.	BS EN 50200:2006
Mineral insulated cables and their terminations with a rated voltage not exceeding 750 V. Cables.	BS EN 60702-1:2002
Modifications to existing electric lifts. (Applicable only to the modernisation of existing lift installations.)	BS 5655-11:2005
Modifications to existing hydraulic lifts. (Applicable only to the modernisation of existing lift installations.)	BS 5655-12:2005
Motor caravans – 12 V direct current extra low voltage electrical installations.	BS EN 1648-2:2005
Non-combustible test for materials.	BS 476-4:1970
Operating conditions and selection of cable type.	BS 7769-3.1:1997 (2005)
Photovoltaic devices. Measurement principles for terrestrial photovoltaic (PV) solar devices with reference spectral irradiance data.	BS EN 60904-3:1993
Physical damage to structures and life hazard.	BS EN 62305-3:2006
Pliable conduit systems.	BS EN 61386-22:2004
Pliable conduit systems. Replaced by BS EN 61386-22:2004 but remains current.	BS EN 50086-2-2:1996
Plugs, socket-outlets and couplers for industrial purposes.	BS EN 60309
Powertrack systems.	BS EN 61534
Powertrack systems intended for wall and ceiling mounting.	BS EN 61534-21:2006
Procedures. Category A.	BS EN 50266-2-2:2001 (2006)
Procedures. Category A F/R.	BS EN 50266-2-1:2001 (2006)
Procedures. Category B.	BS EN 50266-2-3:2001 (2006)
Procedures. Category C.	BS EN 50266-2-4:2001 (2006)
Procedures. Small cables. Category D.	BS EN 50266-2-5:2001 (2006)
Protected-type non-reversible plugs, socket-outlets cable-couplers and appliance-couplers with earthing contacts for single phase a.c. circuits up to 250 volts.	BS 196:1961
Protection against electric shock. Common aspects for installation and equipment.	BS EN 61140:2002
Protection against lightning.	BS EN 62305
Pumps.	BS EN 60335-2-41:2003

(*continued*)

E.2 Listed by subject (*continued*)

Title	BS or EN number
Reciprocating internal combustion engine driven alternating current generating sets. Emergency power supply to safety devices.	BS 7698-12:1998
Requirements for earth-leakage and open-circuit protective devices.	BS EN 50107-2:2005
Requirements for the connection of micro-cogenerators in parallel with public low-voltage distribution networks. This document currently at DPC stage (expired 2004/11/30).	EN 50438
Residual current devices (RCD) in TT, TN and IT systems.	BS EN 61557-6:1998
Residual current operated circuit-breakers without integral overcurrent protection for household and similar uses (RCCBs). General rules. BS EN 61008-1:1995 remains current.	BS EN 61008-1:1995 (2004)
Rigid conduit systems. Replaced by BS EN 61386-21:2004 but remains current.	BS EN 50086-2-1:1996
Rigid conduit systems.	BS EN 61386-21:2004
Risk management.	BS EN 62305-2:2006
Road vehicles. Connectors for the electrical connection of towing and towed vehicles. 13-pole connectors for vehicles with 12V nominal supply voltage.	BS EN ISO 11446:2004
Rom heating (particular requirements for flexible sheet heating elements).	BS EN 60335-2-96:2002
Safety isolating transformers for general use.	BS EN 61558-2-6:1998
Safety of machinery. Electrical equipment of machines.	BS EN 60204
Safety of power transformers, power supply units and similar. General requirements and tests. BS EN 61558-1:1998 remains current.	BS EN 61558-1:1998 (2005)
Safety rules for the construction and installation of electric lifts. (Applicable only to the modernisation of existing lift installations.)	BS 5655-1:1986
Safety rules for the construction and installation of hydraulic lifts. (Applicable only to the modernisation of existing lift installations.)	13S 5655-2:1988
Safety rules for the construction and installation of lifts.	BS EN 81
Sauna heating appliances.	BS EN 60335-2-53:2003
Semiconductor converters. General requirements and line commutated converters. Self-commutated semiconductor converters including direct d.c. converters.	BS EN 60146-2:2000
Shaver transformers and shaver supply units.	BS EN 61558-2-5:1998
Signs and luminous-discharge-tube installations operating from a no-load rated output voltage exceeding 1 kV but not exceeding 10 kV.	BS EN 50107
Slotted cable trunking systems intended for installation in cabinets. Section 3: Slotted in cabinets.	BS EN 50085-2-3:2001
Specification of supplementary requirements for fuses of compact dimensions for use in 240/415V a.c. industrial and commercial electrical installations.	BS 88-6:1988

(*continued*)

E.2 Listed by subject (*continued*)

Title	BS or EN number
Switches for household and similar fixed electrical installations. Specification for general requirements.	BS 3676
Switches for household and similar fixed electrical installations.	BS EN 60669
Switches, disconnectors, switch-disconnectors and fuse-combination units.	BS EN 60947-3:1999
System design, installation, commissioning and maintenance.	BS 5839-1:2002
Telecommunications equipment and telecommunications cabling. Specification for installation, operation and maintenance.	BS 6701:2004
Test methods for cables under fire conditions. Test for vertical flame spread of vertically mounted bunched wires or cables.	BS EN 50266
Test methods. Methods of determining minimum ignition temperatures.	BS EN 50281-2-1:1999
Tests on electric and optical fibre cables under fire conditions. Test for vertical flame propagation for a single insulated wire or cable. Procedure for 1 kW pre-mixed flame.	BS EN 60332-1-2:2004
Thermal insulation for use in pitched roof spaces in dwellings. Specification for installation of man-made mineral fibre and cellulose fibre insulation.	BS 5803-5:1985
Thermal resistance. A method for calculating reduction factors for groups of cables in free air, protected from solar radiation.	BS 7769-2.2:1997 (2005)
Thermal resistance. Section 2.1: Calculation of thermal resistance.	BS 7769-2-2.1:1997 (2006)
Time delay switches (TDS).	BS EN 60669-2-3:2006
Transfer switching equipment.	BS EN 60947-6-1:2005
Transformers for construction sites.	BS EN 61558-2-23:2001
Typetested and partially type-tested assemblies.	BS EN 60439-1:1999
Uninterruptible power systems (UPS).	BS EN 62040
Weather-resistant couplers for household, commercial and light industrial equipment.	BS 6991:1990

E.3 Other Standards to which reference is made in the Regulations

E.3.1 IECa and ISO

IEC 60038-am 2 Ed 6	IEC standard voltages.
IEC 60364	Low-voltage electrical installations.
IEC 60364-5-51	Electrical installations of buildings – Part 5-51: Selection and erection of electrical equipment – Common rules.
IEC 60449-am 1 Ed 1	Voltage bands for electrical installations of buildings.

(*continued*)

E.3.1 IECa and ISO (*continued*)

Title	BS or EN number
IEC 60502-1 Ed 2	Power cables with extruded insulation and their accessories for rated voltages from 1 kV (U_{rn} = 1.2 kV) up to 30 kV (U_{rn} = 36 kV) – Part 1: Cables for rated voltages of 1 kV (U_{rn} = 1.2 kV) and 3 kV (U_{rn} = 3.6 kV).
IEC 60755-am 2	General requirements for residual current operated protective devices.
IEC 60884 Ed 3.1	Plugs and socket-outlets for household and similar purposes. Part 1. General requirements.
IEC 60906	IEC system of plugs and socket-outlets for household and similar purposes.
IEC 61201:1992	Extra-low voltage (ELV). Limit values. Also known as PD 6536.
IEC 61386	Conduit systems for cable management (BS EN 61386 series).
IEC 61386-24 Ed 1	Particular requirements – Conduit systems buried underground.
IEC 61662 TR2 Ed 1	Assessment of the risk of damage due to lightning.
IEC 61936-1 Ed 1	Power installations exceeding 1 kV a.c. – Part 1: Common rules.
IEC 61995-1 Ed 1	Devices for the connection of luminaires for household and similar purposes – Part 1: General requirements.
IEC/TS 62081 Ed 1	Arc welding equipment. Installation and use.
ISO 8820	Road vehicles. Fuse-links.

E.3.2 CENELEC Harmonised Documents

Listed by subject

Agricultural and horticultural premises.	HD 60364-7-705:2007
Application of measures for protection against overcurrent.	HD 384.4.473 Al:1980
Caravan parks, camping parks and similar locations.	HD 384.7.708:2005
Conducting locations with restricted movement.	HD 60364-7-706:2007
Construction and demolition site installations.	HD 60364-7-704:2007
Earthing arrangements, protective conductors and protective bonding conductors.	prHD 60364-5-54:2004
Electrical installations in caravans and motor caravans.	prHD 60364-7-721:2007
Exhibitions, shows and stands.	HD 384.7.711:2003
Extra-low-voltage lighting installations.	HD 60364-7-715:2005
Extra-low-voltage lighting installations.	HD 60364-7-715:2005
Fundamental principles, assessment of general characteristics and definitions.	prHD 60364-:2007
Identification of cores in cables and flexible cords.	HD 308 S2:2001
Initial verification.	HD 384.6.61 S2:2003
Locations containing a bath or shower.	HD 60364-7-701:2007
Marinas and similar locations.	prHD 60364-7-709:2007
Mobile or transportable units.	HD 60364-7-717:2004
Outdoor lighting installations.	HD 384.7.714 S1:2000
Protection against electric shock.	HD 384.4.41 S2/Al:2002

(*continued*)

Listed by subject (*continued*)

Protection against fire where particular risks or danger exist.	HD 384.4.482 S1:1997
Protection against overcurrent.	HD 384.4.43 S2:2001
Protection against overcurrent.	HD 384.4.43 S2:2001
Protection against overvoltages.	HD 384.4.443 S1:2000
Protection against thermal effects.	HD 384.4.42 S1 A2:1994
Rooms and cabins containing sauna heaters.	HD 384.7.703:2005
Selection and erection of equipment – Common rules.	prHD 60364-5-51:2003
Solar photovoltaic (PV) power supply systems.	HD 60364-7-712:2005
Swimming pools and other basins.	HD 384.7.702 S2:2002
Temporary electrical installations for structures, amusement devices and booths at fairgrounds, amusement parks and circuses.	prHD 60364-7-740:2006

Listed by Directive

Fundamental principles, assessment of general characteristics and definitions.	prHD 60364-:2007
Identification of cores in cables and flexible cords.	HD 308 S2:2001
Protection against electric shock.	HD 384.4.41 S2/Al:2002
Protection against thermal effects.	HD 384.4.42 S1 A2:1994
Protection against fire where particular risks or danger exist.	HD 384.4.482 S1:1997
Protection against overcurrent.	HD 384.4.43 S2:2001
Application of measures for protection against overcurrent.	HD 384.4.473 Al:1980
Protection against overvoltages.	HD 384.4.443 S1:2000
Selection and erection of equipment – Common rules.	prHD 60364-5-51:2003
Protection against overcurrent.	HD 384.4.43 S2:2001
Earthing arrangements, protective conductors and protective bonding conductors.	prHD 60364-5-54:2004
Outdoor lighting installations.	HD 384.7.714 S1:2000
Extra-low-voltage lighting installations.	HD 60364-7-715:2005
Initial verification.	HD 384.6.61 S2:2003
Locations containing a bath or shower.	HD 60364-7-701:2007
Swimming pools and other basins.	HD 384.7.702 S2:2002
Rooms and cabins containing sauna heaters.	HD 384.7.703:2005
Construction and demolition site installations.	HD 60364-7-704:2007
Agricultural and horticultural premises.	HD 60364-7-705:2007
Conducting locations with restricted movement.	HD 60364-7-706:2007
Caravan parks, camping parks and similar locations.	HD 384.7.708:2005
Marinas and similar locations.	prHD 60364-7-709:2007
Exhibitions, shows and stands.	HD 384.7.711:2003
Solar photovoltaic (PV) power supply systems.	HD 60364-7-712:2005
Extra-low-voltage lighting installations.	HD 60364-7-715:2005
Mobile or transportable units.	HD 60364-7-717:2004
Electrical installations in caravans and motor caravans.	prHD 60364-7-721:2007
Temporary electrical installations for structures, amusement devices and booths at fairgrounds, amusement parks and circuses.	prHD 60364-7-740:2006

BS 7671 will continue to be amended from time to time to take account of the publication of new or amended CENELEC Standards.

Appendix F

Useful contacts and Further information

Useful contacts

BSI
389 Chiswick High Road
London W4 4AL
Tel: +44 (0)20 8996 9000
email: cservices@bsi-global.com
Fax: +44 (0)20 8996 7001
www.bsi-global.com

As the National Standards Body for the UK, BSI are responsible for developing standards and standardising solutions to meet the needs of business and society.

ECA
ESCA House
34 Palace Court
London W2 4HY
Tel: 020 7313 4800
Fax: 020 7221 7344
email: electricalcontractors@eca.co.uk
Web: www.eca.co.uk

The aims of the Electrical Contractors Association are to:

- Promote quality and safety through:
  - the qualification of companies
  - the training, qualification and reward of individuals
- promote the compliance of all electrical and related installation work to relevant standards
- encourage the adoption of beneficial new technologies and installation practices
- influence the market to ensure that there is an equitable commercial environment

---

ELECSA Limited
44-48 Borough High Street
London SE1 1XB
Tel: +44 (0) 870 749 0080
Fax: +44 (0) 870 749 0085
email: enquiries@elecsa.org.uk
Web: www.elecsa.org.uk

The Fenestration Self-Assessment Scheme is the recognised competent person's scheme for Approved Document P of the Building Regulations.

IEE
Savoy Place
London WC2R 0BL
Tel: +44 (0)20 7240 1871
Fax: +44 (0)20 7240 7735
e-mail: postmaster@iee.org.uk

The IEE (The Institute of Electrical Engineers) is the largest professional engineering society in Europe and has a worldwide membership of 120,000.

NAPIT Suite L4A, Mill 3 Pleasley Vale Business Park Mansfield Nottinghamshire NG19 8RL Tel: 0870 4441392 Fax: 0870 4441427 email: info@napit.org.uk Web: www.napit.org.uk	The National Association of Professional Inspectors and Testers provides an independent professional trade body for electrical inspectors, electrical contractors, electricians and allied trades throughout the UK.
NICEIC Warwick House Houghton Hall Park Houghton Regis Dunstable Bedfordshire LU5 5ZX Tel: 01582 531000 Fax: 01582 531010 email: enquiries@niceic.com Web: niceic.org.uk	The National Inspection Council for Electrical Installation Contracting is the industry's independent, non profit-making, voluntary regulatory body covering the whole of the United Kingdom. The NICEIC's sole purpose is to protect consumers from unsafe and unsound electrical work. We are not a trade association and do not represent the interests of electrical contractors.
Office of the Deputy Prime Minister Eland House Bressenden Place London SW1E 5DU Tel: 020 7944 4400 Fax: 020 7944 9645 email: enquiryodpm@odpm.gsi.gov.uk Web: www.odpm.gov.uk	The job of the Office of the Deputy Prime Minister is to help create sustainable communities, working with other Government departments, local councils, businesses, the voluntary sector, and communities themselves.

Further reading

Copies of the Wiring Regulations may be obtained from either the IET:
P.O. Box 96
Stevenage
SG1 2SD, UK
Tel: +44 (0)1438 767328
Email: sales@theiet.org
or online at www.iee.org/shop

or BSI:
BSI Customer Services
389 Chiswick High Road
London W4 4AL, UK
Tel: +44 (0)20 8996 9001
Email: orders@bsi-global.com
or online at www.bsi-global.com/bsonline

Further assistance

The IET publishes a range of books and runs courses and services for industry to support the use and application of BS 7671. Of particular interest are

the IEE's series of Guidance Notes which offers extensive, industry-endorsed guidance to designers and installers in the effective use of BS 7671.
Seven IEE Guidance Notes are currently available. These are:

1   Selection and Erection of Equipment
2   Isolation and Switching
3   Inspection and Testing
4   Protection against Fire
5   Protection against Electric Shock
6   Protection against Overcurrent
7   Special Locations

Other products

IEE's On-site guide

A convenient, practical guide for electricians which covers domestic installations and smaller industrial and commercial installations up to 100A, 3-line. The most widely used guide in the industry, it removes the need for detailed calculations and outlines a detailed inspection and testing regime for installations. Comprehensive checklists and procedures are provided.

IEE's Commentary on BS 7671: 2001

Written for designers and managers by the IEE's former Principal Engineer, Paul Cook, this title provides clear interpretations of and guidance to the Regulations.

Codes of practice

The *IEE Code of Practice for In-Service Inspection and Testing of Electrical Equipment* offers guidance for the inspection, testing and maintenance of electrical appliances, plus advice on compliance with health and safety legislation.
   The *IEE Electrical Maintenance – Code of Practice* offers guidance on electrical aspects of building maintenance, including electrical installation, fire alarms, emergency lighting and more, plus detailed guidance on legal responsibilities.

CD-ROM of IEE Wiring Regulations

The CD-ROM version of BS 7671 (incorporating all amendments) is a fully structured electronic reference tool, hyperlinked and fully cross-referenced. As well as BS 7671, the CD-ROM contains the *On-Site Guide*, all seven *Guidance Notes* and both *IEE Codes of Practice*.

To order any of these publications, contact The IET:
by telephone: +44 (0)1438 767328
by fax: +44 (0)1438 742792
by email: sales@theiet.org
via their website: www.iee.org/shop

 For free downloads and updates you can join their email list by visiting www.
iee.org/technical

Other publications

IEE On-Site Guide (BS 7671, 16th Edition Wiring Regulations). The Institution
of Electrical Engineers. ISBN 0-85296-987-2, 2002

IEE Guidance Note 1: Selection and erection of equipment. 4th edition. The
Institution of Electrical Engineers. ISBN 0-85296-989-9, 2002

IEE Guidance Note 2: Isolation and switching. 4th edition. The Institution of
Electrical Engineers. ISBN 0-85296-990-2, 2002

IEE Guidance Note 3: Inspection and testing. 4th edition. The Institution of
Electrical Engineers. ISBN 0-85296-991-0, 2002

IEE Guidance Note 4: Protection against fire. 4th edition. The Institution of
Electrical Engineers. ISBN 0-85296-992-9, 2003

IEE Guidance Note 5: Protection against electric shock. 4th edition. The
Institution of Electrical Engineers. ISBN 0-85296-993-7, 2002

IEE Guidance Note 6: Protection against overcurrent. 4th edition. The
Institution of Electrical Engineers. ISBN 0-85296-994-5, 2003

IEE Guidance Note 7: Special locations. 2nd edition (incorporating the 1st
and 2nd amendments). The Institution of Electrical Engineers. ISBN 0-
85296-995-3, 2003

New wiring colours. Leaflet published by the IEE, 2004. Available for downloading from the IEE website at www.iee.org/cablecolours

*ECA comprehensive guide to harmonised cable colours, BS 7671: 2001
Amendment No 2.* Electrical Contractors' Association, March 2004

New fixed wiring colours – A practical guide. National Inspection Council for
Electrical installation Contracting (NICEIC), Spring 2004

Building Regulations in Brief. 4th edition. By Ray Tricker published by
Butterworth-Heinemann. ISBN 0-7506-8058-X (2006)

The Building Regs but not Part P. Article published by the IEE, Spring 2004.
Available for downloading from the IEE website at http://www.iee.org/
Publish/WireRegs/ IEE_Building-Regs.pdf

Electrical Installers' Guide to the Building Regulations. NICEIC and ECA,
August 2004. Available from www.niceic.org.uk and www.eca.co.uk

Appendix G
Books by the same author

Title	Details	Publisher and ISBN
Building Regulations in Brief (fifth edition) 	Handy reference guide to the requirements of the Building Act and its associated Approved Documents. Aimed at experts as well as DIY enthusiasts and those undertaking building projects.	Butterworth–Heinemann ISBN 978-0-7506-8444-6
Scottish Building Standards in Brief 	Takes the highly successful formula of Ray Tricker's Building Regulations in Brief and applies it to the requirements of the Building (Scotland) Regulations 2004. With the same no-nonsense and simple to follow guidance but written specifically for the Scottish Building Standards, it's the ideal book for builders, architects, designers and DIY enthusiasts working in Scotland.	Butterworth–Heinemann ISBN: 978-0-7506-8558-0
Environmental Requirements for Electromechanical and Electronic Equipment 	Definitive reference containing all the background guidance, ranges, test specifications, case studies and regulations worldwide.	Butterworth–Heinemann ISBN: 978-0-7506-3902-6

(*continued*)

Title	Details	Publisher and ISBN
CE Conformity Marking 	Essential information for any manufacturer or distributor wishing to trade in the European Union. Practical and easy to understand.	Butterworth–Heinemann ISBN: 978-0-7506-4813-4
ISO 9001:2000 for Small Businesses (third edition) 	New edition of this top selling quality management handbook. Contains a full description of the ISO 9001:2000 standard plus detailed information on quality control and quality assurance. Fully updated following 4 years practical field experience of the Standard. Includes a sample Quality Manual (that can be customised to suit individual requirements) and assistance on self-certification etc.	Butterworth–Heinemann ISBN: 978-0-7506-6617-6
ISO 9001:2000 Audit Procedures (second edition) 	A complete set of audit check sheets and explanations to assist quality managers and auditors in completing internal, external and third part audits of ISO 9001:2000 Quality Management Systems.	Butterworth–Heinemann ISBN: 978-0-7506-6615-2
ISO 9001:2000 in Brief (second edition) 	Revised and expanded, this new edition of an easy to understand guide provides practical information on how to set up a cost-effective ISO 9001:2000 compliant Quality Management System.	Butterworth–Heinemann ISBN: 978-0-7506-6616-9

(continued)

Title	Details	Publisher and ISBN
ISO 9001:2000 Quality Manual & Audit Checksheets	A CD containing a soft copy of the generic Quality Management System featured in *ISO 9001:2000 for Small Businesses* (3rd edition) plus a soft copy of all the check sheets and example audit forms contained in *ISO 9001:2000 Audit Procedures* (2nd edition).	ISBN 0-9548647-2-7
Quality Management system for ISO 9001:2000	*Quality Management System for ISO 9001:2000* and accompanying CD is probably the most comprehensive set of ISO 9001:2000 compliant documents available world-wide. Fully customisable, it can be used as a basic template for any organisation wishing to work in compliance with, or gain registration to, ISO 9001:2000.	ISBN 0-9548647-4-3
Auditing Management Systems	*Auditing Management Systems* and accompanying CD is the result of 7 years' field experience of the International Standard for Quality Management (i.e. ISO 9001:2000) and is capable of being used to conduct an internal, external or third party audit of ANY Management System.	ISBN 0-9548647-5-1
ISO 9001:2000 The Quality Management Process	*ISO 9001:2000 The Quality Management Process* is unique, being the first publication that is specifically aimed at those professionals who are directly involved in using the Standard. It provides the background to the requirements of the Standard in an easily accessible format and presents the reader with essential, basic answers to the following questions: • What does the Standard actually say? • What needs to be checked? • Where is a particular requirements covered in the Standard? • What will an auditor be looking for?	ISBN 90-77212-77-9

(continued)

Title	Details	Publisher and ISBN
Quality and Standards in Electronics 	Ensures that manufacturers are aware of the all the UK, European and international necessities, knows the current status of these Regulations and Standards, and where to obtain them.	Butterworth–Heinemann ISBN: 978-0-7506-2531-9
Optoelectronic and fiber optic technology 	An introduction to the fascinating technology of fibre optics.	Butterworth–Heinemann ISBN 978-0-7506-5370-1

Index